Game and Decision Theory Applied to Business, Economy and Finance

Game and Decision Theory Applied to Business, Economy and Finance

Guest Editor
Ivan Arraut

MDPI
Basel • Beijing • Wuhan • Barcelona • Belgrade • Novi Sad • Cluj • Manchester

Guest Editor
Ivan Arraut
Laboratory of Applied
Neurosciences
University of Saint Joseph
Macau
China

Editorial Office
MDPI AG
Grosspeteranlage 5
4052 Basel, Switzerland

This is a reprint of the Special Issue, published open access by the journal *Mathematics* (ISSN 2227-7390), freely accessible at: https://www.mdpi.com/journal/mathematics/special_issues/Game_Decis_Theory.

For citation purposes, cite each article independently as indicated on the article page online and as indicated below:

Lastname, A.A.; Lastname, B.B. Article Title. *Journal Name* **Year**, *Volume Number*, Page Range.

ISBN 978-3-7258-5701-2 (Hbk)
ISBN 978-3-7258-5702-9 (PDF)
https://doi.org/10.3390/books978-3-7258-5702-9

Contents

About the Editor

Ivan Arraut

Ivan Arraut obtained his PhD and graduated from the school of Science (Physics) at Osaka University in 2014. He has a Master of Science, Physics, degree from The University of Los Andes (2008) and received a bachelor's degree in Mechanical Engineering from Universidad del Norte in 2004. Between 2008 and 2010 and between 2018 and 2021, he worked as a Professor and as a Lecturer in different universities. Ivan Arraut is one of the best researchers at The University of Saint Joseph.

His research areas include Fundamental Physics (General Relativity and Quantum Field Theory), Engineering Science, Game Theory, Quantum Finance and Machine Learning. Ivan Arraut likes to combine research in different areas. In fact, some of Ivan's best results came to fruition by mapping some ideas from practical areas toward fundamental sciences and vice versa, namely, by mapping ideas from fundamental sciences into ideas related to practical areas. Ivan Arraut is a multidisciplinary researcher, teaching Physics and Mathematics at USJ in Macau. His main research characteristics are originality and simplicity, trying to find the simplest explanation for unexplained phenomena. With this style, he has received 13 academic prizes and/or recognitions from different institutions around the world during his career, including grants, scholarships, commendable thesis awards, fellowships, and recognitions from journals, among others. Among the most special recognitions to note, Ivan Arraut was the third best student in the national exam (ECAES) in Colombia, receiving a recognition from the president of Colombia (Alvaro Uribe). Ivan Arraut received a prize from the journal *Nature* for his presentation at the conference "Horizons of Quantum Physics", organized in Taipei in 2012. He also received the prestigious JSPS fellow and the CAS PIFI fellow for Post-Doctoral studies, among other important recognitions.

mathematics

MDPI

Article

The Mutual Impact of Suppliers' Online Sales Channel Choices and Platform Credit Decisions for Offline Channels

Yangyang Qin

School of Management, University of Science and Technology of China, Hefei 230026, China; qyy22204253@mail.ustc.edu.cn

Abstract: This study examines the strategic decisions and profit dynamics of suppliers marketing their products through both offline and online channels, alongside online e-commerce platforms providing their own consumer credit services. We develop a model that incorporates consumer disposable income, channel preferences, and credit utility. Four supply chain scenarios are analyzed: wholesale and agency models with either private or open credit strategies. Using Stackelberg game theory, we explore suppliers' sales model choices and the conditions under which platforms extend credit to offline channels. Our results show that increasing credit utility generally leads to higher equilibrium prices, while higher service fees compel suppliers to adjust their prices, favoring lower-cost channels. Notably, suppliers are more likely to adopt the wholesale model to secure platform credit for offline sales, especially when credit service fees or credit utility are high. Furthermore, platform credit strategies are strongly influenced by suppliers' sales model choices: In wholesale models, platforms are more inclined to extend credit to offline channels under specific conditions of high disposable income (DPI) and credit utility, whereas in agency models, open credit strategies are only adopted when both DPI proportions and credit utility are low. This research provides new insights into how platforms can tailor credit offerings based on supplier strategies, offering a theoretical foundation for consumer credit policies in multi-channel sales environments and valuable guidance for managers in determining optimal channel strategies and credit service offerings.

Keywords: consumer credit; wholesale model; agency model; game theory

MSC: 91-10; 91A80

1. Introduction

With the rapid growth of global e-commerce, consumer purchasing behaviors and payment habits have undergone significant transformations. Consumer credit refers to a system that allows consumers to borrow money or incur debt with deferred repayment. This encompasses various credit products such as bank credit cards, retail platform consumer credit, and P2P lending. As a new payment method, consumer credit has significantly enhanced consumers' purchasing power. According to the latest industry report, global retail e-commerce sales exceeded USD 6.1 trillion in 2023 and are expected to surpass USD 8 trillion by 2027 (https://worldpay.globalpaymentsreport.com/en, accessed on 11 December 2024). As e-commerce continues to grow, the consumer credit market has expanded rapidly as well. In 2024, the global consumer credit market is expected to reach USD 120 billion, with a projected compound annual growth rate of 3.9% from 2025 to 2033 (https://www.imarcgroup.com/consumer-credit-market, accessed on 15 December

2024). In contrast, traditional credit mechanisms, such as credit card approval processes, are complex and stringent [1], which makes it challenging for younger consumers to access credit. Many platforms, facing significant market potential, have leveraged their technological and data advantages to launch platform-based consumer credit (referred to as "platform credit") online, offering consumers the option of deferred payments with interest-free grace periods. Examples of representative platform credit products include Alibaba's Ant Credit, JD.com's Bai Tiao, and Amazon's Afterpay. Platform credit provides purchasing opportunities for consumers with low disposable income (DPI), thus increasing demand on the platforms and generating additional interest income. In Q3 2022, 43.7% of JD.com's operating cash flow came from accounts receivable from Bai Tiao (https://baijiahao.baidu.com/s?id=1749831670174657381&wfr=spider&for=pc, accessed on 17 December 2024). With continuous innovations in credit payment technology, more platforms have extended credit to various payment scenarios. For instance, Ant Credit expanded its services beyond online channels to include physical stores of certain merchants in 2015.

The literature on consumer credit suggests that credit payments can stimulate greater consumption and are less burdensome for consumers compared to cash payments [2,3]. Consumer credit not only alleviates the issue of insufficient disposable income (DPI) for consumers but also enhances convenience and security through various payment methods such as facial recognition. These advantages have attracted a large number of consumers to use such services. By the end of 2023, Ant Credit had served over 400 million consumers, with more than 60% of them having no prior credit card usage history. For these consumers, Ant Credit is their first encounter with consumer finance (https://news.qq.com/rain/a/20240701A03ITC00, accessed on 11 December 2024). The diversification of payment methods not only reduces consumer payment costs but also enhances their valuation of the platform. Furthermore, consumers who repay on time incur no additional fees.

Online platforms offer significant market opportunities for suppliers, who can choose between two primary sales models: the wholesale model and the agency model. The literature on e-commerce platforms has extensively discussed these two sales models [4,5]. In the wholesale model, the platform acts as a retailer, purchasing products from suppliers and reselling them to consumers. In the agency model, the platform functions as a marketplace, allowing suppliers to sell products directly to consumers through the platform and charging a commission. In practice, e-commerce platforms like Taobao and Amazon allow suppliers to choose between these two sales models and sell their products through traditional offline channels as well. Platform credit not only influences consumer purchasing behavior but also has a profound impact on the profit-sharing dynamics between the platform and suppliers. As both a competitor or collaborator of the supplier and a provider of credit services, the platform can choose to extend credit services to suppliers' offline channels to expand the market for platform credit. Alternatively, as a retailer or online marketplace, the platform can choose not to extend credit services to suppliers' offline channels in order to boost online transactions. The differences in these credit strategies reflect the platform's decision regarding the scope of credit usage, which can influence the supplier's online sales strategy and the consumer's purchasing decisions. Moreover, the extent to which platform credit strategies are open influences the depth of cooperation and market competitiveness between the platform and the supplier, subsequently affecting the operational efficiency and profit distribution across the supply chain. For instance, Apple distributes products via the wholesale model on JD.com, where JD has long provided interest-free credit services for Apple's self-operated flagship stores, but this credit service is not extended to Apple's offline direct retail stores. In contrast, Huawei adopts the agency model to sell products on Taobao, with Taobao offering interest-free credit services for both Huawei's flagship store on the platform and its offline stores (https://www.ithome.com/0/491/699.htm, accessed on 27 October 2024).

To date, no theoretical framework has explained the interaction between platform credit offerings and suppliers' sales model choices in dual-channel systems. Existing studies primarily focus on the intra-platform effects of credit (e.g., online pricing or consumer behavior) but overlook two critical dimensions:

1. Offline Credit Extension: Platforms like Alibaba now offer credit services in physical stores (e.g., Huabei in Apple Stores). However, no model explains how this affects supplier–platform coordination.
2. Sales Model–Credit Interplay: Previous work treats sales models (wholesale/agency) and credit strategies (private/open) as isolated decisions, neglecting their interdependence.

To address these gaps and explain the underlying mechanisms behind the observed phenomena, this study develops a dual-channel supply chain model consisting of suppliers and online platforms. Suppliers sell products through offline channels while simultaneously selling through the platform's online channel. The platform offers two distinct sales formats: the agency model and the wholesale model. As a credit service provider, the platform has two credit strategies: the private credit strategy and the open credit strategy. Under the private credit strategy, platform credit can only be used within the online platform. Under the open credit strategy, the platform also provides credit services for offline channels. This research aims to explore the interplay between platform credit strategies and suppliers' sales model decisions. The study primarily addresses the following three core research questions:

1. How do the equilibrium decisions of suppliers and platforms evolve under different channel structures and consumer credit strategies?
2. Under what market conditions should an online platform extend credit services to suppliers' offline channels?
3. What sales model are suppliers more inclined to choose, and how does the platform respond?

To answer these questions, we construct four scenarios: (1) WP scenario, where the supplier chooses the wholesale model and the platform adopts the private credit strategy; (2) WO scenario, where the supplier chooses the wholesale model and the platform adopts the open credit strategy; (3) AP scenario, where the supplier chooses the agency model and the platform adopts the private credit strategy; and (4) AO scenario, where the supplier chooses the agency model and the platform adopts the open credit strategy. We consider consumer heterogeneity in disposable income (DPI) and the potential for consumer credit to increase consumer valuation. Suppliers and platforms engage in a Stackelberg game, where the supplier first decides on the online sales model, followed by the platform's decision on its credit strategy. In the wholesale model, the platform first sets the offline direct sales price and online wholesale price, then sets the retail price for the wholesale channel. In the agency model, the supplier simultaneously decides the prices for both offline and online channels. By solving these four scenarios, we investigate the interaction between suppliers' sales model decisions and the online retail platform's consumer credit strategy. To the best of our knowledge, this study is the first to consider the interaction between suppliers' online sales model choices and e-commerce platform credit strategies for offline channel expansion. We describe in detail the optimal pricing and strategy decisions of both the supplier and the platform and demonstrate how credit service fees and credit utility affect the pricing decisions of both parties. An increase in credit service fees prompts suppliers to adjust their prices to guide demand, while higher credit utility generally leads to higher equilibrium prices. Platform credit strategies in the agency model facilitate synchronized growth of both online and offline channel sales, whereas in the wholesale model, online sales decrease and offline sales increase. Furthermore, our study shows that the best strategies for both suppliers and platforms depend on credit service fees, the proportion of high-DPI consumers, and credit utility. For the platform, regardless of the sales model, when the proportion of high-DPI consumers is high and credit utility is low, the platform is more likely to offer credit services for offline channels, with a higher likelihood of providing credit services to suppliers' offline channels in the wholesale model. Considering the influence of platform consumer credit, suppliers are likely to adopt the wholesale model only when the proportion of high-DPI consumers is high and credit utility is moderate, or both factors are high. When the proportion of high-DPI consumers is low, suppliers will only consider the agency model, and the platform will not offer credit services for the suppliers' offline channels.

The remainder of this paper is organized as follows. In Section 2, we position this study within the relevant literature on consumer credit, selling formats on online platforms, and interactions between online platforms and offline channels. In Section 3, we establish the problem model and provide the sequence of events along with the basic assumptions. In Section 4, we derive the equilibrium outcomes under different scenarios and analyze the impact of consumer credit on equilibrium decisions. In Section 5, we analyze the decision-making processes of suppliers in selecting online sales models and the platforms' decisions regarding credit provision. In Section 6, we present conclusions and future research directions. Additionally, Appendix A at the end of the paper contains proofs for all results.

2. Literature Review

This study is closely related to three streams of literature: online platform credit, selling formats on online platforms, and interactions between online platforms and offline channels. In this section, we provide an overview of these research streams and highlight our contributions to the field.

2.1. Consumer Credit

Consumer credit has been proven to be an effective payment method for consumers, significantly influencing their purchasing decisions. Credit payments play a key role in enhancing consumers' willingness to pay [6]. Hu et al. [7] found that consumers are increasingly using fintech credit rather than traditional bank credit. Furthermore, consumer credit may have a significant impact on consumers' mental accounting. Xie et al. [8] proposed that product quality and consumer trust significantly affect the willingness of both platform and strategic consumers to use credit services.

Online credit has gained increasing attention in the field of platform supply chain management. Liu et al. [9] demonstrated that credit refunds can lead to price discrimination and demand shaping among consumers, and under certain conditions, they are more profitable than cash refunds. Zhao et al. [10] studied how e-commerce platforms influence consumer behavior through installment payment and price guarantee strategies. They found that, when the proportion of high disposable income consumers is low, interest-free installment strategies are more profitable, while price guarantee strategies limit the use of interest-free installments. Xie et al. [11] examined the impact of e-commerce consumer credit on pricing strategies and the profits of all participants, revealing that credit services benefit sellers, consumers, and the entire supply chain in low- and mid-end product transactions, though the effect on platform profits remains uncertain.

Wu et al. [12] found that the time cost of funds and credit discount rates influence the willingness of suppliers and retailers to open credit payment options. One study presents a model for e-commerce platforms to determine when to introduce consumer credit services and share them with third-party sellers, revealing that sharing services can benefit both parties under certain conditions but may reduce consumer surplus [13]. In another related study, cognitive biases in borrowers, especially over-optimism, are shown to lead to financial mistakes, with the study proposing policy solutions like financial literacy education and reducing default costs to improve welfare, although the results are mixed [14]. Additionally, research on buy now, pay later (BNPL) services highlights their significant impact on consumer financial behavior, influencing the likelihood of adopting other credit options and shaping broader financial decision making [15]. Li et al. [16] explored how the credit strategy of online retail platforms affects interactions with sellers, revealing that offering credit can impact pricing and demand, and that platform credit strategies depend on credit yield and mismatch costs. Wei et al. [17] discussed the dual credit services of e-commerce platforms and their impact on the ecosystem, finding that dual credit can reduce product prices and benefit consumers, though its effects on sellers and platforms remain uncertain. Duan et al. [18] explored the provision strategy of e-commerce consumer credit services in a dual-channel system and their effect on pricing strategies, showing that credit services may generate positive spillover effects for the platform. Li et al. [19] examined the credit strategy and interest issues of online platforms in integrated and independent system platform supply chains.

Although the existing literature provides ample evidence on how consumer credit influences consumer purchasing behavior, there is limited research on how consumer credit affects platform supply chain management decisions. Furthermore, existing studies primarily focus on the impact of credit services within online platforms (such as their effects on platform revenue, consumer behavior, and supply chain pricing strategies). There is a lack of research on whether credit services should extend to external platforms and their potential impacts. While some studies have explored the impact of consumers' disposable income differences and credit utility on credit strategies, their specific effects in online and offline dual-channel supply chains have not been sufficiently analyzed. Our study fills the gap in both areas.

2.2. Selling Formats on Online Platforms

Our study is also related to the literature on selling formats on online platforms. Hagiu et al. [4] found that when marketing activities result in spillover effects or network effects that negatively affect supplier profits, the wholesale model is preferred, while the agency model is favored for long-tail products. Abhishek et al. [5] explored the decision-making process of e-commerce platforms when choosing between agency sales and traditional wholesale models. They discovered that agency sales are more efficient in online channels when traditional channel demand declines, but manufacturers, by controlling retail prices, influence the e-commerce platform's choice. Ha et al. [20] examined how sales efforts impact the channel structure of online retail platforms. Gong et al. [21] pointed out that platforms encourage manufacturers to adopt the agency sales model when information accuracy is high, thereby increasing profits for both parties. Li et al. [22] studied online sales of green products, finding that manufacturers tend to choose the agency sales model at lower agency fees to enhance product greenness, while higher fees suppress online sales, affecting e-commerce platform revenues. Bian et al. [23] found that platforms tend to prefer the wholesale model when offering high-quality services, whereas the agency model may deter suppliers from sharing service information. Chen et al. [24] indicated that a manufacturer's choice of online sales strategy largely depends on consumers' acceptance of online channels and the spillover effects from online to offline sales. Under specific conditions, the agency sales strategy can result in a "win-win-win" situation for the manufacturer, offline retailers, and online platforms. Ke et al. [25] analyzed the selection of three sales models (wholesale, agency sales, and mixed sales) in platform supply chains, concluding that the mixed model typically yields higher profits and that the optimal sales model varies with commission rates. Tian et al. [26] studied the choice of online distribution models between an e-commerce platform and two different suppliers selling mutually substitutable products. Liu et al. [27] proposed that online financial services provided by e-commerce platforms can help capital-constrained suppliers optimize their channel strategies, with their choices being constrained by production costs, financial constraints, platform interest rates, and commission rates. Zhen et al. [28] used a game-theory model to analyze how retailers' sales choices between proprietary channels and third-party platforms are influenced by spillover effects and the intensity of channel competition, showing that under moderate competition, different sales formats may be preferred.

Unlike previous research on selling formats on online platforms, this study aims to delve deeper into the dynamic relationship between vendors' sales model choices and platform credit strategies. We have developed a Stackelberg game model to investigate how platform credit offering strategies influence cooperation or competition between platforms and suppliers in a dual-channel supply chain based on both offline and online channels. As a credit provider, the platform is also a retailer or an online marketplace. It can choose to offer interest-free credit only to consumers purchasing through the supplier's online channel, or it can extend interest-free credit services to consumers in both online and offline channels.

2.3. Interactions Between Online Platforms and Offline Channels

Our work also contributes to the literature on interactions between online platforms and offline channels. In recent years, numerous scholars have explored the reciprocal effects between online and offline channels. Yoo et al. [29] investigated the combination structure of a monopoly manufacturer with both online and offline channels, finding that the introduction of online channels does not always lead to lower retail prices or increased consumer welfare. Chiang et al. [30] discovered that the introduction of a direct sales channel by the supplier can reduce double marginalization effects, thereby increasing profits, while simultaneously benefiting retailers when wholesale prices are reduced. Cattani et al. [31] also examined the situation where manufacturers open offline direct sales channels while cooperating with online wholesale channels, and found that when the convenience of online channels is significantly lower than traditional channels, manufacturers can optimize profits and reduce channel conflicts by implementing strategies aligned with retailer prices. Bhunia and Roy et al. [32] focused on carbon emission reduction in green supply chains and used Stackelberg game theory to explore the balance between profit maximization and carbon reduction for manufacturers and retailers across different sales channels (such as offline and online).

Jiang et al. [33] explored interactions between platforms and independent sellers, showing that platforms' actions to prioritize high-demand products may incentivize sellers to conceal true demand, while platform entry threats may actually benefit sellers. Barman and Roy et al. [34] focused on closed-loop supply chains and introduced a fuzzy optimization model to address the challenges of carbon emission reduction, cost minimization, and policy compliance under uncertain data conditions. This study focuses on the interaction between supplier strategic decisions and platform credit policies in a multi-channel sales environment, particularly examining how this interaction affects the decisions and profits of both parties. Gilbert et al. [35] found that when traditional channel competition is moderate and the substitutability between online platforms and traditional channels is high, suppliers prefer the agency model over wholesale, as it is Pareto superior. Crombrugge et al. [36] found through empirical analysis that the introduction of direct sales channels typically leads to a reduction in the number of SKUs purchased by retailers and an increase in unit wholesale prices. Most previous studies have focused on the impact of online channels on offline channel operations. In contrast, our study considers a different direction of interaction, namely, the interaction between platform credit strategies and the channel strategies of suppliers with both offline and online dual channels. We explore how this unique interaction affects the equilibrium outcomes.

To clearly illustrate the contributions of our work in comparison to recent studies, Table 1 below presents a comparison of the key contributions and research gaps of our study with the following recent works:

Table 1. Comparison of key contributions and research gaps.

Research	Research Focus	Credit Strategy	Sales Models	Channel Interaction	Key Limitations Addressed in Our Work
Liu et al. (2023) [27]	Platform financing for capital-constrained SMEs in dual-channel supply chains.	Debt financing (interest-based loans).	Reselling vs. agency selling.	Focuses on operational–financial integration (e.g., Pareto improvements under high production costs).	Does not study consumer credit (e.g., deferred payment) or offline channel credit extension.
Duan et al. (2024) [18]	Adoption of e-commerce consumer credit (e-CCS) in dual-platform systems.	Platform-specific e-CCS (wholesale vs. agency channels).	Wholesale vs. agency platforms.	Analyzes spillover effects between platforms (e.g., "win-win" under e-CCS).	Limited to intra-platform credit; ignores offline channels and credit utility (τ).
Xie et al. (2023) [11]	Credit impacts on pricing in low/mid-end product transactions.	Single-channel credit (online only).	Wholesale only.	No cross-channel spillovers.	Lacks analysis of credit utility or dual-channel credit coordination.
Li et al. (2022) [19]	Platform credit strategies under yield mismatch costs.	Fixed credit scope (online).	Agency only.	No offline channel integration.	Does not model open credit strategies for offline channels.
Bhunia and Roy et al. [32]	Carbon emissions reduction via green tech, cap-and-trade policies, and dual-channel coordination.	Not studied.	Traditional vs. direct channels (not wholesale/agency).	Focuses on environmental impacts (e.g., emissions, delivery lead time).	Does not address platform credit strategies or financial interactions between channels.
Our Work	Dual-channel coordination; under platform credit strategies.	Private vs. open credit (online/offline).	Wholesale vs. agency with credit fee (τ) and utility (τ).	Credit-driven price synchronization across online/offline channels.	1. First to model credit extension to offline channels. 2. Quantifies profit trade-offs between sales models and credit strategies. 3. Identifies thresholds for platform credit sharing (e.g., α, τ).

3. Model Setup

This study examines a two-tier supply chain consisting of a supplier (S) and an online platform (O), where the supplier sells products directly to consumers through offline channels and via the platform. The platform offers two distinct sales models: wholesale (e.g., JD's self-operated stores) and agency (e.g., the Tmall platform).

3.1. Supplier and Online Platform

We model a dual-channel supply chain in which the supplier sells products directly to consumers through offline channels while simultaneously using the online platform to sell the same products via either a wholesale (W) or agency (A) model. In the wholesale model, the platform functions as a retailer, purchasing products at a wholesale price (w) and selling them at a retail price (p_o). In the agency model, the platform operates as an online marketplace with the supplier acting as a third-party seller and setting the online sales price (p_o). The platform charges the supplier a commission ϕ based on product category. Additionally, if consumers use platform credit to make payments, the platform collects a fee (r) for credit services. If platform credit is allowed in the supplier's offline channel, the supplier must also pay a fee of r to the platform.

Since platform commission rates and credit service fees are typically fixed and disclosed before a supplier joins the platform, we treat ϕ and r as exogenous variables, where $0 < \phi < \frac{1}{4}$ and $0 < r < \frac{1}{4}$. The model ignores the potential risks associated with credit default, simplifying the analysis to focus on credit expansion strategies.

We assume consumers have limited disposable income, and the platform's credit strategy is modeled in two forms: (1) private credit service strategy, where the credit service is exclusive to the platform's online channel; and (2) open credit service strategy, where the credit service is extended to the supplier's offline channels. We assume that the marginal operating costs of both products and credit services are zero, following common practice in the literature, which does not affect the qualitative results of the analysis [16].

3.2. Consumer Choice Model

Consumers have heterogeneous valuations for products, with v uniformly distributed within the interval $[0, 1]$. Since the platform typically offers superior logistics and after-sales services [37], consumers are more likely to accept products purchased via the platform, resulting in a valuation of $(1+\theta)v$, where θ represents the advantage of the platform's channel. The higher the value of θ, the greater the platform's advantage over the supplier's offline channel.

We assume that consumer disposable income (DPI) varies, classifying consumer into two types: high DPI consumers (proportion α) and low DPI consumers (proportion $1-\alpha$). High DPI consumers can afford to purchase products in full, while low DPI consumers rely on platform credit to make purchases. Consumers using credit gain positive utility, represented by τ ($\tau > 0$), such as the psychological benefit of early product access or exclusive discounts.

Therefore, the utility for a consumer with valuation v is as follows:

- For full payment via online platform: $U_{og} = (1+\theta)v - p_o$;
- For credit payment via online platform: $U_{oc} = (1+\theta)v + \tau - p_o$;
- For full payment via offline channel: $U_{sg} = v - p_s$;
- For credit payment via offline channel: $U_{sc} = v - p_s$.
- When the platform does not offer credit services for the supplier's offline channel, high DPI consumers will choose to buy using credit on the platform if $U_{sg} \geq U_{oc}$. In contrast, low DPI consumers will always use platform credit. The sales volumes of online and offline channels are as follows:
- Online channel: $D_o = \alpha(1 - \frac{-\tau+p_o-p_s}{\theta}) + (1-\alpha)(1 - \frac{-\tau+p_o}{1+\theta})$;
- Offline channel: $D_s = \alpha(\frac{-\tau+p_o-p_s}{\theta} - p_s)$.
- When the platform offers credit services for offline channels, low DPI consumers can choose to buy via credit in either channel. The demand for online and offline channels are as follows:
- Online channel: $D_o = 1 - \frac{p_o-p_s}{\theta}$;
- Offline channel: $D_s = \frac{p_o-p_s}{\theta} - (-\tau + p_s)$.

In the wholesale model, the profit functions for the supplier and the platform are as follows:

- $\Pi_S = p_s(D_{sg} + D_{sc}) + w(D_{og} + D_{oc}) - rp_sD_{sc}$;
- $\Pi_O = (p_o - w)(D_{og} + D_{oc}) + rp_sD_{sc}$.

In the agency model, the profit functions for the supplier and the platform are as follows:

- $\Pi_S = p_s(D_{sg} + D_{sc}) + (1-\phi)p_o(D_{og} + D_{oc}) - r(p_s D_{sc} + p_o D_{oc})$;
- $\Pi_O = \phi p_o(D_{og} + D_{oc}) + r(p_s D_{sc} + p_o D_{oc})$.

If the platform only provides credit for the online channel, the demand for offline credit purchases is $D_{sc} = 0$.

The meanings of the relevant parameters and variables are summarized in Table 2.

Table 2. Summary of notation.

Notations	Description
v	Consumer's valuation of the product, uniformly distributed between 0 and 1
τ	Utility derived from the platform's credit service ($\tau > 0$)
α	Proportion of high DPI consumers
r	Platform credit service fee
ϕ	Platform commission rate
θ	Platform channel advantage factor
U_{ij}	Utility of consumer with valuation v for j-payment via i-channel
D_{ij}	Demand from consumers via i-channel (j-payment)
π_S	Supplier's profit
π_O	Platform's profit
w	Wholesale price in the wholesale model
p_i	Retail price via i-channel
$i = s,\ o$	Subscripts representing offline channel (s) and online channel (o)
$j = g, c$	Subscripts representing regular purchase (g) and credit purchase (c)

The sequence of events is as follows: In the first stage, the supplier chooses the sales model for the online platform channel, either reselling or acting as an agent. In the second stage, the platform decides whether to offer consumer credit services through offline channels. In the third stage, if the supplier selects the reselling model, the supplier first sets the wholesale price and the offline retail price, after which the platform determines the online retail price. If the supplier chooses the agency model, the supplier simultaneously sets both the offline and online retail prices. Finally, consumers choose their preferred purchase channel and payment method, and market demand is realized. The timing of these events is illustrated in Figure 1.

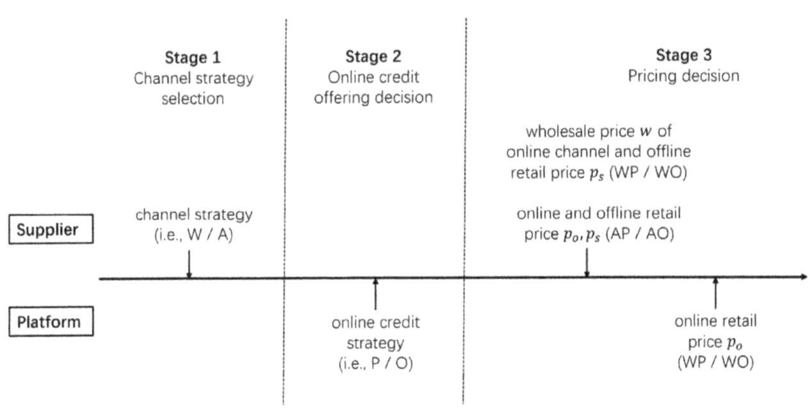

Figure 1. Timing of events.

4. Equilibrium Results

This section derives the equilibrium results for four scenarios: WP, WO, AP, and AO. The first letter represents the supplier's decision regarding the online sales model in the first stage, while the second letter indicates the online platform's decision regarding the provision of platform credit in the second stage. The equilibrium results for these four scenarios are derived using backward induction.

4.1. Case WP

In the WP case, the supplier selects the wholesale model for the online platform, while the platform adopts a private credit strategy, making credit available only for online purchases. A representative example of the WP scenario is Apple's wholesale partnership with JD.com. Apple supplies products to JD at a fixed wholesale price, while JD operates the "Apple JD Self-operated Flagship Store", offering exclusive "JD Baitiao" installment services for online purchases. Notably, Apple's offline flagship stores do not integrate JD's credit system, which aligns with the definition of the private credit strategy. Using backward induction, we derive the optimal retail price for the online platform by solving the following optimization problem: $\underset{p_o}{\text{Max}}\ [(p_o - w)\left(\alpha\left(1 - \frac{-\tau+p_o-p_s}{\theta}\right) + (1-\alpha)\left(1 - \frac{-\tau+p_o}{1+\theta}\right)\right)]$. Solving this optimization problem, we obtain the platform's optimal reaction price: $\hat{p}_o(w, p_s) = \frac{w(\alpha+\theta)+\alpha\tau+\theta(1+\theta+\tau)+\alpha(1+\theta)p_s}{2(\alpha+\theta)}$. Given this, the supplier anticipates the platform's optimal price and selects the wholesale price w and offline retail price p_s to maximize profit:

$$\underset{w,p_s}{\text{Max}}\ [p_s\alpha(\tfrac{-\tau+\hat{p}_o-p_s}{\theta} - p_s) + w(\alpha(1 - \tfrac{-\tau+\hat{p}_o-p_s}{\theta}) + (1-\alpha)(1-\tfrac{-\tau+\hat{p}_o}{1+\theta}))]$$

Solving the above optimization problem, we derive the supplier's optimal wholesale price and offline retail price:

$$w^{WP} = \frac{1}{2}(1+\theta+\tau), p_s^{WP} = \frac{1}{2}, p_o^{WP} = \frac{1}{4}(\frac{(1+\theta)(2\alpha+3\theta)}{\alpha+\theta} + 3\tau)$$

Substituting these prices into the profit functions for the supplier and the platform, we obtain the optimal profits:

$$\Pi_s^{WO} = \tfrac{1}{8}\left(\tfrac{(1+\theta)(2\alpha+\theta)}{\alpha+\theta} + 2\tau + \tfrac{(\alpha+\theta)\tau^2}{\theta(1+\theta)}\right)$$
$$\Pi_o^{WO} = \tfrac{(\alpha\tau+\theta(1+\theta+\tau))^2}{16\theta(1+\theta)(\alpha+\theta)}$$

4.2. Case WO

In the WO case, the supplier selects the wholesale model for the online channel, and the platform extends credit services to the supplier's offline channel. The WO scenario is exemplified by LEGO's wholesale collaboration with Amazon. Through Amazon Vendor Central, LEGO provides products at wholesale prices, with Amazon setting retail prices and offering "Amazon Pay Later" installments (five-period interest-free) both on its platform and in LEGO's physical stores (e.g., New York Fifth Avenue flagship). This dual-channel credit extension embodies the open credit strategy. Using backward induction, we derive the optimal online retail price p_o for the platform by solving the following optimization problem:

$$\underset{p_o}{\text{Max}}\ [(p_o - w)\left(1 - \frac{p_o - p_s}{\theta}\right) + rp_s\left(\frac{p_o - p_s}{\theta} + \tau - p_s\right)]$$

Solving this gives the platform's optimal reaction price:

$$\hat{p}_o(w, p_s) = \frac{1}{2}(w + \theta + (1 + r)p_s)$$

Substituting this into the supplier's profit function, the supplier chooses w and p_s to maximize their profit:

$$\underset{w,p_s}{\text{Max}}(1 - r)p_s(\frac{\hat{p}_o - p_s}{\theta} + \tau - p_s) + w(1 - \frac{\hat{p}_o - p_s}{\theta})$$

The optimal pricing decisions for the supplier in the WO case are

$$w^{WO} = \frac{1}{2}(1 + \theta + \tau - r(1 + \tau)), p_s^{WO} = \frac{1 + \tau}{2}$$

Substituting these into $\hat{p}_o(w, p_s)$, we obtain the platform's optimal pricing decision:

$$p_o^{WO} = \frac{1}{4}(2 + 3\theta + 2\tau)$$

Substituting these optimal prices into the profit functions of the supplier and the platform, we obtain

$$\Pi_s^{WO} = \tfrac{1}{8}(\theta + 2(1 + \tau)^2 - 2r(1 + \tau)^2)$$
$$\Pi_o^{WO} = \tfrac{1}{16}(\theta + 4r(1 + \tau)^2)$$

4.3. Case AP

In the AP case, when the supplier sells through the agency model on an online platform, and the platform adopts a private credit strategy (where the platform serves as both an online marketplace and a provider of consumer credit services), the supplier simultaneously determines the online (p_o) and offline (p_s) retail prices. A representative AP case is DJI's agency model partnership with Pinduoduo. DJI operates an official store on Pinduoduo using the agency model, where "Duoduofenqi" installment services are exclusively available online. Crucially, DJI's offline flagship stores in China reject Pinduoduo's credit system, relying instead on Alipay Huabei/bank installments—consistent with private credit constraints. The optimization problem is formulated as follows:

$$\underset{p_s,p_o}{\text{Max}}[p_s\alpha(\tfrac{-\tau+p_o-p_s}{\theta} - p_s) + (1 - \phi - r)p_o(\alpha * (1 - \tfrac{-\tau+p_o-p_s}{\theta}) + (1 - \alpha)(1 - \tfrac{-\tau+p_o}{1+\theta}))]$$

By solving the above optimization equation, we obtain the equilibrium results:

$$p_s^{AP} = \frac{(1 - r - \phi)\left(\theta^2(-2 + r + \phi) + \alpha\tau(r + \phi) + \theta(-2 + r + r\tau + \phi + \tau\phi)\right)}{(1 + \theta)(-4\theta + (r + \phi)(4\theta + \alpha(r + \phi)))}$$
$$p_o^{AP} = \frac{2\theta^2(-1 + r + \phi) + 2\theta(1 + \tau)(-1 + r + \phi) + \alpha\tau(r + \phi)}{-4\theta + (r + \phi)(4\theta + \alpha(r + \phi))}$$

By substituting p_s^{AP} and p_o^{AP} into the profit functions of the supplier and the platform, we derive their optimal profits:

$$\Pi_s^{AP} = \frac{(1-r-\phi)\left(\theta^3(-1+r+\phi)+2\theta^2(1+\tau)(-1+r+\phi)+\alpha\tau(r(1+\tau)+\tau(-1+\phi)+\phi)+\theta\left(r-(1+\tau)^2+r\tau(2+\alpha+\tau)+\phi+\tau(2+\alpha+\tau)\phi\right)\right)}{(1+\theta)(-4\theta+(r+\phi)(4\theta+\alpha(r+\phi)))}$$
$$\Pi_o^{AP} = \frac{1}{(1+\theta)(-4\theta+(r+\phi)(4\theta+\alpha(r+\phi)))^2}((r + \phi)(r(1 + \theta)(\alpha + 2\theta) - 2\alpha\tau + 2r(\alpha + \theta)\tau - 2\theta(1 + \theta + \tau) + 2\theta(1 + \theta + \tau)\phi + \alpha(1 + \theta + 2\tau)\phi)(2\theta^2(-1 + r + \phi) + 2\theta(1 + \tau)(-1 + r + \phi) + \alpha\tau(r + \phi)))$$

4.4. Case AO

In the AO case, when the supplier sells through the agency model on the online platform, and the platform adopts an open credit strategy (where the platform serves as both an online marketplace and a credit service provider for both online and offline channels), the supplier simultaneously determines the online (p_o) and offline (p_s) retail prices. The AO model finds perfect implementation in Apple's Tmall partnership. Through Tmall's agency model, Apple directly controls pricing in its flagship store (paying ~5% commission), with "Huabei Installment" services (3/6/12-period interest-free) synchronized across both Tmall online store and Apple's physical stores in China—a textbook example of open credit strategy under agency sales. The optimization problem is formulated as follows:

$$\underset{p_s,p_o}{\text{Max}}(1-r)p_s(\frac{p_o-p_s}{\theta}-(-\tau+p_s))+(1-\phi-r)p_o(1-\frac{p_o-p_s}{\theta})$$

By solving this optimization problem, we obtain the supplier's optimal pricing strategy:

$$p_s^{AO}=\frac{\theta(-1+r+\phi)(2(-1+r)(1+\tau)+\phi)}{-\phi^2+4(-1+r)\theta(-1+r+\phi)}$$

$$p_o^{AO}=\frac{(-1+r)\theta(2(-1+r)(1+\theta+\tau)+(2+2\theta+\tau)\phi)}{-\phi^2+4(-1+r)\theta(-1+r+\phi)}$$

Substituting p_s^{AO} and p_o^{AO} into the profit functions of the supplier and the platform, we derive their optimal profits:

$$\Pi_s^{AO}=-\frac{(-1+r)\theta(-1+r+\phi)\left((-1+r)\left(\theta+(1+\tau)^2\right)+(1+\theta+\tau)\phi\right)}{-\phi^2+4(-1+r)\theta(-1+r+\phi)}$$

$$\begin{aligned}\Pi_o^{AO}=&\frac{1}{(\phi^2-4(-1+r)\theta(-1+r+\phi))^2}(r\theta(-1+r+\phi)^2(2(-1+r)(1+\tau)\\&+\phi)(2(-1+r)\theta\tau-(1+\theta+\tau)\phi)+(-1+r)^2\theta(r+\phi)(2(-1\\&+r)(1+\theta+\tau)+(2+2\theta+\tau)\phi)(\phi+\tau\phi+2\theta(-1+r+\phi)))\end{aligned}$$

4.5. Sensitivity Analysis

In the wholesale model, we conducted a sensitivity analysis of equilibrium prices and profits with respect to the credit service fee rate r and credit utility τ. By differentiating with respect to r and τ, we derived Proposition 1, which reveals how pricing strategies and profit distribution adjust in response to changes in the parameters of consumer credit.

Proposition 1. *When the supplier uses the wholesale model for the online channel, the equilibrium outcomes change as follows with increasing credit service fee rates r and credit utility τ:*

(a) *For the WP case:*

i. $\frac{\partial w^{WP}}{\partial r}=0,\ \frac{\partial p_S^{WP}}{\partial r}=0,\ \frac{\partial p_O^{WP}}{\partial r}=0,\ \frac{\partial \Pi_S^{WP}}{\partial r}=0,\ \frac{\partial \Pi_O^{WP}}{\partial r}=0;$

ii. $\frac{\partial w^{WP}}{\partial \tau}>0,\ \frac{\partial p_S^{WP}}{\partial \tau}=0,\ \frac{\partial p_O^{WP}}{\partial \tau}>0,\ \frac{\partial \Pi_S^{WP}}{\partial \tau}>0,\ \frac{\partial \Pi_O^{WP}}{\partial \tau}>0;$

(b) *For the WO case:*

i. $\frac{\partial w^{WO}}{\partial r}<0,\ \frac{\partial p_S^{WO}}{\partial r}=0,\ \frac{\partial p_O^{WO}}{\partial r}=0,\ \frac{\partial \Pi_S^{WO}}{\partial r}<0,\ \frac{\partial \Pi_O^{WO}}{\partial r}>0;$

ii. $\frac{\partial w^{WO}}{\partial \tau}>0,\ \frac{\partial p_S^{WO}}{\partial \tau}>0,\ \frac{\partial p_O^{WO}}{\partial \tau}>0,\ \frac{\partial \Pi_S^{WO}}{\partial \tau}>0,\ \frac{\partial \Pi_O^{WO}}{\partial \tau}>0.$

Proposition 1 indicates that when the supplier adopts the wholesale model, an increase in the credit service fee rate r will lead to a decrease in the supplier's wholesale price, unless under the open credit strategy. On the other hand, an increase in credit utility τ will lead to an increase in the equilibrium price, except for the offline retail price of the supplier under the private credit strategy. In the wholesale model, the response of profits to credit parameters depends on the power structure: Under the private credit strategy (WP), the supplier dominates the profit allocation, while under the open credit strategy (WO), the platform captures incremental profits through service fees.

Specifically, under the resale model, both parties' profits increase as τ rises. The enhancement of credit utility stimulates consumers' preference for credit payments, resulting in a demand expansion effect. Although the online (or offline) retail prices increase, the reduced demand elasticity makes the simultaneous increase in both quantity and price possible, thus driving profit growth. Under the WP case, both parties' profits are unaffected by the unit credit service fee rate r. In contrast, under the WO case, as r increases, the supplier's profit decreases, while the platform's profit rises. As the provider of credit services, the platform directly gains from the service fee, while the supplier is forced to lower the wholesale price to compensate for channel costs, thereby squeezing their profit.

Through differentiation of the equilibrium outcomes with respect to the credit service fee rate r and credit utility τ, we derive the following proposition:

Proposition 2. *When the supplier adopts the agency model for online sales, an increase in the credit service fee rate and credit utility leads to the following equilibrium changes for the supplier and platform:*

(a) *For the AP case:*

i. $\frac{\partial p_S^{AP}}{\partial r} < 0,\ \frac{\partial p_O^{AP}}{\partial r} > 0,\ \frac{\partial \Pi_S^{AP}}{\partial r} < 0,\ \frac{\partial \Pi_O^{AP}}{\partial r} > 0$

ii. $\frac{\partial p_S^{AP}}{\partial \tau} < 0,\ \frac{\partial p_O^{AP}}{\partial \tau} > 0,\ \frac{\partial \Pi_S^{AP}}{\partial \tau} > 0,\ \frac{\partial \Pi_O^{AP}}{\partial \tau} > 0$

(b) *For the AO case:*

i. $\frac{\partial p_S^{AO}}{\partial r} < 0,\ \frac{\partial p_O^{AO}}{\partial r} > 0,\ \frac{\partial \Pi_S^{AO}}{\partial r} < 0,\ \frac{\partial \Pi_O^{AO}}{\partial r} > 0$

ii. $\frac{\partial p_S^{AO}}{\partial \tau} > 0,\ \frac{\partial p_O^{AO}}{\partial \tau} > 0,\ \frac{\partial \Pi_S^{AO}}{\partial \tau} > 0,\ \frac{\partial \Pi_O^{AO}}{\partial \tau} > 0$

Proposition 2 suggests that when suppliers adopt the agency model on an online platform, the equilibrium price of the online agency channel will consistently increase with the rise in both the platform's consumer credit service fee rate r and credit utility τ. However, the equilibrium price of the offline direct sales channel does not necessarily decrease. When the platform adopts an open credit strategy, an increase in credit utility generally leads to a rise in the supplier's offline direct sales price.

The platform transfers the credit costs to the supplier through a revenue-sharing mechanism. However, the demand-creation effect of credit utility provides a win–win profit space for both parties, especially under the open credit strategy (AO), where price and profit both rise simultaneously. Specifically, the supplier's profit is negatively affected by the increase in r, while the platform's profit benefits. As r increases, the supplier is forced to lower offline prices to sustain demand, while the platform raises online prices and captures service fee revenue, creating a "scissors gap" in profit distribution. On the other hand, when credit utility τ increases, both parties' profits improve. This is because, under the private credit strategy, the demand expansion effect driven by credit utility dominates the profit change: The revenue from increased online sales offsets the losses from offline discounts, creating a "price-for-quantity" synergy effect. Under the open credit strategy, the platform and the supplier jointly leverage the brand premium capacity of credit utility to achieve cross-channel price coordination, prevent demand substitution, and maximize joint profits.

5. Decision Analyses

In this section, we examine the impact of platform credit strategies under two channel models and provide the conditions under which the platform should choose different credit strategies. The decision-making process for selecting a credit strategy is influenced by several factors, including the proportion of low-DPI consumers, credit utility, and the platform's overall pricing strategy. Through a comparative analysis of optimal pricing decisions and profitability under private and open credit strategies in the online wholesale model, we derive Proposition 3.

Proposition 3. *(a)* $w^{WP} > w^{WO}, p_s^{WP} < p_s^{WO}, p_o^{WP} > p_o^{WO}$; *(b)* $D_s^{WP} < D_s^{WO}$, $D_o^{WP} > D_o^{WO}$ *and* $D_s^{WP} + D_o^{WP} < D_s^{WO} + D_o^{WO}$ *always hold.*

Proposition 3 indicates that in the online wholesale model, compared to private credit strategies, an open credit strategy reduces the optimal wholesale and retail prices for online channels, leading to a decrease in the online platform's sales. This decision is primarily driven by the platform's strategy to balance the benefits of increased offline sales with the potential loss in online sales. Conversely, the retail price and sales in offline channels increase. This is because, under a private credit strategy, offline sales are constrained by consumers' disposable income, limiting overall sales. However, with open credit strategies, this constraint is alleviated, enabling more sales in offline channels, which also allows for higher prices.

Regarding the platform, the open credit strategy may weaken its competitiveness, especially among price-sensitive consumers with low disposable income, who may switch to offline channels, reducing online sales. However, as the usage of credit expands, the total market demand increases, as more consumers can now afford the products. Therefore, the platform's decision is not just about short-term pricing but also about long-term market expansion, balancing the trade-off between competitive pressure and demand growth.

By comparing the unit profit of the platform in the wholesale channel under different credit strategies, we derived Proposition 4.

Proposition 4. *Marginal unit profit for the platform in wholesale model: When* $Max[0, \tau_1] < \tau < Min[\tau_2, \tau_5]$, *the unit profit increases* ($p_o^{WP} - w^{WP} < p_o^{WO} - w^{WO}$)*; otherwise, it decreases* ($p_o^{WP} - w^{WP} > p_o^{WO} - w^{WO}$)*.*

Proposition 4 indicates that in the wholesale model, when credit utility is low, the open credit strategy positively impacts the platform's marginal profit by reducing the double marginalization effect in the wholesale channel. However, when credit utility is high, opening credit to offline channels reduces the platform's marginal profit, although additional revenue from credit service fees may offset this loss. The platform's decision is based on maximizing its profit, where it has to weigh the immediate gains from credit service fees against the long-term impact on sales and pricing.

Next, we compare the differences in online and offline prices under private and open credit strategies in the agency model, with results detailed in Proposition 5.

Proposition 5. *(a)* $p_s^{AP} < p_s^{AO}$ *always holds;* $p_o^{AP} < p_o^{AO}$ *holds only when* $0 < \alpha < \alpha_1$, $0 < \tau < \tau_3$ *or* $\alpha_1 < \alpha < 1$, $\tau_6 < \tau < \tau_3$*; otherwise,* $p_o^{AP} > p_o^{AO}$*; (b)* $D_s^{AP} > D_s^{AO}$ *holds only when* $\alpha_2 < \alpha < 1$, $0 < \tau < \tau_7$*; otherwise,* $D_s^{AP} < D_s^{AO}$*; similarly,* $D_o^{AP} < D_o^{AO}$ *holds only when* $\alpha_3 < \alpha < 1$, $0 < \tau < \tau_8$*; otherwise,* $D_o^{AP} > D_o^{AO}$*.*

Proposition 5 indicates that in the agency model, when the platform adopts the open credit strategy, suppliers raise offline prices. This is because with open credit, consumers can use credit to increase their valuation of products, attracting low-income consumers, and allowing suppliers to engage in price discrimination without reducing demand.

For the online channel, the availability of platform credit may change the substitution relationship between online and offline channels. Since online channels already offer credit, prices may remain stable or slightly increase. Specifically, when the proportion of low-DPI consumers is high or credit utility is large, online prices may rise, whereas when low-DPI consumers are fewer or credit utility is small, prices may decrease. Offline channels, enhanced by credit payment options, gain competitiveness, which allows them to match or even exceed online pricing, particularly when offering better customer experiences.

In contrast to the wholesale model, the effect of open credit strategies on equilibrium in the agency model is more complex. In the wholesale model, the availability of credit usually leads to a decrease in online sales and an increase in offline sales. However, in the agency model, when the proportion of low-DPI consumers is high and credit utility is low (i.e., when $1-\alpha$ is large and τ is small), the opening of platform credit may lead to increased online sales and reduced offline sales. This difference arises from the supplier's greater control over pricing in the agency model, where they can adjust online and offline retail prices directly to manage sales. The decision-making process here is more nuanced, as it involves strategic pricing by both the platform and the suppliers, taking into account not only credit utility but also the elasticity of demand.

By comparing platform profits under private and open credit strategies, we derive the conditions for platform decisions, as shown in Proposition 6.

Proposition 6. *The conditions under which the platform opens consumer credit services in different models are as follows. (a) When the supplier adopts the wholesale model on the online channel, (i) when $0<\alpha<\alpha_4$, and $max[0,\tau_1]<\tau<\tau_9$ or $\tau_{10}<\tau<\tau_2$, $\Pi_o^{WP}<\Pi_o^{WO}$, otherwise, $\Pi_o^{WP}>\Pi_o^{WO}$; and (ii) when $\alpha_4<\alpha<1$, and $max[0,\tau_1]<\tau<Min[\tau_{10},\tau_2]$, $\Pi_o^{WP}<\Pi_o^{WO}$, otherwise, $\Pi_o^{WP}>\Pi_o^{WO}$. (b) When the supplier adopts the agency model on the online channel, the platform chooses the open credit strategy only when $\alpha_5<\alpha<1$ and $0<\tau<\tau_{12}$, otherwise, the platform chooses the private credit strategy.*

As illustrated in Figure 2, Proposition 6 shows that in the wholesale model, when the proportion of low-DPI consumers is high (i.e., $0<\alpha<\alpha_4$), the platform does not benefit from the open credit strategy unless the credit utility is moderate, in which case the platform selects the private credit strategy. Conversely, when the proportion of low-DPI consumers is low and credit utility is small, the platform chooses the open credit strategy.

In the agency model, when the proportion of low-DPI consumers is high, the platform does not offer credit services to the supplier's offline channels. However, when the proportion of low-DPI consumers is low, and credit utility is below threshold τ_{12}, the platform chooses the open credit strategy, otherwise the private credit strategy is selected.

We further analyze that when the proportion of low-DPI consumers is low and credit utility is small, regardless of the supplier's sales model, the platform will always choose the open credit strategy. The platform's decision to expand credit services to offline channels reflects a strategic move to increase overall market reach and combat price competition. This is because the higher credit utility and low-DPI consumer ratio make it beneficial for the platform to expand credit services to offline channels, alleviating price competition and generating additional income. However, when credit utility is high, if the platform does not offer credit to offline channels, it gains a competitive advantage over suppliers, forcing them to increase wholesale prices, thereby reducing the platform's marginal profit. Finally, by comparing the conditions for providing credit services to offline channels in the wholesale and agency models, we find that the conditions for extending credit services to offline channels are more relaxed in the wholesale model.

We continue to analyze the first-stage supplier's strategy equilibrium as presented in Proposition 7.

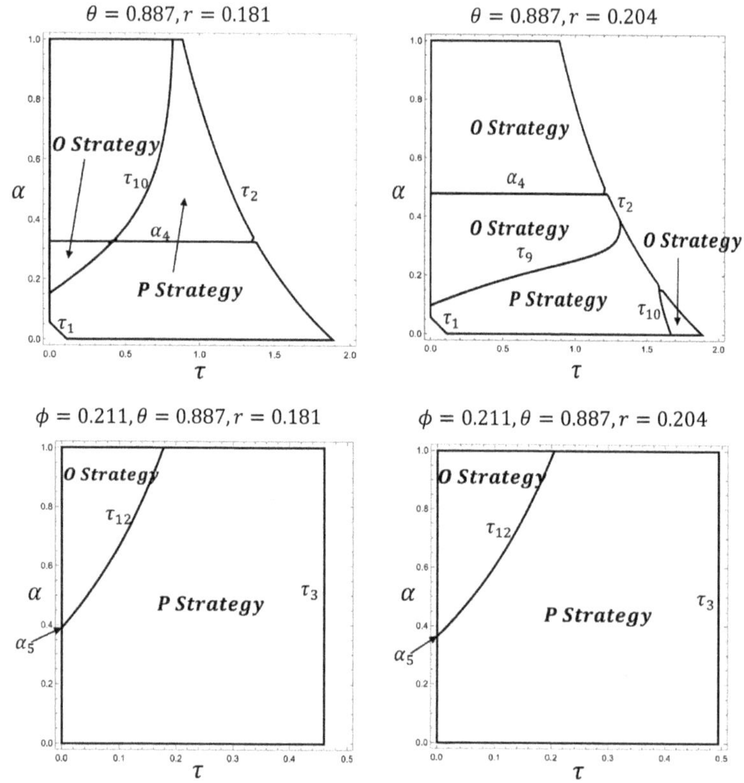

Figure 2. The platform's equilibrium consumer credit strategy under the wholesale and agency models.

Proposition 7. *(a) WP is the equilibrium strategy when* $\alpha_6 < \alpha < 1$, $\tau_{10} < \tau < Min[\tau_3, \tau_{14}]$; *(b) WO is the equilibrium strategy when* $r_2 < r < \phi$ *and* $\alpha_8 < \alpha < 1$, $Max[\tau_{12}, \tau_{17}] < \tau < \tau_{10}$ *or* $\alpha_5 < \alpha < 1$, $Max[0, \tau_{15}] < \tau < \tau_{12}$; *(c) AP is the equilibrium strategy when* $\alpha_8 < \alpha < 1$, $Max[0, \tau_{12}] < \tau < Min[\tau_{10}, \tau_{17}]$, *or* $0 < \alpha < \alpha_6$, $Max[0, \tau_1] < \tau < \tau_3$, *or* $\alpha_6 < \alpha < 1$, $Max[\tau_{10}, \tau_{14}] < \tau < \tau_3$; *and (d) AO is the equilibrium strategy when* $0 < r < r_1$, $\alpha_5 < \alpha < 1$, *and* $0 < \tau < Min[\tau_{12}, \tau_{15}]$.

Proposition 7 analyzes key parameters, including the credit service fee rate r, the proportion of high-DPI consumers α, and credit utility τ, to determine the optimal strategy for suppliers and platforms under different market conditions. As shown in Figure 3, when the proportion of high-DPI consumers is high and credit utility is large, suppliers choose the wholesale model, and the platform chooses the private credit strategy. When the credit service fee rate is high and credit utility is small or moderate, suppliers choose the wholesale model, and the platform chooses the open credit strategy. When the proportion of high-DPI consumers is low or credit utility is very small or large, suppliers choose the agency model, and the platform chooses the private credit strategy. The platform chooses the open credit strategy only when the credit service fee rate is low, and the proportion of high-DPI consumers is high with small credit utility.

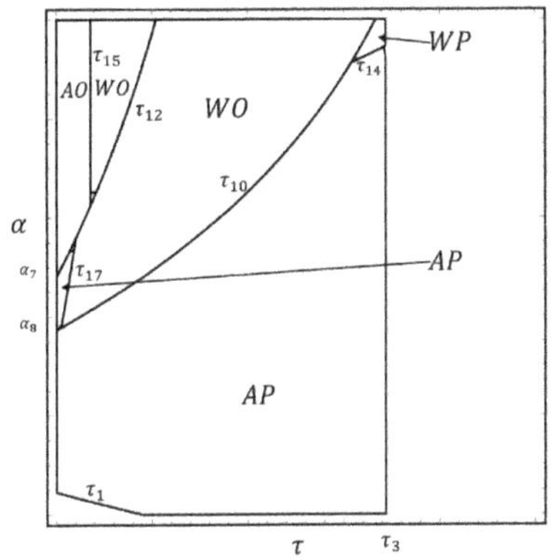

Figure 3. Supplier's equilibrium channel strategy under consumer credit.

6. Conclusions and Discussion

6.1. Conclusions

This study explores the complex relationship between channel strategies and platform credit strategies within a dual-channel supply chain consisting of a supplier and an online platform. The supplier sells products through both offline and online channels, while the platform provides interest-free credit services to consumers. The platform has two distinct credit offering strategies: private and open credit strategies. These strategies are influenced by the supplier's online channel sales model. We analyze four specific scenarios: the wholesale model with a private credit strategy (WP), the wholesale model with an open credit strategy (WO), the agency model with a private credit strategy (AP), and the agency model with an open credit strategy (AO). Using a game theory model, we derive the optimal pricing and strategic decisions for both the supplier and platform under different channel strategies. Through sensitivity analysis, we examine the impact of credit service fees and credit utility on the equilibrium decisions of both the supplier and platform. In particular, the credit service fee does not affect the platform's equilibrium decision. However, under the open credit strategy or the agency model, as the per-unit credit service fee increases, the supplier is incentivized to adjust online and offline prices—raising the online price and lowering the offline price. This helps redirect demand toward the lower-cost channel, reducing overall expenses. An increase in credit utility generally leads to higher equilibrium prices. However, in the agency model with a private credit strategy, an increase in credit utility results in a decrease in the offline sales price. Importantly, the profit allocation mechanism exhibits distinct patterns across channel strategies. Under the wholesale model, the private credit strategy stabilizes supplier profits through fixed contractual terms, while the open credit strategy enables the platform to capture additional gains through service fee extraction. When adopting the agency model, platforms strategically transfer credit costs to suppliers while leveraging credit utility to create mutual profit growth opportunities—a phenomenon most evident in AO scenarios where coordinated price adjustments drive synchronized profit escalation for both parties.

Subsequently, we conducted a comparative analysis, which revealed that the platform's credit strategy significantly influences the allocation of pricing power between the platform and the supplier. In the agency model, suppliers can effectively control channel sales by flexibly adjusting online and offline retail prices, potentially achieving synchronized sales growth across both channels. This contrasts with the wholesale model, where the platform's open credit service leads to decreased online sales and increased offline sales. Additionally, we find that the proportion of high-DPI consumers, denoted as α, and the positive credit utility, denoted as τ, play crucial roles in the platform's willingness to provide credit services to the supplier's offline channels. Specifically, regardless of the vendor's online sales model, when the proportion of low-DPI consumers and the credit utility are both low, the platform will choose to offer credit services through the vendor's offline channels. However, when the proportion of low-DPI consumers is high, the platform will only provide consumer credit services to offline channels under a wholesale model, and only when the credit utility is moderate. Under the wholesale model, the platform's conditions for offering credit services to offline channels are more lenient.

Our study further analyzes the supplier's optimal channel strategy and the platform's best response. The optimal channel and credit strategy are determined by the credit service fee, the proportion of high-DPI consumers, and the credit utility. For suppliers, they are more likely to adopt the wholesale model when the proportion of high-DPI consumers is relatively high and credit utility is moderate, or when both the proportion of high-DPI consumers and credit utility are high. In such cases, the platform decides whether to offer credit services to the offline channel to maximize its own benefit. Both suppliers and platforms need to adjust their sales models and credit strategies flexibly based on changes in the credit service fee, the proportion of high-DPI consumers, and credit utility. This provides a theoretical basis for suppliers and e-commerce platforms to make strategic decisions that consider channel preferences and consumer DPI heterogeneity.

6.2. Managerial Insights

The findings of this study offer valuable practical insights for both platforms and suppliers. As the utility of credit increases, the equilibrium prices for both parties tend to rise. Therefore, platforms should focus on enhancing the quality of their credit services to foster mutually beneficial profit growth. Additionally, changes in the per-unit credit service fee significantly influence pricing and profits. Platforms must adapt their credit strategies according to different models to achieve the optimal equilibrium. Pricing strategies for suppliers and platforms vary across sales models. Platforms should be flexible in adjusting product pricing based on whether the supplier adopts a wholesale or agency sales model to maximize overall profit. For suppliers, choosing an online sales model is influenced by the platform's credit policy. Platforms should design their credit policies thoughtfully, considering factors such as the supplier's product positioning, to encourage suppliers to select sales models that align with the platform's long-term development goals. Under certain conditions, the platform's unit profit will vary depending on the supplier's choice of sales model. Therefore, platforms should closely monitor key parameters like credit service fees and credit utility, adjusting strategies in response to market feedback to ensure optimal profit maximization.

6.3. Limitations of the Work and Future Research

While this study provides valuable insights, several limitations warrant further investigation. Our study has several limitations. First, we do not account for the potential negative utility of consumer credit, such as the failure of credit providers to remind consumers about repayments or the possibility of consumers forgetting to repay, which could damage their credit. Future research could explore these aspects and their impact on platform and supplier strategies. Second, high-credit-risk consumers could significantly impact both the supplier's and platform's channel and credit strategies. Future studies should investigate how credit risk influences strategic decision making, providing a more comprehensive understanding of the dynamics at play. Third, we assume that the platform's consumer credit is the sole option for the supplier's offline channels, without considering alternative credit products such as credit cards. However, a variety of competing consumer credit products could influence the strategic equilibrium between the supplier and platform. Future research could explore how the presence of alternative credit options affects platform–supplier dynamics, providing further insights into optimal credit strategy development.

Funding: This research received no external funding.

Data Availability Statement: Data are contained within the article.

Conflicts of Interest: The author declares no conflicts of interest.

Appendix A

Proof of Equilibrium Results. In the agency model (AP), the supplier simultaneously decides both the online and offline retail prices to maximize its profit. The profit function is

$$\begin{aligned}\max_{p_s,p_o}\Pi_s = p_s\alpha\left(\tfrac{-\tau+p_o-p_s}{\theta} - p_s\right) + (1-\phi)p_o\Big(\alpha\left(1-\tfrac{-\tau+p_o-p_s}{\theta}\right) + (1\\ -\alpha)\left(1-\tfrac{-\tau+p_o}{1+\theta}\right)\Big)\end{aligned}$$

The second-order Hessian matrix AP_1 is

$$AP_1 = \begin{pmatrix} -2\alpha\left(1+\frac{1}{\theta}\right) & \frac{\alpha}{\theta}+\frac{\alpha\phi}{\theta} \\ \frac{\alpha}{\theta}+\frac{\alpha\phi}{\theta} & -2\left(\frac{\alpha}{\theta}+\frac{1-\alpha}{1+\theta}\right)\phi \end{pmatrix}$$

To ensure negative definiteness, the following conditions must be satisfied:

$$\theta_1 < \theta < 1$$

We solve the first-order conditions:

$$\frac{\partial \Pi_s^{AP}}{\partial p_s} = 0, \frac{\partial \Pi_s^{AP}}{\partial p_o} = 0$$

These yield the optimal supplier price decisions:

$$p_s^{AP} = \frac{(1-r-\phi)\left(\theta^2(-2+r+\phi)+\alpha\tau(r+\phi)+\theta(-2+r+r\tau+\phi+\tau\phi)\right)}{(1+\theta)(-4\theta+(r+\phi)(4\theta+\alpha(r+\phi)))}$$

$$p_o^{AP} = \frac{2\theta^2(-1+r+\phi)+2\theta(1+\tau)(-1+r+\phi)+\alpha\tau(r+\phi)}{-4\theta+(r+\phi)(4\theta+\alpha(r+\phi))}$$

To avoid negative demand, we need to satisfy the condition $0 < \tau < \tau_3$, where $\theta_1 = \dfrac{2\phi-3\phi^2}{1-4\phi+4\phi^2}, \tau_3 = \dfrac{r+r\theta+\phi+\theta\phi}{2-r-\phi}$. Similarly, the prices, demand, and profits under the WP, WO, and AO modes can be derived. □

Proof of Proposition 1. When the supplier adopts the wholesale model, and the platform does not provide consumer credit to offline channels, it is evident that the equilibrium prices w^{WP}, p_s^{WP} and p_o^{WP} do not depend on r. This implies that $\dfrac{\partial w^{WP}}{\partial r} = 0$, $\frac{\partial p_s^{WP}}{\partial r} = 0$, $\dfrac{\partial p_o^{WP}}{\partial r} = 0$.

The derivative of w^{WP}, p_s^{WP}, and p_o^{WP} with respect to τ is as follows:

$$\frac{\partial w^{WP}}{\partial \tau} = \frac{1}{2} > 0$$
$$\frac{\partial p_s^{WP}}{\partial \tau} = 0$$
$$\frac{\partial p_o^{WP}}{\partial \tau} = \frac{3}{4} > 0$$

When suppliers adopt a wholesale model and the platform chooses to provide consumer credit to offline channels, by taking the derivative of the equilibrium outcomes with respect to r and τ, we obtain

$$\frac{\partial w^{WO}}{\partial r} = -\frac{1}{2}(1+\tau) < 0$$
$$\frac{\partial p_s^{WO}}{\partial r} = 0$$
$$\frac{\partial p_o^{WO}}{\partial r} = 0$$
$$\frac{\partial w^{WO}}{\partial \tau} = \frac{r}{2} > 0$$
$$\frac{\partial p_s^{WO}}{\partial \tau} = \frac{1}{2} > 0$$
$$\frac{\partial p_o^{WO}}{\partial \tau} = \frac{1}{2} > 0$$

This completes the proof. □

Proof of Proposition 2. When supplier adopt an agency model and the platform chooses not to provide consumer credit to offline channels, by taking the derivative of the equilibrium outcomes with respect to r and τ, we obtain

$$\begin{aligned}\frac{\partial p_s^{AP}}{\partial r} &= -\frac{(1-r-\phi)(4\theta+2\alpha(r+\phi))\left(\theta^2(-2+r+\phi)+\alpha\tau(r+\phi)+\theta(-2+r+r\tau+\phi+\tau\phi)\right)}{(1+\theta)(-4\theta+(r+\phi)(4\theta+\alpha(r+\phi)))^2} \\ &\quad -\frac{(\alpha\tau+\theta(1+\theta+\tau))(1-r-\phi)(4\theta-(r+\phi)(4\theta+\alpha(r+\phi)))}{(1+\theta)(-4\theta+(r+\phi)(4\theta+\alpha(r+\phi)))^2} \\ &\quad -\frac{(\theta^2(2-r-\phi)-\alpha\tau(r+\phi)+\theta(2-r-r\tau-\phi-\tau\phi))(4\theta-(r+\phi)(4\theta+\alpha(r+\phi)))}{(1+\theta)(-4\theta+(r+\phi)(4\theta+\alpha(r+\phi)))^2} \\ &< 0\end{aligned}$$

$$\begin{aligned}\frac{\partial p_o^{AP}}{\partial r} &= \frac{\alpha(-4\theta\tau+4r\theta(1+\theta+\tau)+r^2(\alpha\tau+2\theta(1+\theta+\tau))}{(-4\theta+(r+\phi)(4\theta+\alpha(r+\phi)))^2} \\ &\quad +\frac{4\theta(1+\theta+\tau)\phi-2r(\alpha\tau+2\theta(1+\theta+\tau))\phi-(\alpha\tau+2\theta(1+\theta+\tau))\phi^2)}{(-4\theta+(r+\phi)(4\theta+\alpha(r+\phi)))^2} > 0\end{aligned}$$

$$\frac{\partial p_s^{AP}}{\partial \tau} = -\frac{(\alpha+\theta)(1-r-\phi)(r+\phi)}{(1+\theta)(4\theta-(r+\phi)(4\theta+\alpha(r+\phi)))} < 0$$

$$\frac{\partial p_o^{AP}}{\partial \tau} = \frac{2\theta(1-r-\phi)-\alpha(r+\phi)}{4\theta-(r+\phi)(4\theta+\alpha(r+\phi))} > 0$$

When suppliers adopt the agency model and the platform chooses to provide consumer credit to offline channels, by taking the derivative of the equilibrium outcomes with respect to r and τ, we obtain

$$\frac{\partial p_s^{AO}}{\partial r} = -\frac{\theta\phi\left(4\theta(-1+r+\phi)^2+\phi(4(-1+r)(1+\tau)+(3+2\tau)\phi)\right)}{(\phi^2-4(-1+r)\theta(-1+r+\phi))^2} < 0$$

$$\frac{\partial p_o^{AO}}{\partial r} = \frac{\theta\phi\left(4(-1+r)^2\theta\tau-4(-1+r)(1+\theta+\tau)\phi-(2+2\theta+\tau)\phi^2\right)}{(\phi^2-4(-1+r)\theta(-1+r+\phi))^2} > 0$$

$$\frac{\partial p_s^{AO}}{\partial \tau} = \frac{2(1-r)\theta(1-r-\phi)}{-\phi^2+4(1-r)\theta(1-r-\phi)} > 0$$

$$\frac{\partial p_o^{AO}}{\partial \tau} = \frac{(1-r)\theta(2-2r-\phi)}{-\phi^2+4(1-r)\theta(1-r-\phi)} > 0$$

This completes the proof. □

Proof of Proposition 3. The difference between the wholesale price under the WO and WP models is

$$w^{WO} - w^{WP} = -\frac{1}{2}r(1+\tau) < 0$$

This shows that the wholesale price in the WO model is lower than in the WP model for $r > 0, \tau > 0$.

The difference between the offline price in the WO and WP models is

$$p_s^{WO} - p_s^{WP} = \frac{\tau}{2} > 0$$

The difference between the online price in the WO and WP models is

$$p_o^{WO} - p_o^{WP} = -\frac{\alpha\tau + \theta(1+\theta+\tau)}{4(\alpha+\theta)} < 0$$

For demand,

$$D_s^{WP} = \tfrac{1}{4}\alpha(\tfrac{1+\theta}{\alpha+\theta} - \tfrac{\tau}{\theta}),\ D_o^{WP} = \tfrac{1}{4}(1 + \tfrac{(\alpha+\theta)\tau}{\theta(1+\theta)})$$
$$D_s^{WO} = \tfrac{1}{4}(1+2\tau),\ D_o^{WO} = \tfrac{1}{4}$$

The difference in demand between the WO and WP models is

$$D_s^{WO} - D_s^{WP} = \frac{\alpha^2\tau + 3\alpha\theta\tau + \theta^2(1-\alpha+2\tau)}{\theta(\alpha+\theta)} > 0$$
$$D_o^{WO} - D_o^{WP} = -\frac{(\alpha+\theta)\tau}{4\theta(1+\theta)} < 0$$

This completes the proof. □

Proof of Proposition 4. We define $g_1(\tau)$ as the difference between $p_o^{WO} - w^{WO}$ and $p_o^{WP} - w^{WP}$: $g_1(\tau) = (p_o^{WO} - w^{WO}) - (p_o^{WP} - w^{WP})$.

Substituting the expressions for the prices,

$$g_1(\tau) = \frac{1}{4}\left(\frac{(-1+\alpha)\theta}{\alpha+\theta} - \tau + 2r(1+\tau)\right)$$

To analyze the behavior of $g_1(\tau)$, we compute the derivative of $g_1(\tau)$ with respect to τ:

$$\frac{\partial g_1(\tau)}{\partial \tau} = -\frac{1}{4}(1-2r) < 0$$

This is a constant, indicating that $g_1(\tau)$ is a linear function of τ. Since $\frac{\partial g_1(\tau)}{\partial \tau} < 0$, we know that $g_1(\tau)$ is decreasing in τ.

To find the value of τ where $g_1(\tau) = 0$, solving for τ,

$$\tau_5 = \frac{\theta - \alpha\theta - 2r(\alpha+\theta)}{(-1+2r)(\alpha+\theta)}$$

Thus, for τ values between 0 and τ_5, we have

$$g_1(\tau) > 0$$

For $\tau > \tau_5$, we have

$$g_1(\tau) < 0$$

From the above analysis, we conclude that when τ is between $Max[0, \tau_1] < \tau < Min[\tau_2, \tau_5]$, we have $g_1(\tau) > 0$, meaning $(p_o^{WO} - w^{WO}) - (p_o^{WP} - w^{WP}) > 0$; otherwise, $g_1(\tau) < 0$.

This completes the proof of Proposition 4. □

Proof of Proposition 5. The equilibrium difference in the offline prices is defined as

$$p_s^{AO} - p_s^{AP} = (1 - r - \phi)\left(\frac{\theta(2(1-r)(1+\tau)+\phi)}{-\phi^2 + 4(1-r)\theta(1-r-\phi)} - \frac{\theta^2(2-r-\phi) - \alpha\tau(r+\phi) + \theta(2-r-r\tau-\phi-\tau\phi)}{(1+\theta)(4\theta - (r+\phi)(4\theta+\alpha(r+\phi)))}\right) > 0$$

We define $g_2(\tau)$ as the difference in the online prices for the AO and AP models:

$$g_2(\tau) = p_o^{AO} - p_o^{AP} = \frac{(-1+r)\theta(2(-1+r)(1+\theta+\tau) + (2+2\theta+\tau)\phi)}{-\phi^2 + 4(-1+r)\theta(-1+r+\phi)} - \frac{2\theta^2(-1+r+\phi) + 2\theta(1+\tau)(-1+r+\phi) + \alpha\tau(r+\phi)}{-4\theta + (r+\phi)(4\theta+\alpha(r+\phi))}$$

We now compute the derivative of $g_2(\tau)$ with respect to τ:

$$\frac{\partial g_2(\tau)}{\partial \tau} = \frac{(1-r)\theta(2-2r-\phi)}{-\phi^2 + 4(1-r)\theta(1-r-\phi)} + \frac{2\theta - (\alpha+2\theta)(r+\phi)}{-4\theta + (r+\phi)(4\theta+\alpha(r+\phi))} > 0$$

Since the partial derivative is positive, this implies that $g_2(\tau)$ is increasing in τ.

When $0 < \alpha < \alpha_1$, $g_2(0) > 0$; otherwise, $g_2(0) < 0$. Let $g_2(\tau) = 0$, which gives τ_6. Therefore, when $0 < \alpha < \alpha_1$ or $\alpha_1 < \alpha < 1, \tau_6 < \tau < \tau_3$, $g_2(\tau) > 0$; otherwise, $g_2(\tau) < 0$.

Consider the function $g_3(\tau) = D_s^{AO} - D_s^{AP} = (-1+r+\phi)(\frac{2(-1+r)\theta\tau-(1+\theta+\tau)\phi}{-\phi^2+4(-1+r)\theta(-1+r+\phi)} - \frac{\alpha(-2\tau+r(1+\theta+\tau)+(1+\theta+\tau)\phi)}{-4\theta+(r+\phi)(4\theta+\alpha(r+\phi))})$, where $\frac{\partial g_3(\tau)}{\partial \tau} = (1-r-\phi)\left(\frac{2(1-r)\theta+\phi}{-\phi^2+4(1-r)\theta(1-r-\phi)} + \frac{\alpha(2-r-\phi)}{4\theta-(r+\phi)(4\theta+\alpha(r+\phi))}\right) > 0$ implies that $g_3(\tau)$ is an increasing function of τ. It is given that $g_3(\tau_3) = \frac{2(1+\theta)(-1+r+\phi)(-\phi+(-1+r)\theta(r+\phi))}{(2-r-\phi)(-\phi^2+4(-1+r)\theta(-1+r+\phi))} > 0$. When $0 < \alpha < \alpha_2$, we have $g_3(0) = (1+\theta)(-1+r+\phi)(\frac{\phi}{\phi^2-4(-1+r)\theta(-1+r+\phi)} - \frac{\alpha(r+\phi)}{-4\theta+(r+\phi)(4\theta+\alpha(r+\phi))}) > 0$; otherwise, $g_3(0) < 0$. By setting $g_3(\tau) = 0$, we determine τ_7. Therefore, $g_3(\tau) > 0$ for $0 < \alpha < \alpha_2$ or $\alpha_2 < \alpha < 1, \tau_7 < \tau < \tau_3$; otherwise, $g_3(\tau) > 0$. Where $\alpha_2 = \frac{4\theta\phi(1-r-\phi)}{(r+\phi)(r\phi+4(-1+r)\theta(-1+r+\phi))}$,
$\tau_7 = \frac{(1+\theta)\left(4(-1+r)^2 r\alpha\theta+\left(r^2\alpha+4(-1+r)(1+(-1+2r)\alpha)\theta\right)\phi+(r\alpha+4(1+(-1+r)\alpha)\theta)\phi^2\right)}{(4\theta(2(-1+r)\theta-\phi)(-1+r+\phi)+\alpha(-\phi(r^2+(2+r)\phi)-2(-1+r)\theta(4+r^2+2r(-3+\phi)+(-6+\phi)\phi)))}$.

Finally, we consider the function $g_4(\tau) = D_o^{AO} - D_o^{AP} = \frac{(-1+r)(\phi+\tau\phi+2\theta(-1+r+\phi))}{-\phi^2+4(-1+r)\theta(-1+r+\phi)} - \frac{r(1+\theta)(\alpha+2\theta)-2\alpha\tau+2r(\alpha+\theta)\tau-2\theta(1+\theta+\tau)+2\theta(1+\theta+\tau)\phi+\alpha(1+\theta+2\tau)\phi}{(1+\theta)\left(-4\theta+4\alpha(r+\phi)^2\right)}$. It holds that $g_4(\tau_3) = \frac{(-2+2r+\phi)(-\phi+(-1+r)\theta(r+\phi))}{(-2+r+\phi)(-\phi^2+4(-1+r)\theta(-1+r+\phi))} < 0$ and $\frac{\partial g_4(\tau)}{\partial \tau} = \frac{(-1+r)\phi}{-\phi^2+4(-1+r)\theta(-1+r+\phi)} - \frac{(\alpha+\theta)(-1+r+\phi)}{2(1+\theta)\left(-\theta+\alpha(r+\phi)^2\right)} < 0$ for all τ in the interval $(0, \tau_3)$. When $0 < \alpha < \alpha_3$, we have $g_4(0) = \frac{(-1+r)(\phi+2\theta(-1+r+\phi))}{-\phi^2+4(-1+r)\theta(-1+r+\phi)} + \frac{2\theta-(\alpha+2\theta)(r+\phi)}{-4\theta+(r+\phi)(4\theta+\alpha(r+\phi))} > 0$; otherwise, $g_3(0) < 0$. By setting $g_4(\tau) = 0$, we determine τ_8. Therefore, $g_4(\tau) > 0$ for $\alpha_3 < \alpha < 1$ and $0 < \tau < \tau_8$; otherwise, $g_3(\tau) > 0$.

The proof of proposition 5 is completed. □

Proof of Proposition 6.

Part 1: Wholesale Model (WO and WP)

Let us define the difference in profits between the online platform under the WO and WP models as $f_1(\tau) = \Pi_o^{WO} - \Pi_o^{WP}$.

We are given that

$$f_1(\tau) = A_1\tau^2 + B_1\tau + C_1$$

where the coefficients A_1, B_1 and C_1 are given as

$$\begin{aligned} A_1 &= \tfrac{-\alpha+\theta(-1+4r(1+\theta))}{16\theta(1+\theta)} \\ B_1 &= -\tfrac{1}{8}(1-4r) \\ C_1 &= \tfrac{1}{16}\left(4r+\tfrac{(-1+\alpha)\theta}{\alpha+\theta}\right) \end{aligned}$$

We also know that $f_1(\tau)$ is a quadratic function, and thus, the number of roots of this equation (i.e., the values of τ) depends on the discriminant of the quadratic.

The roots of $f_1(\tau)$ are denoted by τ_9 and τ_{10}.

When $0 < \alpha < \alpha_4$, we have $A_1 > 0$, and the roots satisfy $\tau_9 < \tau_{10}$. In this case, if τ lies between τ_9 and τ_{10}, then $f_1(\tau) < 0$; otherwise, $f_1(\tau) > 0$.

When $\alpha_4 < \alpha < 1$, $A_1 < 0$, and the roots satisfy $\tau_9 < 0 < \tau_{10}$. Here, if τ lies between τ_{10} and τ_2, then $f_1(\tau) < 0$; otherwise, $f_1(\tau) > 0$.

The roots are given by $\tau_{9,10} = \dfrac{\theta(1+\theta)\left(1-4r\pm\sqrt{\dfrac{\theta(\alpha+\theta)^2+4r\left(\alpha^2-3\alpha\theta^2-(1+\alpha)\theta^3\right)}{\theta(1+\theta)(\alpha+\theta)}}\right)}{4r\theta(1+\theta)-\alpha-\theta}$,

where $\alpha_4 = -\theta+4r\theta+4r\theta^2$.

Part 2: Agency Model (AO and AP)

Let us define the profit difference function between the online platform under the AO and AP models as $f_2(\tau) = \Pi_o^{AO} - \Pi_o^{AP}$.

Profit Difference Function $f2(\tau)f2(\tau)$

For the agency model, the profit difference function is

$$f_2(\tau) = A_2\tau^2 + B_2\tau + C_2$$

We also know that

$$\frac{\partial A_2}{\partial \alpha} = \frac{2\theta\phi(-1+\phi^2)\left(4\theta\phi^2+\alpha(-1+\phi)(1+3\phi)\right)}{(1+\theta)\left(-\alpha(-1+\phi)^2+4\theta\phi\right)^3}$$

is negative, indicating that A_2 is decreasing with respect to α.

$A_2(\alpha = 0) < 0$, so, we conclude that $A_2 < 0$, given that $A_2 < 0$, $f_2(\tau_3) < 0$, $f_2(\tau)$ is a downward-opening parabola. Now, we investigate the behavior of $f_2(\tau)$ under different values of α:

When $0 < \alpha < \alpha_5$, $f_2(0) < 0$, we find that $f_2(\tau) < 0$ holds for τ in the range $0 < \tau < \tau_3$.
When $\alpha_5 < \alpha < 1$, $f_2(0) > 0$, and we solve $f_2(\tau) = 0$ to find the roots τ_{11} and τ_{12}.
In this case, $f_2(\tau) > 0$ for $0 < \tau < \tau_{12}$, and $f_2(\tau) < 0$ otherwise.

where $\alpha_5 = \frac{\phi^4+4r\theta\phi^2(-1+r+\phi)^2-16(-1+r)^2\theta^2(-1+r+\phi)^3}{(r+\phi)^2\left(4(-1+r)^3r\theta+4(-1+r)^2(-1+2r)\theta\phi+(-1+r)(-2-r+4(-1+r)\theta)\phi^2-r\phi^3\right)} + \sqrt{\frac{\left((\phi^2-4(-1+r)\theta(-1+r+\phi))^2\left(\phi^4+16(-1+r)^2\theta^2(-1+r+\phi)^4-4\theta\phi^2(-1+r+\phi)^2(-2+r(r+\phi))\right)\right)}{\left((r+\phi)^4\left(4(-1+r)^3r\theta+4(-1+r)^2(-1+2r)\theta\phi+(-1+r)(-2-r+4(-1+r)\theta)\phi^2-r\phi^3\right)^2\right)}}$,

$A_2 = \frac{2(\alpha+\theta)(-1+\phi)\phi(\alpha(-1+\phi)+2\theta\phi)}{(1+\theta)\left(\alpha(-1+\phi)^2-4\theta\phi\right)^2} + \frac{r\theta\left(-r^3+r^3\phi+r(3+4\theta-2r(1+2\theta))\phi^2+(-2+r)\phi^3\right)}{(r^2+\phi^2-2r(\phi+2\theta\phi))^2}$,

$B_2 = \frac{(-1+\phi)\left(\alpha^2(-1+\phi)^2+8\theta^2\phi^2+4\alpha\theta\phi(-1+2\phi)\right)}{\left(\alpha(-1+\phi)^2-4\theta\phi\right)^2} - \frac{\theta\left(r^4+(-2+r)r^3\phi-r^2(-6+r+4(-2+r)\theta)\phi^2-r(2+r+4r\theta)\phi^3+(-1+r)\phi^4\right)}{(r^2+\phi^2-2r(\phi+2\theta\phi))^2}$,

and $C_2 = \frac{\theta(1+\theta)(-1+r+\phi)\left(4(-1+r)^3r\theta+4(-1+r)^2(-1+2r)\theta\phi+(-1+r)(-2-r+4(-1+r)\theta)\phi^2-r\phi^3\right)}{(\phi^2-4(-1+r)\theta(-1+r+\phi))^2} - \frac{2\theta(1+\theta)(-1+r+\phi)(r+\phi)(-2\theta+(\alpha+2\theta)(r+\phi))}{(-4\theta+(r+\phi)(4\theta+\alpha(r+\phi)))^2}$.

This completes the proof of Proposition 6. □

Proof of Proposition 7.

Case 1: WP and AP Region

Since $\alpha_4 < \alpha_5$, the intersection region where both WP and AP can coexist is

$$\alpha_5 < \alpha < 1, Max[\tau_{10}, \tau_{12}] < \tau < \tau_3$$

Also, when $\alpha > \frac{\theta - 4r\theta}{4r+\theta}$, we have $\tau_{10} > 0$.

We define the profit difference function for the supplier:

$$f_3(\tau) = \Pi_s^{WP} - \Pi_s^{AP} = A_3\tau^2 + B_3\tau + C_3$$

where

$$A_3 = \frac{(\alpha+\theta)(4\theta + (r+\phi)(-12\theta + (\alpha+8\theta)(r+\phi)))}{8\theta(1+\theta)(-4\theta + (r+\phi)(4\theta + \alpha(r+\phi)))} < 0$$

This means that $f_3(\tau)$ is a quadratic function opening downward. The roots of this function, τ_{13} and τ_{14}, are derived from solving $f_3(\tau) = 0$.

When $\alpha_6 < \alpha < 1$, $\tau_{14} > 0$, and $f_3(\tau) > 0$ when $0 < \tau < \tau_{14}$. Thus, in this region, $\Pi_s^{WP} > \Pi_s^{AP}$, and when $\tau > \tau_{14}$, $\Pi_s^{WP} < \Pi_s^{AP}$.

Thus, the equilibrium strategy for supplier in the WP and AP region is WP when $\alpha_6 < \alpha < 1, 0 < \tau < \tau_{14}$, where $\alpha_6 = \frac{8\theta - 9r\theta - 9\theta\phi}{4(r+\phi)} - \frac{1}{4}\sqrt{\frac{1}{(r+\phi)^2}(32\theta^2 - 48r\theta^2 + 17r^2\theta^2 - 48\theta^2\phi + 34r\theta^2\phi + 17\theta^2\phi^2)}$

Case 2: WP and AO Region

Since $\alpha_4 < \alpha_5$, and $\tau_{12} < \tau_{10}$, the intersection between WP and AO is empty, meaning that the WP and AO strategies do not overlap in this region.

Thus, there is no equilibrium in the WP and AO region.

Case 3: WO and AP Region

When $0 < \alpha < \alpha_8$, we have $\tau_9 < 0$ and $\tau_{10} > \tau_3$, meaning that the WO and AP regions only overlap in areas with lower DPI consumer proportions.

Let the profit difference function be $f_5(\tau) = \Pi_s^{WO} - \Pi_s^{AP} = \frac{1}{8}\left(\theta + 2(1+\tau)^2 - 2r(1+\tau)^2\right) + \frac{(-1+r+\phi)\left(\theta^3(-1+r+\phi) + 2\theta^2(1+\tau)(-1+r+\phi) + \alpha\tau(r(1+\tau) + \tau(-1+\phi) + \phi) + \theta\left(r - (1+\tau)^2 + r\tau(2+\alpha+\tau) + \phi + \tau(2+\alpha+\tau)\phi\right)\right)}{(1+\theta)(-4\theta + (r+\phi)(4\theta + \alpha(r+\phi)))}$.

When $0 < \alpha < \alpha_8$, $f_5(0) = \frac{1}{8}(2 - 2r + \theta) + \frac{\theta(1+\theta)(-1+r+\phi)^2}{-4\theta + (r+\phi)(4\theta + \alpha(r+\phi))} > 0$, and $f_5(\tau)$ is monotonically increasing in the interval $(0, \tau_3)$, so WO is always the equilibrium strategy in this region.

When $\alpha_8 < \alpha < 1$, if $0 < r < r_2$, then $f_5(\tau_3) < 0$; if $r_2 < r < \phi$, then $f_5(\tau_3) > 0$. In this case, for $r_2 < r < \phi$ and $\alpha_8 < \alpha < 1$, there exists a point τ_{17} in the interval $(0, \tau_3)$ such that $f_5(\tau_{17}) = 0$.

When $Max[\tau_{12}, \tau_{17}] < \tau < \tau_{10}$, $f_5(\tau) > 0$, meaning WO is the equilibrium strategy. Otherwise, if the condition is not met, AP is the equilibrium strategy.

Where $r_2 = Root\left[4\theta - 8\phi - 12\theta\phi - \theta\phi^2 - 2\theta^2\phi^2 + (-12\theta + 6\theta\phi - 4\theta^2\phi + 2\theta^2\phi^2)\#1 + (7\theta - 2\theta^2 + 4\theta^2\phi)\#1^2 + 2\theta^2\#1^3\&, 2\right]$, r_2 is the solution of $f_5(\tau_3) = 0$ in the interval $(0, \phi)$, $\alpha_8 = \frac{4\theta(-1+r+\phi)(2\phi + \theta(-1+2r+2\phi))}{(-2+2r-\theta)(r+\phi)^2}$

Case 4: WO and AO Region

We define the profit difference function in the WO and AO region as

$$f_4(\tau) = \Pi_s^{WO} - \Pi_s^{AO} = A_4\tau^2 + B_4\tau + C_4$$

where $A_4 = \frac{(1-r)\phi^2}{4\phi^2-16(1-r)\theta(1-r-\phi)} < 0$. The roots of $f_4(\tau) = 0$ are τ_{15} and τ_{16}, with $\tau_{16} > \tau_3$ always. If $0 < r < r_1$, then $\tau_{15} > 0$, and $\Pi_s^{WO} > \Pi_s^{AO}$ holds when τ is between τ_{15} and τ_3. Thus, when $\alpha_5 < \alpha < 1$ and $\max[0, \tau_{15}] < \tau < \tau_{12}$, the equilibrium strategy is WO. WP is the equilibrium strategy only when $0 < r < r_1$, $\alpha_5 < \alpha < 1$, and $0 < \tau < \tau_{15}$, where $\tau_{15,16} = \frac{2\theta-2r\theta-\phi-2\theta\phi}{\phi} \mp \frac{\sqrt{\frac{-4\theta^2+8r\theta^2-4r^2\theta^2+4\theta^2\phi-4r\theta^2\phi+\theta\phi^2}{(-1+r)\phi^2}}}{\sqrt{2}}$, $r_1 = Root\left[-4\theta^2+8\theta\phi+12\theta^2\phi-2\phi^2-9\theta\phi^2-8\theta^2\phi^2+\left(16\theta^2-16\theta\phi-28\theta^2\phi+2\phi^2+8\theta\phi^2+8\theta^2\phi^2\right)\#1+\left(-20\theta^2+8\theta\phi+16\theta^2\phi\right)\#1^2+8\theta^2\#1^3\&,1\right]$, and r_1 is the solution of $\tau_{15} = 0$ in the interval $(0, \phi)$.

This completes the proof of Proposition 7. □

References

1. Allen, F.; Gale, D. Competition and Financial Stability. *J. Money Credit Bank.* **2004**, *36*, 453–480. [CrossRef]
2. Feinberg, R.A. Credit Cards as Spending Facilitating Stimuli: A Conditioning Interpretation. *J. Consum. Res.* **1986**, *13*, 348–356. [CrossRef]
3. Soman, D. Effects of Payment Mechanism on Spending Behavior: The Role of Rehearsal and Immediacy of Payments. *J. Consum. Res.* **2001**, *27*, 460–474. [CrossRef]
4. Hagiu, A.; Wright, J. Marketplace or Reseller? *Manag. Sci.* **2015**, *61*, 184–203. [CrossRef]
5. Abhishek, V.; Jerath, K.; Zhang, Z.J. Agency Selling or Reselling? Channel Structures in Electronic Retailing. *Manag. Sci.* **2016**, *62*, 2259–2280. [CrossRef]
6. Prelec, D.; Simester, D. Always Leave Home Without It: A Further Investigation of the Credit-Card Effect on Willingness to Pay. *Mark. Lett.* **2001**, *12*, 5–12. [CrossRef]
7. Hu, J.; Huang, Y.; Liu, J. The changing face of consumer credit: Evidence from a big tech platform in China. *Pac.-Basin Finance J.* **2024**, *83*, 102254. [CrossRef]
8. Xie, X.; Hu, P.; Yu, J.; Dai, B. Impact of capacity on the supplier's distribution channel selection in facing a retail platform. *Nav. Res. Logist. NRL* **2021**, *68*, 837–854. [CrossRef]
9. Liu, Y.; Zhang, D. Intraconsumer Price Discrimination with Credit Refund Policies. *Manag. Sci.* **2024**, *70*, 6835–6851. [CrossRef]
10. Zhao, C.; Wu, J. Online installment payments and price guarantees under randomized pricing. *Decis. Support Syst.* **2024**, *178*, 114127. [CrossRef]
11. Xie, J.; Wei, L.; Zhu, W.; Zhang, W. Platform supply chain pricing and financing: Who benefits from e-commerce consumer credit? *Int. J. Prod. Econ.* **2021**, *242*, 108283. [CrossRef]
12. Wu, H.; Zheng, H.; Zhang, M. Credit payment services and pricing strategy in the digital economy era. *Ann. Oper. Res.* **2022**, 1–18. [CrossRef]
13. Feng, Y.; Zhang, Z.; Yao, Y.; Chen, Y.; Liu, J. To Share or Not? Hybrid e-Commerce Platform's Provision and Sharing Strategies of Consumer Credit Service. *Manag. Decis. Econ.* **2025**, *46*, 910–927. [CrossRef]
14. Exler, F.; Livshits, I.; MacGee, J.; Tertilt, M. Consumer Credit with Over-Optimistic Borrowers. *J. Eur. Econ. Assoc.* **2024**, jvae057. [CrossRef]
15. Cervellati, E.M.; Filotto, U.; Sgrulletti, D.; Stella, G.P. Buy now, pay later consumer credit behavior: Impacts on financing decisions. *Qual. Res. Financ. Mark.* 2025; *ahead-of-print*. [CrossRef]
16. Proprietarily or Open? Online Credit Offering Strategy for a Platform-Based Retailer. Available online: https://wvpn.ustc.edu.cn/https/77726476706e69737468656265737421f9f244993f20645f6c0dc7a59d50267b1ab4a9/document/10292498 (accessed on 18 December 2024).
17. Wei, L.; Xie, J.; Zhu, W.; Li, Q. Pricing of platform service supply chain with dual credit: Can you have the cake and eat it? *Ann. Oper. Res.* **2023**, *321*, 589–661. [CrossRef]
18. Adoption strategies of E-commerce consumer credit services: A dual-channel competition perspective. *Int. J. Prod. Econ.* **2024**, *277*, 109387. [CrossRef]
19. Li, Q.; Zha, Y.; Li, L.; Yu, Y. The Platform's Credit-Offering Strategy in the Presence of Integrated and Independent Systems. *IEEE Trans. Syst. Man Cybern. Syst.* **2022**, *52*, 2933–2944. [CrossRef]
20. Ha, A.Y.; Tong, S.; Wang, Y. Channel Structures of Online Retail Platforms. *Manuf. Serv. Oper. Manag.* **2022**, *24*, 1547–1561. [CrossRef]
21. Gong, C.; Ignatius, J.; Song, H.; Chai, J.; Day, S.J. The Impact of Platform's Information Sharing on Manufacturer Encroachment and Selling Format Decision. *Eur. J. Oper. Res.* **2024**, *317*, 141–155. [CrossRef]

22. Li, J.; Wang, H.; Shi, V.; Sun, Q. Manufacturer's choice of online selling format in a dual-channel supply chain with green products. *Eur. J. Oper. Res.* **2024**, *318*, 131–142. [CrossRef]
23. Bian, Y.; Wang, H.; Yan, S.; Han, X. Platform channel strategy of selling format: Effect of asymmetric service information. *Transp. Res. Part E Logist. Transp. Rev.* **2024**, *185*, 103508. [CrossRef]
24. Chen, L.; Nan, G.; Li, M.; Feng, B.; Liu, Q. Manufacturer's online selling strategies under spillovers from online to offline sales. *J. Oper. Res. Soc.* **2023**, *74*, 157–180. [CrossRef]
25. Ke, W.; Zhou, X.; Lev, B.; Zhang, K. Agency, reselling, or hybrid mode? Competing with store brand. *Transp. Res. Part E Logist. Transp. Rev.* **2024**, *184*, 103487. [CrossRef]
26. Tian, L.; Vakharia, A.J.; Tan, Y.; Xu, Y. Marketplace, Reseller, or Hybrid: Strategic Analysis of an Emerging E-Commerce Model. *Prod. Oper. Manag.* **2018**, *27*, 1595–1610. [CrossRef]
27. Liu, Y.; Lu, J.; Yan, N. Agency or reselling? Supplier's online channel strategies with platform financing. *Eur. J. Oper. Res.* **2024**, *318*, 1014–1027. [CrossRef]
28. Zhen, X.; Xu, S.; Li, Y.; Shi, D. When and how should a retailer use third-party platform channels? The Impact of spillover effects. *Eur. J. Oper. Res.* **2022**, *301*, 624–637. [CrossRef]
29. Yoo, W.S.; Lee, E. Internet Channel Entry: A Strategic Analysis of Mixed Channel Structures. *Mark. Sci.* **2011**, *30*, 29–41. [CrossRef]
30. Chiang, W.K.; Chhajed, D.; Hess, J.D. Direct Marketing, Indirect Profits: A Strategic Analysis of Dual-Channel Supply-Chain Design. *Manag. Sci.* **2003**, *49*, 1–20. [CrossRef]
31. Cattani, K.; Gilland, W.; Heese, H.S.; Swaminathan, J. Boiling Frogs: Pricing Strategies for a Manufacturer Adding a Direct Channel that Competes with the Traditional Channel. *Prod. Oper. Manag.* **2006**, *15*, 40–56. [CrossRef]
32. Bhunia, S.; Das, S.K.; Jablonsky, J.; Roy, S.K. Evaluating carbon cap and trade policy effects on a multi-period bi-objective closed-loop supply chain in retail management under mixed uncertainty: Towards greener horizons. *Expert Syst. Appl.* **2024**, *250*, 123889. [CrossRef]
33. Jiang, B.; Jerath, K.; Srinivasan, K. Firm Strategies in the "Mid Tail" of Platform-Based Retailing. *Mark. Sci.* **2011**, *30*, 757–775. [CrossRef]
34. Barman, H.; Pervin, M.; Roy, S.K.; Weber, G.-W. Analysis of a dual-channel green supply chain game-theoretical model under carbon policy. *Int. J. Syst. Sci. Oper. Logist.* **2023**, *10*, 2242770. [CrossRef]
35. Gilbert, S.M.; Hotkar, P.; Liu, C. Channel Choice via On-Line Platform. *Prod. Oper. Manag.* **2024**, *33*, 1373–1392. [CrossRef]
36. Van Crombrugge, M.; Breugelmans, E.; Gryseels, F.; Cleeren, K. How Retailers Change Ordering Strategies When Suppliers Go Direct. *J. Mark.* **2024**, *89*, 00222429241266576. [CrossRef]
37. Song, W.; Chen, J.; Li, W. Spillover Effect of Consumer Awareness on Third Parties' Selling Strategies and Retailers' Platform Openness. *Inf. Syst. Res.* **2021**, *32*, 172–193. [CrossRef]

mathematics

MDPI

Article

Blockchain for Mass Customization: The Value of Information Sharing Through Data Accuracy by Contract Coordination

Zhening Ye [1,2], Jie Wang [1] and Huida Zhao [3,4,*]

1 Transportation Engineering College, Dalian Maritime University, Dalian 116026, China; yehchening@gmail.com (Z.Y.); dlwjie@163.com (J.W.)
2 Department of Maritime Sciences, Kobe University, 5-1-1 Fukaeminami-machi, Higashinada-ku, Kobe 658-0022, Japan
3 College of Management, Shenzhen University, Shenzhen 518061, China
4 Shenzhen International Maritime Institute, Shenzhen 518083, China
* Correspondence: lskjd003@gmail.com

Abstract: This study provides the game-theoretical framework to investigate the relationship between the blockchain service and mass customization in the environment of information sharing and contract coordination. Specifically, we construct the game-theoretical models of the manufacturer and the retailer to discuss the optimal strategy of information sharing by the retailer in the case of mass customization. The result explores the conditions of information sharing for the retailer because she understands the end market information of nearby consumers. This discussion helps us to understand that the motivation of the manufacturer pays for the retailer's construction of a blockchain system in the case of two types of products, such as a standard product and a customization product. Finally, we use the method of contract coordination to obtain the optimal strategy. Results reveal that information costs significantly impact sharing decisions, and cost-sharing contracts can incentivize retailers to share market data. This study has two main contributions. On the one hand, this study adopts the blockchain service for mass customization by supporting contract coordination, showing the technical value of avoiding false information and tampering-proof. On the other hand, although big data has the same information sharing function, this technology can't play the role of secure data transmission. In order to increase the accuracy of information sharing, we analyze the fusion results of two technologies in the aspect of increasing the accuracy of data sharing, which better reveals the technical value.

Keywords: mass customization; information sharing; blockchain service; contract coordination; relevant stakeholders

MSC: 90B06

1. Introduction

1.1. Background and Motivation

The emergence of blockchain services has promoted changes in the production mode, increased the flexibility of the production process, and reduced the problem of limited production quantity of the product line [1,2]. Crucially, this change can enable mass customization in supply chain activities [3]. As one of the production modes, mass customization allows the customers to design a product according to their needs while keeping costs closer to those of the mass-produced products [4–7]. Therefore, a key distinction

 https://doi.org/10.3390/math13030404

between mass customization and traditional production lies in its dependence on personalized consumer data, which guide production and marketing decisions after customer needs are specified [8].

Retailers at the forefront of consumer interaction have a natural advantage in collecting and understanding detailed market information, such as preferences, purchasing behavior, and demand trends. However, despite the strategic importance of such data, information sharing between retailers and manufacturers often faces significant challenges [9]. These challenges arise from differences in perceptions of data ownership, concerns about data privacy, and the costs associated with data sharing. Retailers may also fear losing their competitive advantage if critical market insights are disclosed to manufacturers. Consequently, this lack of trust and alignment between supply chain participants undermines the efficiency of mass customization processes.

Blockchain technology offers a robust solution to these challenges. It has key features that can enhance data security and reliability, including immutability, transparency, and decentralized trust, thereby fostering trust in information sharing [10]. By leveraging blockchain services, manufacturers can create a secure and tamper-proof environment for sharing sensitive consumer data, addressing retailers' concerns about data misuse or accuracy. Moreover, blockchain-enabled smart contracts can be used to design innovative cost-sharing mechanisms and align the interests of both parties, ensuring that retailers are adequately incentivized to share information, such as mixed technologies.

In traditional customized production, there are some problems related to the security and reliability of information transmission. Furthermore, data transmission between humans and machines is insufficient [11]. Compared with this, blockchain services can increase the security and credibility of information sharing. However, the manufacturer may be hesitant to obtain customized data from retailers to gain a market advantage [12]. Rather, by providing blockchain services to retailers, manufacturers can gain information superiority and, thus increase market efficiency. Therefore, due to the dynamic and profitable information, they need a reasonable contract to realize information sharing between both parties. This study investigates the potential of blockchain technology to enable more effective collaboration in mass customization supply chains. Specifically, this study seeks to answer the following critical questions:

Question 1: What is the impact of the information cost on the sharing behavior via blockchain services under mass customization?

Question 2: How can a manufacturer achieve reasonable contract coordination using the retailer's market information to obtain benefits?

Question 3: How can a manufacturer's positive interests be realized through new technology (such as big data)?

1.2. Major Findings

To answer the aforementioned questions, we construct a game-theoretical model to explore the optimal strategy for information sharing via supporting blockchain services in a mass customization environment. Our main results are as follows:

First, this study investigates the influence of information costs on sharing behavior. The results show that the retailer must pay an information cost for each product when sharing information. In other words, nothing can be carried out too early. The retailer's information sharing behavior increases the cost of the unit product.

Second, if the retailer does not receive any benefits from the manufacturer, it will not share the information on the terminal market. Instead, through contract coordination between themselves, the manufacturer and the retailer can achieve the goal of information sharing and obtain benefits from the cost-sharing contract, discount coupons, and wholesale price discounts.

Finally, we analyze the stability of the results from different aspects, including unreal product quantity in information sharing, changes in the utility function in the baseline analysis, optimal strategy in the case of an uncovered market, optimal strategy in the case of a product with a delay strategy, and optimal strategy with additional new technology. By supporting blockchain services and big data technology, these discussions can increase our understanding of mass customization information sharing.

1.3. Contribution Statement, Feature, and Structure

The contributions of this study are twofold: On the one hand, this study adopts the blockchain service for mass customization by supporting contract coordination, which shows the technical value of avoiding false information and tampering. This highlights the relationship between blockchain services and mass customization. On the other hand, although big data has the same function as information sharing, this technology cannot play a role in data security transmission. To improve the accuracy of information sharing, this study analyzes the effects of two technologies on improving the accuracy of data sharing in a mass customization environment, thus better revealing the value of these two technologies.

This study has two main features. First, this study considers blockchain services in mass customization to increase the accuracy of information sharing, thus highlighting the technical value of these services. Specifically, we demonstrate that this service mode improves the customer experience by helping firms to better meet customers' shopping needs. Second, this study considers supply chain cooperation through contract design. We find that blockchain technology can improve information accuracy while reducing the proportion of false information, thereby highlighting the value of blockchain in avoiding false problems.

The structure of this paper is as follows: Section 2 presents the literature review. Section 3 introduces the model setting. Section 4 constructs our basic analysis from different aspects of the equilibrium, including the influence of blockchain technology on different factors, the influence of information cost on sharing behavior, profits for different shareholders, and contract coordination for the equilibrium strategy. Section 5 makes some extensions by considering the cases of unreal product quantity in information sharing, changes in the utility function in the basic analysis, optimal strategy in the case of the uncovered market, optimal strategy in the case of a product with a delay strategy, and optimal strategy in the case of additional new technology. Some results are discussed in Section 6. Finally, the relevant conclusions and management implications are presented in Section 7. To show the results, we offer brief proofs of the propositions in the Appendix B.

2. Literature Review

We review the related literature, including mass customization, information sharing, and contract coordination, and then discuss the research gap.

2.1. Mass Customization in Supply Chain Management

Here, we discuss mass customization in the literature. Specifically, we review the literature to find related content that describes mass customization in supply chain management. Among extant studies, Kouhpayeh et al. investigate the role on the trust-building element in the marketing activity by the support of blockchain technology [13]. Qu et al. propose a framework of asynchronous federated learning with blockchain technology in order to promote markets' shift toward mass customization [14]. Guo et al. explore the application of 3D printing in mass customization projects [15]. Yetis et al. propose a reliable and optimized framework for mass customization to fully exploit the positive role of blockchain, the Internet of Things, and cyber-physical systems in adopting a personalized customization production mode [1]. Jain et al. attempt to establish a theory of mass customization in manufacturing units [16]. Longo et al. propose a two-stage design process based on a platform to support clothing brands in implementing mass customization strategies [17]. Qi et al. empirically examine the influence of lean and agile practices, mass customization, and product innovation capability on service implementation [18]. Shao develops an analytical model to study the optimal production strategy for mass customization in a market in which customers have different preferences for product types [19]. Wu et al. reveal the important responsibility and role of information-sharing practices in coordinating suppliers' modular practices [20].

Overall, our study is most closely related to that of [1], who analyze the relationship between mass customization and blockchain services. However, few quantitative studies have examined this relationship. Here, we consider the information-sharing value of blockchain services in mass customization, similar to [20]. Furthermore, in the aspect of mass customization, blockchain could increase the trust foundation of trusted customers. Our research is more about the trust of blockchain under information sharing.

2.2. Information Sharing with Blockchain Service

Next, the literature demonstrates the technical value of information sharing via blockchain services. For example, Qu et al. study a new framework of information sharing with an asymmetric encryption transmission method by the support of blockchain technology [21]. Xu et al. use the game-theoretical model to analyze the role of information sharing by blockchain technology in the case of offline and e-platform channels under the cap-and-trade scheme, which shows the technical value of traceability and high transparency [22]. Zheng et al. study the risk decision-making problem faced by spacecraft supply chain participants based on information sharing via blockchain services [23].

Taking advantage of blockchain services in information recording and sharing, Du et al. propose a new business process and construction scheme for a medical information-sharing platform based on blockchain services [24]. Dwivedi et al. design a scheme based on blockchain services to securely share information in a pharmaceutical supply chain system using an intelligent contract and consensus mechanism [25]. Wang et al. propose an information management framework for prefabrication supply chains based on blockchain services, which is helpful for the timely delivery of prefabricated components and tracking the causes of disputes related to these components [26].

Based on theoretical research on supply chain information asymmetry and collaboration, Si et al. analyze the lightweight security framework for information sharing via the Internet of Things based on blockchain technology and a dynamic game method of node cooperation to prevent malicious acts of local domination [27].

Although some studies analyze the necessity of blockchain technology in information sharing, few studies highlight the useful conditions for blockchain services in an environment of information sharing for supply chain members, especially the tamper-proof and accuracy characteristics of blockchain services. Our study explores the condition of using blockchain technology in the case of information sharing, which could show the value of blockchain technology on the environment of supply chain activity.

2.3. Contract Coordination in Supply Chain Management

The literature discusses various types of contracts, including cost sharing, delay, and revenue-sharing contracts. For example, Crettez et al. study how to coordinate the benefits of vertical supply chain members through a revenue sharing contract with wholesale price and fixed sales revenue share [28]. Quadir et al. investigate the result that sharing demands information in competing supply chains with greening efforts [29]. Gago-Rodríguez et al. (2021) discuss the extent to which asymmetric bargaining power among supply chain members mitigates the impact of environmental delay costs on negotiation results [30]. Hosseini et al. (2021) investigate the effects of discounts and advertisements on aggregate demand and profits in the ecotourism supply chain [31]. Zhang et al. compare the optimal green decision and profit under a single- or dual-channel strategy with and without green investment, and determine the optimal channel strategy for the supply chain under prepayment financing [32]. He et al. use cost allocation to explain that consumers' low-carbon preference, the marginal profit of chain members, and corporate social responsibility behavior significantly influence the optimal solution [33].

To optimize the revenue of supply chain members under decentralized decision making, Wu et al. construct a revenue-sharing contract between a recycling center and a third-party recycler and, finally, realize the overall coordination of the supply chain [34]. Considering the reference emissions and cost-learning effects, Yu et al. study Stackelberg differential games with manufacturers as leaders and retailers as followers [35]. Wang et al. indicate that implementing a cost-sharing contract can improve the carbon emission reduction level, product quantity, and supply chain profit [36]. Fan et al. study the impact of the product liability cost allocation mechanism caused by quality defects on product quality and pricing decisions, and the equilibrium profitability of supply chain members and the entire supply chain [37]. Gilotra et al. discuss the impact of human error checks on emissions costs, transportation costs, and deferred payments in inventory management [38]. Raza et al. propose a supply chain coordination scheme for pricing inventories and a corporate social responsibility investment decision for a single manufacturer–retailer supply chain [39].

Yet, there is little discussion on contract coordination in the case of information sharing via supporting blockchain services. Here, we reveal the value of contract coordination for information sharing via the use of blockchain services in supply chain management.

2.4. Discussion of the Existing Literature

The literature review shows that, first, mass customization helps improve customer experiences rather than standard production. However, it is difficult to obtain timely and accurate market information for mass customization, such as customer habits and shopping experiences. Here, the emergence of blockchain services can alleviate this situation by providing supporting technical features, such as tamper prevention and timeliness. Indeed, extant studies rarely consider the relationship between blockchain services and mass customization based on quantitative research. Here, we explore the conditions for using blockchain services to improve the usefulness of information sharing.

Second, while studies have considered contract coordination in supply chain management, few have investigated the impact of contract coordination on blockchain service applications, particularly for different types of contract coordination. Meanwhile, we discuss contract coordination between mass customization and blockchain services in information sharing. This provides a reasonable strategy for applying blockchain services. In summary, we provide specific conditions for implementing blockchain services in information sharing. The results reveal the practical value of blockchain services in mass customization

3. Model

3.1. Scenario Description

Consider a supply chain with a manufacturer and retailer each. The retailer enjoys market superiority due to it being near the end market. Thus, it can capture more information about customers, such as their shopping preferences and search habits. Correspondingly, the manufacturer can obtain market information from the retailer's information sharing via contract coordination. However, the retailer has the right to decide whether to share the market information with the manufacturer. If the retailer chooses exclusive market information, the scenario is marked N as a strategy. Otherwise, the scenario is marked Y, representing the strategy when the retailer shares market information. To obtain the optimal strategy of information sharing by firms, we consider the following types of contracts: cost-sharing contract (strategy Y-C), discount coupons (strategy Y-S), and discount of wholesale price (strategy Y-D).

Next, we also consider two types of products: standard product A and customized product B. Specifically, the manufacturer produces two types of products. For a standard product, the manufacturer can use their own market data to determine the production plan. Meanwhile, for customized products, they need data from customers to determine how to produce products to satisfy customer requirements in a mass customization environment. Finally, based on the product characteristics, the retailer sells these two types of products in the market.

3.2. Model Assumptions

We make some assumptions in our analysis:

(1) According to the literature [40,41], the retailer offers two types of products to end consumers. Specifically, the consumers' perceived value is V; their utility from standard product A and customized product B offerings are $U_A = V - p_A$ and $U_B = \theta V - p_B + t$, respectively. Note that the decision variables p_A and p_B are the two types of product prices in the environment of standard production and mass customization, respectively. The parameter $\theta \in (0,1)$ represents the valuation difference between products A and B. This design is based on product A to analyze the competitive relationship between customized products and standard products. Finally, the parameter t represents the accuracy level of data sharing by the blockchain system supporting retailer.

(2) The retailer has access to a demand signal Γ_i, which is an unbiased estimator Y, and decides whether to share it with the manufacturer before the demand is observed. According to [42,43], the signal accuracy is defined as t and demand is forecasted as follows: $E[Y|\Gamma_i] = E\left[\Gamma_j|\Gamma_i\right] = \frac{t_i\sigma^2}{1+t_i\sigma^2}\Gamma_i$. σ is the variance of market information. This information structure is common knowledge.

(3) Following the literature, we consider that the service cost involves an operational cost of the unit product (kt) and an investment cost (at^2) [44–46]. The parameters k and a belong to the cost parameter, which satisfy with $k, a \in (0,1)$. Specifically, we set $a = 1$ to show the main discussion. Moreover, retailers choose to invest directly in blockchain technology to ensure the transparency of the supply chain, reduce the problem of counterfeit goods in the supply chain, and improve the trust of consumers. For example, JD company has also achieved a lot of explorations in the application of blockchain technology, especially in supply chain management, commodity traceability, and anti-counterfeiting. It has developed its own blockchain system to strengthen its control and traceability of commodity quality, avoid the circulation of counterfeit goods, and ensure that consumers can obtain real information.

(4) We assume rational economic agents who maximize their benefits to obtain maximum benefits or maximum consumption experience.

3.3. Utility and Profit Functions

We consider two scenarios where in the retailer does (N) and does not share information (Y). In the basic model, the model settings of the two scenarios are the same. Different from non-information sharing, information sharing is influenced by market signals Γ_i. To express the model clearly, the utility functions of two products are designed as follows:

$$U_A = V - p_A \tag{1}$$

$$U_B = \theta V - p_B + t \tag{2}$$

The profit function expressions of the retailer and manufacturer are as follows:

$$\pi_R = (p_A - w_A)q_A + (p_B - w_B - kt)q_B - t^2 \tag{3}$$

$$\pi_M = w_A q_A + w_B q_B \tag{4}$$

$$\pi_{SC} = \pi_M + \pi_R \tag{5}$$

The wholesale prices of the two products are w_A and w_B. Correspondingly, the product quantities are q_A and q_B, respectively. Finally, according to whether information sharing is introduced, the corresponding model superscript is marked in the model, such as π_R^N and π_M^N.

3.4. Decision Sequence

The decision sequence is as follows: First, the manufacturer determines the wholesale price for the two types of products. Second, the retailer considers whether to adopt information sharing. Moreover, according to the retailer's information-sharing behavior, the retail prices of the two types of products can be considered. Finally, the market is cleared.

4. Basic Analysis

Here, we discuss four aspects: the influence of blockchain technology on different factors, the influence of information costs on sharing behavior, profits of different shareholders, and contract coordination of the equilibrium strategy. These discussions reveal the value of blockchain services in information sharing and contract coordination under mass customization.

4.1. The Influence of Blockchain Technology on Different Factors

We first examine the influence of blockchain services on various factors under mass customization, including the wholesale price, retail price, and product quantity. The results are shown as the following Lemma 1.

Lemma 1 (influence of blockchain services on different factors)**.**

(1) *The impact of blockchain services on the wholesale price:*

$$\frac{\partial w_B^N}{\partial t} > 0; \ \frac{\partial w_B^Y}{\partial t} > 0; \ \frac{\partial w_B^N}{\partial \Gamma} = 0; \ \frac{\partial w_B^Y}{\partial \Gamma} > 0$$

(2) *The impact of blockchain services on the retail price:*

$$\frac{\partial p_B^N}{\partial t} > 0;\ \frac{\partial p_B^Y}{\partial t} > 0;\ \frac{\partial p_B^N}{\partial \Gamma} > 0;\ \frac{\partial p_B^Y}{\partial \Gamma} > 0$$

(3) *The impact of blockchain services on the product quantity:*

$$\frac{\partial q_B^N}{\partial t} > 0;\ \frac{\partial q_B^Y}{\partial t} > 0;\ \frac{\partial q_B^N}{\partial \Gamma} = 0;\ \frac{\partial q_B^Y}{\partial \Gamma} > 0.$$

The Lemma 1 illustrates the value of a blockchain services. Specifically, the wholesale price increases with the accuracy of data activity regardless of information sharing in the supply chain system. Moreover, the demand signal has a positive effect in the case of information sharing, increasing the wholesale price. These changes show the value of the blockchain services through a change in the unit price. Specifically, it can result in a high wholesale price without considering the manufacturer's unit cost. Similarly, it shows the results for the retail price and product quantity with blockchain services. Furthermore, these results show similar trends, revealing the value of blockchain services in terms of data accuracy and demand signals. Thus, blockchain services increase the factor values of mass customization through the wholesale price, retail price, and product quantity.

4.2. The Influence of Information Cost for Sharing Behavior

Next, we consider the influence of information costs on sharing behaviors. Specifically, we define a term $c_B^j (j = N, Y)$, which can be written as $c_B^j = t^2 \big/ q_B^j, (j = N, Y)$. This term describes the information cost of a unit product under a blockchain service considering the impact of data accuracy and demand signals. The results are presented in Proposition 1.

Proposition 1 (information cost of unit product).

(1) $c_B^N = \frac{4t(1-\theta)\theta\left(1+t\sigma^2\right)}{1-k+((1-k)t+2\Gamma(1-\theta)\theta)\sigma^2}$; $c_B^Y = \frac{4t(1-\theta)\theta\left(1+t\sigma^2\right)}{1-k-((1-k)t+\Gamma(1-\theta)\theta)\sigma^2}$.

(2) $c_B^N < c_B^Y$.

Proposition 1 shows the information cost of a unit of product in the two cases. Based on the definition of information cost per unit product, we can obtain the results of two information-sharing models in the supply chain: with and without information sharing. The result shows that $c_B^N < c_B^Y$, which can be referred to in Appendix B.1. This indicates the higher cost with information sharing. This may explain why firms do not want to share information with others, as it can increase the unit product cost. This also indirectly decreases the benefits of the unit product and product price. Therefore, from the perspective of the unit cost of information sharing, firms prefer not to share information to obtain superiority in terms of product price. To obtain a potential strategy for information sharing via blockchain services, we can further consider the changes in firms' profits.

4.3. Profits for Different Shareholders

Next, we discuss the changes in the profit for firms to find the optimal strategy. The result is shown as Proposition 2 by analyzing the profits of firms and total the profit for the supply chain.

Proposition 2 (comparison of the profit for firms). $\pi_R^N > \pi_R^Y$; $\pi_M^N < \pi_M^Y$; $\pi_{SC}^N > \pi_{SC}^Y$.

Proposition 2 presents the results of the analysis for profit changes. Specifically, $\pi_R^N > \pi_R^Y$ and $\pi_M^N < \pi_M^Y$ indicate the firms' different choices, which can be referred to in Appendix B.2. The manufacturer would like to share information to obtain greater profits and consumer quantity. However, the retailer does not prefer information sharing. Thus, the retailer can obtain superior information from consumers in the process of a mass customization promotion.

Additionally, from the perspective of the equilibrium strategy, the results show two cases. On the one hand, the profits of the retailer and supply chain remain consistent in the optimal strategy. On the other hand, the profits of the manufacturer are inconsistent. Thus, a feasible strategy needs to explore new solutions to meet the interests of the manufacturer, retailer, and supply chain. Therefore, in Section 4.4, we further explore a reasonable solution through the three types of contract coordination.

4.4. Contract Coordination for the Equilibrium Strategy

Here, we consider three types of contract coordination to find the solution for the equilibrium: the cost-sharing contract, discount coupons, and wholesale price discount.

4.4.1. The Cost-Sharing Contract

Here, we change the profit function using a cost-sharing contract. Specifically, contract design is based on how information transmission happens. For example, the retailer has its own blockchain information system to support business needs, such as customer and process management, by closing the end market. The manufacturer needs to pay some costs to the retailer to obtain market information or customer information to stay away from the market, which is especially unfavorable in a mass customization environment, such as price design [40]. Therefore, to design the production plan, the manufacturer pays the retailer the cost of acquiring correlative information from the retailer's information system, which shares information in the blockchain system with the permitted supply chain.

Based on the above, we have changed the profit function with the cost parameter m_B^{Y-C} to share the technical cost based on the existing models. This cost parameter changes in the technical investment cost $m_B^{Y-C}t^2$ for the retailer in the case of information sharing (Y). Moreover, the manufacturer incurs the cost $\left(1-m_B^{Y-C}\right)t^2$ for the profit function. Therefore, the new profit functions are as follows:

$$\pi_R^{Y-C} = \left(p_A^{Y-C} - w_A^{Y-C}\right)q_A^{Y-C} + \left(p_B^{Y-C} - w_B^{Y-C} - kt\right)q_B^{Y-C} - m_B^{Y-C}t^2 \quad (6)$$

$$\pi_M^{Y-C} = w_A^{Y-C}q_A^{Y-C} + w_B^{Y-C}q_B^{Y-C} - \left(1-m_B^{Y-C}\right)t^2 \quad (7)$$

Moreover, we compare the profits of the retailer, manufacturer, and supply chain. We also draw the results from the information cost of a unit product according to the above analysis. These results are helpful in improving our understanding of the equilibrium strategy. The results are presented in Proposition 3.

Proposition 3 (comparison of the profits for the firms under a cost-sharing contract).

(1) $\pi_R^Y < \pi_R^{Y-C}$; $\pi_M^Y > \pi_M^{Y-C}$; $\pi_{SC}^Y = \pi_{SC}^{Y-C}$; *and* $c_B^Y > c_B^{Y-C}$.

(2) $m_B^{Y-C} \in \left(0, 1 - \frac{1+t\sigma^2(1+\theta\sigma^2)}{8t^2(1+t\sigma^2)} - \frac{(1-k)^2}{8(1-\theta)\theta}\right)$.

Proposition 3 presents the optimal strategy for a cost-sharing contract. Specifically, the equilibrium strategy for supply chain members is strategy Y-C, which refers to information sharing through a cost-sharing contract despite the unfavorable profit to the manufacturer. In this case, the manufacturer must pay the cost proportion

$m_B^{Y-C} \in \left(0, 1 - \frac{1+t\sigma^2(1+\theta\sigma^2)}{8t^2(1+t\sigma^2)} - \frac{(1-k)^2}{8(1-\theta)\theta}\right)$. From the perspective of the unit product information cost, the manufacturer incurs the cost of supporting information sharing with the blockchain system. This strategy avoids the manufacturer's free-riding. Moreover, this strategy implies that the unit information cost is lower than that in the basic model. Therefore, although the manufacturer must pay the cost of capturing customer information, it also obtains market information sharing from the retailer and exceeds the cost expenditure. In other words, a contract is an effective scheme for the supply chain members. Moreover, it can provide management implications for determining a reasonable cost-payment scheme within the effective cost ratio (m_B^{Y-C}) for firms.

4.4.2. The Discount Coupons

Next, we discuss a new contract to obtain the new scheme for mass customization. Specifically, we design the discount coupon plan for the manufacturer to determine the optimal strategy. As the leader of the supply chain, the manufacturer expects to obtain benefits that will pay the cost of discount coupons to customers and the retailer. It indirectly participates in the construction of a retailer's blockchain system under mass customization, which increases its market advantage. In other words, discount coupons can capture information from consumers and retailers to support the accuracy of their product plans for customized products.

Based on the above, we add discount coupon value s to retailers and consumers by the manufacturer, respectively. The new model seeks to obtain the market information required by the manufacturer from the information of the terminal market and retailer in the case of mass customization; it can guide the manufacturer in making reasonable production plans, including consumer habits, consumption preferences, and consumption behaviors. Therefore, we change the utility and profit functions based on the basic model as follows:

$$U_A^{Y-S} = V - p_A^{Y-S} \tag{8}$$

$$U_B^{Y-S} = V - p_B^{Y-S} + t + s \tag{9}$$

$$\pi_M^{Y-S} = w_A^{Y-S} q_A^{Y-S} + \left(w_B^{Y-S} - 2s\right) q_B^{Y-S} \tag{10}$$

$$\pi_R^{Y-S} = \left(p_A^{Y-S} - w_A^{Y-S}\right) q_A^{Y-S} + \left(p_B^{Y-S} - w_B^{Y-S} - kt + s\right) q_B^{Y-S} \tag{11}$$

In the case of mass customization, the manufacturer must pay both types of channel costs to indirectly support the construction of a retailer-sharing system. Moreover, we only consider that the discount coupons of customers and retailers have the same value, mainly focusing on changes in the optimal strategy. By comparing the profits of all the parties in the new and basic models, we obtain the analytical results of Proposition 4.

Proposition 4 (comparison of firms' profits under discount coupons). $\pi_R^Y < \pi_R^{Y-S}$; $\pi_M^Y < \pi_M^{Y-S}$; $\pi_{SC}^Y < \pi_{SC}^{Y-S}$; *and* $c_B^Y > c_B^{Y-S}$.

Proposition 4 compares the firms' profits under discount coupons. Specifically, the equilibrium strategy is Y-S, which is an effective way to introduce a new strategy. Consequently, the manufacturer should use discount coupons to increase consumer surplus and guide consumers toward using mass customization. In other words, this strategy indirectly increases benefits to consumers through mass customization. Moreover, this helps increase the retailer's profit. Combined with the information cost per unit of product, strategy Y-S can reduce the information cost. An increase in the quantity of products dilutes the input cost of technology sharing in mass customization. Therefore, from the management's perspective for firms, the manufacturer can adopt a special scheme of discount coupons to satisfy the requirement of mass customization, thereby increasing the benefit of participating firms.

4.4.3. The Discount of the Wholesale Price

Next, we consider the wholesale price discount. Specifically, we designed a cost parameter to describe the change in wholesale prices in the new strategy. Firms use discount prices to obtain market resources such as enterprise relationship management. Thus, the firm incurs a relatively low price to retain business relationships. In other words, this strategy is called discount behavior. It has some discount types, such as wholesale and retail prices. We discuss the discount in wholesale prices for manufacturers, which can increase the product quantity for mass customization. In addition, the manufacturer attempts to indirectly reduce the wholesale price to obtain better results for mass customization under information sharing.

Based on the above-mentioned factors, we adopt the parameter of wholesale price in mass customization to capture the change in the wholesale price. Specifically, we add the parameter of the wholesale price ϕ into the basic model, resulting in the wholesale price being ϕw_B^{Y-D}. Moreover, by rewriting the profit function of the retailer and combining the exiting model design, the result of the new strategy Y-D can be obtained, which is shown as follows:

$$\pi_R^{Y-D} = \left(p_A^{Y-D} - w_A^{Y-D}\right)q_A^{Y-S} + \left(p_B^{Y-D} - \phi w_B^{Y-D} - kt\right)q_B^{Y-D} \tag{12}$$

Furthermore, we compare the profit of firms and information costs in the new and basic models, which provides the result in Proposition 5 and Figure 1. This result reveals the equilibrium strategy of firms, and provides a reference for the decision making of firms.

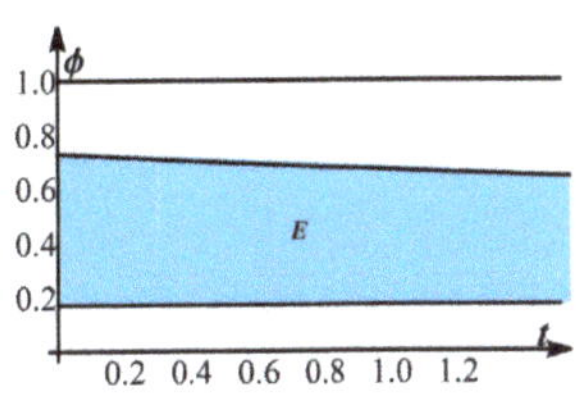

Figure 1. The result of equilibrium (E) ($k = 1, \theta = 1/2, \sigma = 1/2$).

Proposition 5 (comparison of the profit for firms under the condition of wholesale price discount). $\pi_R^Y > (<)\pi_R^{Y-D}; \pi_M^Y > (<)\pi_M^{Y-D}; \pi_{SC}^Y > (<)\pi_{SC}^{Y-D}; c_B^Y > (<)c_B^{Y-D}$.

Proposition 5 and Figure 1 show the equilibrium results between the basic and new models. Specifically, Proposition 5 compares firm profits under the condition of a wholesale price discount. These results indicate different cases. To show the result clearly, we have further drawn the results graph so that the changes in the equilibrium results can be visually displayed in Figure 1. This figure shows the results for equilibrium E. This shows that firms can reach an equilibrium when the discount factor must be maintained in the middle and low ranges. Moreover, this change indicates that discount factor ϕ must maintain a specific range to achieve equilibrium. The discount factor is further discussed, which shows that the discount factor designed by the manufacturer is not a bottomless change to balance the relationship between maintaining one's own profits and maintaining a going concern. Additionally, the higher the accuracy of the retailer's blockchain system, the more attractive the discount ratio required by the manufacturer to achieve an equilibrium strategy under extreme conditions. Therefore, from the management perspective, the manufacturer can adopt a reasonable discount wholesale price to satisfy the requirement of mass customization, thus increasing the benefit of participating firms.

In summary, this study discusses the basic model and three types of contract designs to satisfy the requirement of an equilibrium strategy, including a cost-sharing contract, discount coupons, and discount on the wholesale price. Moreover, we provide management implications to support firms' decision making in the business activity process. To further explore additional schemes to support the optimal strategy, this study also analyzed the new results under different conditions.

5. Extensions

We further discuss the new conditions for finding a new equilibrium which can provide some new management implications: unreal product quantity in information sharing, the changes in the utility function for the basic analysis, the optimal strategy in the case of an uncovered market, the optimal strategy in the case of a product with a delay strategy, and the optimal strategy in the case of additional new technology.

5.1. The Case of Unreal Product Quantity in Information Sharing

Here, we consider the value of information sharing. Specifically, we design the case of unreal product quantity in information sharing owing to commercial factors. For example, a retailer wishes to obtain more products and profit in an environment of mass customization that may require more product quantity higher than the actual quantity. The manufacturer cannot control the business strategy because of the independent business relationship between the manufacturer and retailer.

Therefore, we design a new model to describe the changes in product quantity using y^{Y-R}. This parameter describes the proportion of quantity that may increase and is shown as $(1+y^{Y-R})q_B^{Y-R}$ in the new product quantity. Based on this new quantity, we can change the profit functions of the manufacturer and retailer as follows:

$$\pi_M^{Y-R} = w_A^{Y-R}q_A^{Y-R} + \left(1+y^{Y-R}\right)w_B^{Y-R}q_B^{Y-R} \tag{13}$$

$$\pi_R^{Y-R} = \left(p_A^{Y-R} - w_A^{Y-R}\right)q_A^{Y-R} + \left(p_B^{Y-R} - w_B^{Y-R} - kt\right)q_B^{Y-R} \tag{14}$$

This is called strategy Y-R. Moreover, we compare the new strategy with the basic strategy to obtain the firms' equilibrium. In addition, we seek the relationship between the parameters of untrue product quantity ($\left(y_{MAX}^{Y-R}\right)^*$) and accuracy ($t$) of the supply chain system. Propositions 6 and Figure 2 present the results.

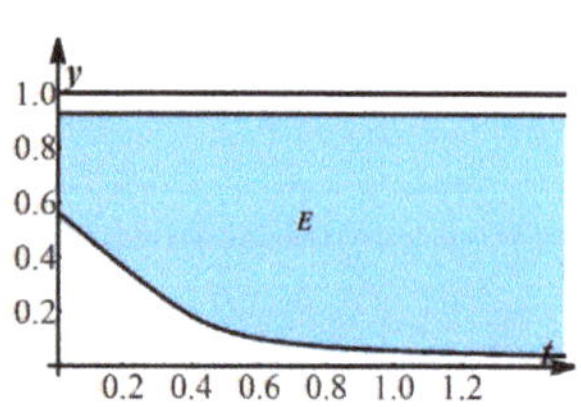

Figure 2. The result of equilibrium (E) ($k = 1/2, \theta = 9/10, \sigma = 1/2$).

Proposition 6 (false quantity of products in the information-sharing environment).

(1) $\pi_R^Y > (<)\pi_R^{Y-R}$; $\pi_M^Y > (<)\pi_M^{Y-R}$; $\pi_{SC}^Y > (<)\pi_{SC}^{Y-R}$; *and* $c_B^Y > (<)c_B^{Y-R}$.

(2) *An optimal increment of false output* $\left(y_{MAX}^{Y-R}\right)^*$ *exists such that* $\frac{\partial\left(y_{MAX}^{Y-R}\right)^*}{\partial t} < 0$ *is satisfied when information is shared.*

Proposition 6 presents two aspects of this discussion. Specifically, Proposition 6 (1) shows the analytical results for firm profits. This result is consistent with the basic discussion. To clearly show the equilibrium result, we further visualize the results, as shown in Figure 2.

Figure 2 shows the result of the equilibrium as E, in which the system accuracy changes (t) and the product quantity changes falsely (y^{Y-R}). This can be explained by the fact that y^{Y-R} has both upper and lower boundaries. In other words, although profit is high, the product quantity that the retailer can obtain is not limited to mass customization.

Furthermore, to reduce the influence of the optimal increment of false output $\left(y_{MAX}^{Y-R}\right)^*$, we seek the relationship between $\left(y_{MAX}^{Y-R}\right)^*$ and system accuracy (t). Proposition 6 (2) shows the relationship changes for both factors. It can be found that the higher the accuracy (t) of the blockchain system, the lower the proportion the optimal unreal product quantity ($\left(y_{MAX}^{Y-R}\right)^*$) will be. In other words, the blockchain system helps improve the distortion of market quantity, which also improves the technical value of commercial activity. Thus, the management implications can be derived. On the one hand, firms can use the unreal quantity of products to realize a new equilibrium strategy, which is helpful in increasing the firm's profit. In other words, the blockchain system in the supply chain can play the role of reducing unreasonable quantities, which reflects the technical value of resisting unreal market demand.

5.2. The Changes of Utility Function for Basic Analysis

In this section, we discuss the result of utility function changes. To be specific, we analyze the new strategy as X' ($X' = \pi_j^{i'}$, $i = M, R, SC$; $j = N, Y$). In the basic model, the utility function as $U_A > U_B > 0$ changes to the new utility function as $U_B > U_A > 0$ to obtain the new equilibrium strategy. The purpose of this model is to check the stability of the basic result. The profit functions are the same as the basic model. In other words, it takes into account the influence of the demand function changes caused by the change in utility function on the strategy result. Note that the utility of mass customization is higher than standard production product, which captures the potential changes in reality. Therefore, we analyze new conditions of the model to show the stability of the result, as shown in Proposition 7.

Proposition 7 (the stability of the result owing to the change in the utility of the function). $\pi_R^{N'} > \pi_R^{Y'}$; $\pi_M^{N'} < \pi_M^{Y'}$; $\pi_{SC}^{N'} > \pi_{SC}^{Y'}$.

Proposition 7 indicates that the result is a profit analysis after the utility function is changed. Specifically, this result is similar to that of the basic model and points to the manufacturer's sharing preferences. Moreover, this result is stable and shows the same effect on information-sharing preferences. The management implications are as follows: Although the structure of market demand has changed, retailers have not changed their existing manufacturer choices. Therefore, the manufacturer must consider new contracts such as cost-sharing contracts to achieve the equilibrium strategy.

5.3. The Optimal Strategy in the Case of an Uncovering Market

In this section, we expand the existing discussion to find new results for information sharing among firms. Specifically, in the basic analysis, we designed a complete market at the firm scale. In this subsection, we discuss the optimal strategy for an uncovered market. In reality, some firms have a part of the market scope for complex reasons, such as the influence of COVID-19. These reasons may lead to the shortening of the supply chain and make the market smaller, thus opening up all markets for enterprises.

In order to capture these changes in the market, we call this strategy Y-N. Specifically, the model is adopted as the distance cost of the consumer as x and cost parameter as ρ, which depicts the cost that consumers need to pay. Moreover, the market capacities of the standard product and mass customization are M. Accordingly, these market capacities are matched with the parameters of two market shares as λ_A^{Y-N} and λ_B^{Y-N}, respectively. In addition, we also provide the parameter l to capture the influence of system accuracy in the uncovered market. Furthermore, in order to find the optimal strategy, we design the cost transfer value F from manufacturer to retailer due to information sharing by the blockchain system of the retailer, which is called the strategy Y-N-C. In addition, we also analyze the firms' alliance, which is called strategy Y-N-CO. Therefore, we sort the new model settings which are shown as follows. The analytical result is shown as Proposition 8.

$$U_A^{N-C} = V - p_A - \rho x \tag{15}$$

$$U_B^{N-C} = \theta V - p_B - \rho x + lt \tag{16}$$

$$\pi_R^{Y-N} = \left(p_A^{Y-N} - w_A^{Y-N}\right)\lambda_A^{Y-N}M + \left(p_B^{Y-N} - w_B^{Y-N} - kt\right)\lambda_B^{Y-N}M - t^2 \tag{17}$$

$$\pi_M^{Y-N} = w_A^{Y-N}\lambda_A^{Y-N}M + w_B^{Y-N}\lambda_B^{Y-N}M \tag{18}$$

$$\pi_R^{Y-N-C} = \left(p_A^{Y-N} - w_A^{Y-N}\right)\lambda_A^{Y-N}M + \left(p_B^{Y-N} - w_B^{Y-N} - kt\right)\lambda_B^{Y-N}M - t^2 + F \tag{19}$$

$$\pi_M^{Y-N-C} = w_A^{Y-N}\lambda_A^{Y-N}M + w_B^{Y-N}\lambda_B^{Y-N}M - F \tag{20}$$

Proposition 8 (optimal strategy of incompletely covering the market). *$\pi_R^{Y-N} < \pi_R^{Y-N-C}$; $\pi_M^{Y-N} > \pi_M^{Y-N-C}$; $\pi_{SC}^{Y-N} = \pi_{SC}^{Y-N-C}$; and $\pi_R^{Y-N} < \pi_R^{Y-N-CO}$.*

Proposition 8 shows two types of optimal strategies that do not completely cover the market. Specifically, on the one hand, the retailer's cost-sharing plan helps to increase the profit. Although the manufacturer needs to pay cost F, it can obtain the benefit of sharing information, which is called an effective scheme. This is because the manufacturer obtains superior information, but this advantage does not result in actual profit. Therefore, they choose a strategy that is different from that of the retailer. However, the adoption of strategic alliances can also help narrow the profit gap, which is a feasible choice for firms. Based on the above, management implications were obtained. First, a firm can use shared information by adding it into the manufacturer's fixed costs, which is the best choice for both parties. Firms can also use strategic alliances to achieve market superiority when uncovering a market.

5.4. The Optimal Strategy in the Case of a Product with a Delay Strategy

In this subsection, we present a new strategy for mass customization. Specifically, we use a delay strategy to achieve equilibrium. In fact, the delay strategy in mass customization can reduce the impact of the price increase in customized products, thereby increasing the product quantity in the market. In other words, this strategy produces a promotional effect at a relatively low price, which is more common for products with high price sensitivity.

Therefore, in mass customization, we adopt the delay parameter δ to describe the price changes in the delay-state environment. This delay strategy is known as the Y-DE strategy. In this model, we change the utility function based on the basis model to $U_B^{Y-DE} = \theta V - \delta p_B^{Y-S} + t$. Moreover, this study discusses two cases of the utility function: $U_A^{Y-DE} > U_B^{Y-DE}$ and $U_A^{Y-DE} < U_B^{Y-DE}$. By analyzing these two types of models, we show the optimal delay strategy in Propositions 9 and Figure 3.

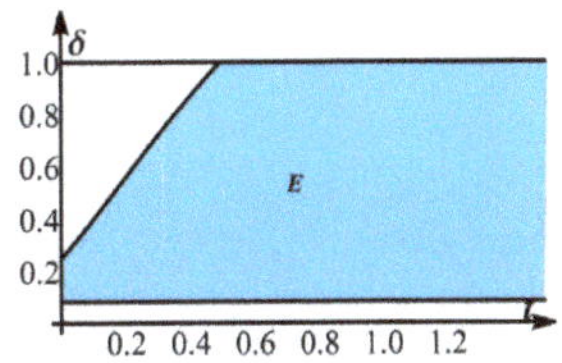

Figure 3. The result of equilibrium (E) ($k = 1/2, \theta = 1/2, \sigma = 1/2$).

Proposition 9 (optimal strategy of delay strategy).

(1) *When* $U_A^{Y-DE} > U_B^{Y-DE}$ *and* $\frac{2-\theta-2\sqrt{1-\theta}}{\theta} < \delta < 1$: $\pi_R^Y < \pi_R^{Y-DE}$; $\pi_M^Y < \pi_M^{Y-DE}$; $\pi_{SC}^Y < \pi_{SC}^{Y-DE}$.

(2) *When* $U_A^{Y-DE} < U_B^{Y-DE}$: $\pi_R^Y > (<)\pi_R^{Y-DE}$; $\pi_M^Y > (<)\pi_M^{Y-DE}$; $\pi_{SC}^Y > (<)\pi_{SC}^{Y-DE}$.

Proposition 9 shows the optimal delay strategy for firms. Specifically, this strategy shows two cases of utility change. In the case of $U_A^{Y-DE} > U_B^{Y-DE}$, the equilibrium strategy is Y-DE when the $\frac{2-\theta-2\sqrt{1-\theta}}{\theta} < \delta < 1$. This means that the delay strategy is effective but cannot be too long, which needs to meet the requirements of the firms' profits. Moreover, in the case of $U_A^{Y-DE} < U_B^{Y-DE}$, the firms' equilibrium strategy is E. To show the results clearly, we have used Figure 3.

The results in Figure 3 show that system accuracy helps improve the selection possibility of equilibrium results and reduces the impact of delayed activity. In other words, the delay strategy increases firms' profits. Therefore, the meaning of the management is obtained. For the mass customization of firms, they choose a reasonable delay time and method to help improve the profit of firms and benefit customers to activate the market.

5.5. The Optimal Strategy in the Case of Additional New Technology

Finally, we examine the influence of the adoption of big data technology. Specifically, we use a blockchain system to ensure information security during the information release process. Moreover, big data technology in an information system can capture large amounts of business information. In other words, both technologies can achieve secure information sharing. In reality, Walmart uses blockchain in combination with big data analytics to enhance transparency and traceability in its supply chain. By integrating blockchain with real-time data from IoT sensors, Walmart tracks the origin and journey of food products, ensuring safety and reducing spoilage. Big data analytics helps in predicting trends and optimizing inventory management.

Specifically, we adopt big data technology for the retailer because it is close to the end market. The cost of big data technology is c per product of mass customization. Moreover, to achieve a high level of information sharing, the manufacturer provides a fixed-cost F to the retailer. Finally, customers can benefit from two technologies—the blockchain system and big data—which show changes in the utility function. To capture the technology benefit-to-cost ratio (BCB) in big data, we adopted $BCB = b/c$ to demonstrate the value of big data technology. $BCB > 1$, $BCB = 1$, and $BCB < 1$ represent beneficial, neutral, and harmful effects, respectively [47]. Correspondingly, this new strategy is referred to as Y-NEW. The new utility and profit functions are as follows: Proposition 10 presents these results.

$$U_A^{Y-NEW} = V - p_A^{Y-NEW} \tag{21}$$

$$U_B^{Y-NEW} = \theta V - p_B^{Y-NEW} + t + b \tag{22}$$

$$\pi_R^{Y-NEW} = \left(p_A^{Y-NEW} - w_A^{Y-NEW}\right)q_A^{Y-NEW} + \left(p_B^{Y-NEW} - w_B^{Y-NEW} - kt - c\right)q_B^{Y-NEW} - t^2 + F \tag{23}$$

$$\pi_M^{Y-NEW} = w_A^{Y-NEW} q_A^{Y-NEW} + w_B^{Y-NEW} q_B^{Y-NEW} - F \tag{24}$$

Proposition 10 (the optimal strategy of new technology enhancement).

(1) *When* $BCB > 1$ *and* $0 < F < \frac{((b-c)+2t(1-k))(b-c)}{8\theta(1-\theta)}$: $\pi_R^Y < \pi_R^{Y-NEW}$; $\pi_M^Y < \pi_M^{Y-NEW}$; $\pi_{SC}^Y < \pi_{SC}^{Y-NEW}$.

(2) *When* $BCB \leq 1$, $0 < t < \frac{c-b}{2(1-k)}$, *and* $0 < F < \frac{((b-c)+2t(1-k))(b-c)}{8\theta(1-\theta)}$: $\pi_R^Y < \pi_R^{Y-NEW}$; $\pi_M^Y < \pi_M^{Y-NEW}$; $\pi_{SC}^Y < \pi_{SC}^{Y-NEW}$.

Proposition 10 presents the optimal strategy for enhancing the new technology. Specifically, on the one hand, when the technology is beneficial ($BCB > 1$), as long as the investment cost of technical compensation does not exceed the specific range ($0 < F < \frac{((b-c)+2t(1-k))(b-c)}{8\theta(1-\theta)}$), the strategy of technology enhancement can reach equilibrium. This result is helpful in realizing the Y-NEW equilibrium strategy. However, when technology is harmful ($BCB \leq 1$), the equilibrium strategy conditions are $0 < F < \frac{((b-c)+2t(1-k))(b-c)}{8\theta(1-\theta)}$ and $0 < t < \frac{c-b}{2(1-k)}$. In this case, the equilibrium strategy is Y-NEW. These results are common in the Y-NEW equilibrium strategy. In other words, big data is effective as an enhanced sharing technology, particularly when the accuracy of blockchain technology is not high. Big data can provide full play to its technological advantages. Management implications are obtained based on the results of the discussion. A firm can consider adopting a variety of technologies to gain market superiority, especially in a mass-customized environment. In other words, a variety of technologies can quickly capture consumers' habits and behaviors and realize the technical value of sharing relevant data with firms.

6. Discussion

This study discusses strategies based on changes in contracts to seek the optimal strategy. Therefore, for a deeper discussion, we sort the strategies to show the potential development laws. Specifically, in the basic model, an optimal strategy for the manufacturer and retailer does not exist. In other words, although the manufacturer prefers to obtain information from the retailer through the blockchain system, the retailer cannot share the information because of construction cost limitations. Furthermore, to find a reasonable solution, we discuss some new types of contracts to obtain the equilibrium strategy, including the cost-sharing contract, discount coupons, wholesale price discounts, unreal product quantity in information sharing, the optimal strategy in the case of an uncovered market, optimal strategy in the case of a product with a delay strategy, and the optimal strategy in the case of additional new technology. Specifically, these contracts under different conditions find the optimal strategy that guides the information benefit of the manufacturer from the retailer in mass customization. Among them, the basic idea is that the cost of obtaining information for the manufacturer should be similar to the cost of technical investment. In other words, the firm expects to obtain benefits from information sharing and must pay the cost as well.

7. Conclusions, Managerial Implications, and Limitations

7.1. Conclusions

We develop game-theoretical models involving manufacturers and retailers to explore the retailer's optimal information-sharing strategy in a mass customization environment. Specifically, we focus on the retailer's ability to share terminal market information about nearby consumers, which is essential for mass customization. This analysis helps to clarify the manufacturer's incentives to support the retailer in establishing blockchain systems for two types of products: standard and customized products. Our investigation reveals the following preliminary conclusions:

First, retailers cannot share market information without incurring costs to manufacturers, meaning that no information can be exchanged "early" without some form of compensation. Second, by coordinating cost-sharing contracts, discount coupons, and wholesale price discounts, the manufacturer and retailer can achieve an equilibrium strategy under specific conditions. Lastly, the results hold even when the underlying assumptions are adjusted, such as altering the utility function or finding new market conditions. We also identify new equilibrium strategies under various scenarios, including handling unreal product quantities in information sharing, optimizing strategies for products with delayed delivery policies, and strategies for incorporating new technologies.

7.2. Managerial Implications

From the manufacturer's perspective, it is crucial to consider the market conditions in a mass customization environment when evaluating the retailer's information-sharing strategy. Manufacturers should implement blockchain systems to facilitate secure information sharing with retailers, leveraging the technology's ability to prevent data tampering. Moreover, manufacturers should adopt optimal strategies under well-structured contract conditions, such as cost-sharing contracts, discount coupons, wholesale price discounts, unreal product quantities in information sharing, strategies for uncovered markets, delayed product strategies, and incorporating new technologies. These contract mechanisms are vital for ensuring the quality and reliability of information shared by the retailer.

From the retailer's perspective, implementing a blockchain system can protect the integrity of data within the supply chain, ensuring that relevant data are not tampered with. Retailers should evaluate optimal strategies for sharing information with the manufacturer to enhance the benefits derived from data sharing. Additionally, retailers can leverage big data technologies in conjunction with blockchain to maximize technical value, creating a more efficient and transparent information-sharing ecosystem. Together, these technologies can significantly improve the retailer's operational capabilities and overall business strategy.

7.3. Limitations and Future Work

First, we use information sharing to improve the market value of mass customization. Future research should consider the competitive environment of mass customization. Second, the government should also be considered in the model. We did not consider the government's role in smart sales, especially the power of technology, such as oversight or technological subsidies. Future research could explore competitive dynamics in mass customization markets, incorporate government roles like subsidies or oversight. Moreover, it also could analyze consumer behavior's impact on information sharing. Additionally, investigating the integration of emerging technologies and their ethical implications could provide further insights.

Author Contributions: Conceptualization, Z.Y., J.W. and H.Z.; Software, Z.Y. and H.Z.; Investigation, J.W.; Resources, H.Z.; Writing—original draft, Z.Y., J.W. and H.Z.; Writing—review & editing, Z.Y. and J.W.; Visualization, Z.Y. and H.Z.; Supervision, J.W. and H.Z.; Project administration, H.Z.; Funding acquisition, H.Z. All authors have read and agreed to the published version of the manuscript.

Funding: This research is supported by the National Natural Science Foundation of China (72404190); Guangdong Basic and Applied Basic Research Foundation (2023A1515110302); Guangdong Planning Office of Philosophy and Social Science (GD24YGL09); China Society of Logistics (2024CSLKT3-419); Shenzhen Peacock Pengcheng Talents Funding; Shenzhen Key Research Base of Humanities and Social Sciences; and Shenzhen University Young Teachers Research Initiation Project (000001032153, GY00007).

Data Availability Statement: The original contributions presented in this study are included in the article.

Acknowledgments: We appreciate the editors and reviewers that spent time on the original manuscript.

Conflicts of Interest: The authors declare no conflicts of interest.

Appendix A. Basic Results

1. The supply chain does not use the solution of information sharing (N):

$$p_A^N = \frac{3}{4} + \frac{t\Gamma\theta\sigma^2}{2+2t\sigma^2}$$

$$p_B^N = \frac{1}{4}\left(3\theta + t\left(3+k+\frac{2\Gamma\theta\sigma^2}{1+t\sigma^2}\right)\right)$$

$$q_A^N = \frac{1-\theta-(1-k)t}{4(1-\theta)}$$

$$q_B^N = \frac{1}{4}\left(\frac{(1-k)t}{(1-\theta)\theta} + \frac{2t\Gamma\sigma^2}{1+t\sigma^2}\right)$$

$$w_A^N = \frac{1}{2}$$

$$w_B^N = \frac{(1-k)t+\theta}{2}$$

$$\pi_R^N = \frac{1}{16}\left(1 + \frac{t^2(1-(2-k)k-16(1-\theta)\theta)}{(1-\theta)\theta} + \frac{4t\theta\sigma^4}{1+t\sigma^2}\right)$$

$$\pi_M^N = \frac{1}{8}\left(1 + \frac{(1-k)^2t^2}{(1-\theta)\theta}\right)$$

$$\pi_{SC}^N = \frac{1}{16}\left(3 + \frac{t^2\left(3(1-k)^2-16(1-\theta)\theta\right)}{(1-\theta)\theta} + \frac{4t\theta\sigma^4}{1+t\sigma^2}\right)$$

2. The supply chain uses the solution of information sharing (Y):

$$p_A^Y = \frac{3}{4}\left(1 + \frac{t\Gamma\theta\sigma^2}{1+t\sigma^2}\right)$$

$$p_B^Y = \frac{1}{4}\left(3\theta + t\left(3+k+\frac{3\Gamma\theta\sigma^2}{1+t\sigma^2}\right)\right)$$

$$q_A^Y = \frac{1-\theta-(1-k)t}{4(1-\theta)}$$

$$q_B^Y = \frac{1}{4}\left(\frac{(1-k)t}{(1-\theta)\theta} + \frac{t\Gamma\sigma^2}{1+t\sigma^2}\right)$$

$$w_A^Y = \frac{1}{2}\left(1 + \frac{t\Gamma\theta\sigma^2}{1+t\sigma^2}\right)$$

$$w_B^Y = \frac{1}{2}\left((1-k)t + \theta + \frac{t\Gamma\theta\sigma^2}{1+t\sigma^2}\right)$$

$$\pi_R^Y = \frac{1}{16}\left(1 + \frac{t^2(1-(2-k)k-16(1-\theta)\theta)}{(1-\theta)\theta} + \frac{t\theta\sigma^4}{1+t\sigma^2}\right)$$

$$\pi_M^Y = \frac{1}{8}\left(1 + \frac{(1-k)^2t^2}{(1-\theta)\theta} + \frac{t\theta\sigma^4}{1+t\sigma^2}\right)$$

$$\pi_{SC}^Y = \frac{1}{16}\left(3 + \frac{t^2\left(3(1-k)^2 - 16(1-\theta)\theta\right)}{(1-\theta)\theta} + \frac{3t\theta\sigma^4}{1+t\sigma^2}\right)$$

3. The supply chain uses the solution of information sharing in the case of cost-sharing (Y-C):

$$p_A^{Y-C} = \frac{3}{4}\left(1 + \frac{t\Gamma\theta\sigma^2}{1+t\sigma^2}\right)$$

$$p_B^{Y-C} = \frac{1}{4}\left(3\theta + t\left(3 + k + \frac{3\Gamma\theta\sigma^2}{1+t\sigma^2}\right)\right)$$

$$q_A^{Y-C} = \frac{1-\theta-(1-k)t}{4(1-\theta)}$$

$$q_B^{Y-C} = \frac{1}{4}\left(\frac{(1-k)t}{(1-\theta)\theta} + \frac{t\Gamma\sigma^2}{1+t\sigma^2}\right)$$

$$w_A^{Y-C} = \frac{1}{2}\left(1 + \frac{t\Gamma\theta\sigma^2}{1+t\sigma^2}\right)$$

$$w_B^{Y-C} = \frac{1}{2}\left((1-k)t + \theta + \frac{t\Gamma\theta\sigma^2}{1+t\sigma^2}\right)$$

$$\pi_R^{Y-C} = \frac{1}{16}\left(1 - 16m_B^{Y-C}t^2 - \frac{(-1+k)^2t^2}{(-1+\theta)\theta} + \frac{t\theta\sigma^4}{1+t\sigma^2}\right)$$

$$\pi_M^{Y-C} = \frac{1}{8}\left(1 + \frac{t^2\left(1-(2-k)k-8\theta\left(1-m_B^{Y-C}\right)(1-\theta)\right)}{(1-\theta)\theta} + \frac{t\theta\sigma^4}{1+t\sigma^2}\right)$$

$$\pi_{SC}^{Y-C} = \frac{1}{16}\left(3 + \frac{t^2\left(3(1-k)^2 - 16(1-\theta)\theta\right)}{(1-\theta)\theta} + \frac{3t\theta\sigma^4}{1+t\sigma^2}\right)$$

4. The supply chain uses the solution of information sharing in the case of discount coupons (Y-S):

$$p_A^{Y-S} = \frac{3}{4}\left(1 - s + \frac{t\Gamma\theta\sigma^2}{1+t\sigma^2}\right)$$

$$p_B^{Y-S} = \frac{1}{4}\left(3\theta + s + t\left(3 + k + \frac{3\Gamma\theta\sigma^2}{1+t\sigma^2}\right)\right)$$

$$q_A^{Y-S} = \frac{1-\theta-(1-k)t}{4(1-\theta)}$$

$$q_B^{Y-S} = \frac{(1-k)t-(1-\theta)s}{4(1-\theta)\theta} + \frac{t\Gamma\sigma^2}{4(1+t\sigma^2)}$$

$$w_A^{Y-S} = \frac{1}{2}\left(1-s+\frac{t\Gamma\theta\sigma^2}{1+t\sigma^2}\right)$$

$$w_B^{Y-S} = \frac{1}{2}\left((1-k)t+3s+\theta+\frac{t\Gamma\theta\sigma^2}{1+t\sigma^2}\right)$$

$$\pi_R^{Y-S} = \frac{1-16t^2(1+t\sigma^2)}{16(1+t\sigma^2)} + \frac{(1-k)^2t^2}{16(1-\theta)\theta} + \frac{s^2-2s((1-k)t+\theta)}{16\theta} + \frac{t\sigma^2(1-\theta)(1+\sigma^2\theta)}{16(1+\theta)(1+t\sigma^2)}$$

$$\pi_M^{Y-S} = \frac{1+t\sigma^2(1+\theta\sigma^2)}{8(1+t\sigma^2)} + \frac{(1-k)^2t^2}{8(1-\theta)\theta} + \frac{s^2-2s((1-k)t+\theta)}{8\theta}$$

$$\pi_{SC}^{Y-S} = \frac{3(1+t\sigma^2(1+\theta\sigma^2))}{16(1+t\sigma^2)} + \frac{3(1-k)^2t^2}{16(1-\theta)\theta} + \frac{3s^2-6s((1-k)t+\theta)}{16\theta} - t^2$$

5. The supply chain uses the solution of information sharing in the case of discount of wholesale price (Y-D):

$$p_A^{Y-D} = \frac{1}{2}\left(1+\frac{t\Gamma\theta\sigma^2}{1+t\sigma^2}+\frac{t(1-k)(1-\phi)}{4\phi-\theta(1+\phi)^2}+\frac{(1-\theta)}{4\phi-\theta(1+\phi)^2}\left(2\phi+\frac{t\Gamma\theta\sigma^2(1+\phi)}{1+t\sigma^2}\right)\right)$$

$$p_B^{Y-D} = \frac{1}{2}\left((1+k)t+\theta+\frac{t\Gamma\theta\sigma^2}{1+t\sigma^2}+\frac{(1-k)(2-(1+\phi)\theta)t\phi}{4\phi-\theta(1+\phi)^2}+\frac{\phi(1-\theta)\theta}{4\phi-\theta(1+\phi)^2}\left(1+\frac{2t\Gamma\sigma^2}{1+t\sigma^2}+\phi\right)\right)$$

$$q_A^{Y-D} = \frac{2\phi-(1-k)(1+\phi)t}{2\left(4\phi-\theta(1+\phi)^2\right)} - \frac{\theta(1+\phi+t\sigma^2(1+\Gamma+(1-\Gamma)\phi))}{2(1+t\sigma^2)\left(4\phi-\theta(1+\phi)^2\right)}$$

$$q_B^{Y-D} = \frac{(1-k)t\phi}{\theta\left(4\phi-\theta(1+\phi)^2\right)} + \frac{\phi(1-\phi)}{2\left(4\phi-\theta(1+\phi)^2\right)} + \frac{\phi(t\Gamma\sigma^2(2-\theta(1+\phi)))}{2(1+t\sigma^2)\left(4\phi-\theta(1+\phi)^2\right)}$$

$$w_A^{Y-D} = \frac{t(1-k)(1-\phi)+2\phi(1-\theta)}{4\phi-\theta(1+\phi)^2} + \frac{t\Gamma\theta\sigma^2(1+\phi)(1-\theta)}{(1+t\sigma^2)\left(4\phi-\theta(1+\phi)^2\right)}$$

$$w_B^{Y-D} = \frac{(1-k)t(2-(1+\phi)\theta)+(1-\theta)\theta(1+\phi)}{4\phi-\theta(1+\phi)^2} + \frac{2t\Gamma\sigma^2(1-\theta)\theta}{(1+t\sigma^2)\left(4\phi-\theta(1+\phi)^2\right)}$$

$$
\pi_R^{Y-D} = \frac{\left(\begin{array}{l}
\theta(\theta^2 - \phi((4\theta - (4+\theta+\theta^2)\phi) + \theta\phi^2(2(3-\theta)+\phi)))+ \\
t^3\sigma^2\left(4(1-k)^2\phi^2 + \theta\left(4(1+\phi)^2\theta\left(8\phi - \theta(1+\phi)^2\right) + \left((1-k)^2(1-2\phi) - (67-3(2-k)k)\phi^2\right)\right)\right)+ \\
+t^2\left(\begin{array}{l} 4(1-k)^2\phi^2 - 2\theta^2(1+\phi)\left(2\theta(1+\phi)^3 + \left((1-k)(1-\phi)^2\sigma^2 - 16\phi(1+\phi)\right)\right) \\ +\theta\left(1+k^2(1+\phi)(1-3\phi) - \phi\left(2+4\sigma^2(1-\phi)^2+67\phi\right) - k\left(2-2\phi\left(2+2\sigma^2(1-\phi)^2+3\phi\right)\right)\right)\end{array}\right)+ \\
\left(t\theta\left(\begin{array}{l} 4\phi\left(k(1-\phi)^2 - 1 + (2+\sigma^2-\phi)\phi\right) + \theta^3\sigma^4\phi(2+\phi(1+\phi^2))+ \\ \theta^2\sigma^2(1+(1+2\phi)\phi^2 + \sigma^2(1-\phi(6-\phi(1-4\phi))))+ \\ \theta\left(2-2k(1-\phi)^2(1+\phi) - 2\phi(1-\phi(2\sigma^4-(1-\phi)) + \sigma^2(4-\phi+(6-\phi)\phi^2))\right)\end{array}\right)\right)
\end{array}\right)}{4(\theta+t\theta\sigma^2)\left(4\phi-\theta(1+\phi)^2\right)^2}
$$

$$
\pi_M^{Y-D} = \frac{(1-k)^2t^2 + (1-k)t\theta - (1-k)t\theta\phi + (1-\theta)\theta\phi}{2\theta\left(4\phi - \theta(1+\phi)^2\right)} + \frac{t(1-\theta)\theta^2\sigma^4}{2\theta(1+t\sigma^2)\left(4\phi-\theta(1+\phi)^2\right)}
$$

$$
\pi_{SC}^{Y-D} = \frac{\left(\begin{array}{l}
\theta(\theta^2 - \phi((4\theta - (4+\theta+\theta^2)\phi) + \theta\phi^2(2(3-\theta)+\phi)))+ \\
t^3\sigma^2\left(4(1-k)^2\phi^2 + \theta\left(4(1+\phi)^2\theta\left(8\phi - \theta(1+\phi)^2\right) + \left((1-k)^2(1-2\phi) - (67-3(2-k)k)\phi^2\right)\right)\right)+ \\
+t^2\left(\begin{array}{l} 4(1-k)^2\phi^2 - 2\theta^2(1+\phi)\left(2\theta(1+\phi)^3 + \left((1-k)(1-\phi)^2\sigma^2 - 16\phi(1+\phi)\right)\right) \\ +\theta\left(1+k^2(1+\phi)(1-3\phi) - \phi\left(2+4\sigma^2(1-\phi)^2+67\phi\right) - k\left(2-2\phi\left(2+2\sigma^2(1-\phi)^2+3\phi\right)\right)\right)\end{array}\right)+ \\
\left(t\theta\left(\begin{array}{l} 4\phi\left(k(1-\phi)^2 - 1 + (2+\sigma^2-\phi)\phi\right) + \theta^3\sigma^4\phi(2+\phi(1+\phi^2))+ \\ \theta^2\sigma^2(1+(1+2\phi)\phi^2 + \sigma^2(1-\phi(6-\phi(1-4\phi))))+ \\ \theta\left(2-2k(1-\phi)^2(1+\phi) - 2\phi(1-\phi(2\sigma^4-(1-\phi)) + \sigma^2(4-\phi+(6-\phi)\phi^2))\right)\end{array}\right)\right)
\end{array}\right)}{4(\theta+t\theta\sigma^2)\left(4\phi-\theta(1+\phi)^2\right)^2} + \frac{(1-k)^2t^2 + (1-k)t\theta - (1-k)t\theta\phi + (1-\theta)\theta\phi}{2\theta\left(4\phi - \theta(1+\phi)^2\right)} + \frac{t(1-\theta)\theta^2\sigma^4}{2\theta(1+t\sigma^2)\left(4\phi-\theta(1+\phi)^2\right)}
$$

6. The supply chain uses the solution of information sharing in the case of unreal product quantity (Y-R):

$$
p_A^{Y-R} = \frac{1}{2}\left(1 + \frac{t\Gamma\theta\sigma^2}{1+t\sigma^2} + \frac{(1+y)}{4(1+y)-(2+y)^2\theta}\left((1-k)ty + (1-\theta)\left(2+\frac{t(2+y)\Gamma\theta\sigma^2}{1+t\sigma^2}\right)\right)\right)
$$

$$
p_B^{Y-R} = \frac{1}{2}\left(\begin{array}{c} t+kt+\theta+\dfrac{t\Gamma\theta\sigma^2}{1+t\sigma^2} + \dfrac{((1-k)t(2(1+y)-(2+y)\theta))}{4(1+y)-(2+y)^2\theta} - \\ \dfrac{(1-\theta)\theta}{4(1+y)-(2+y)^2\theta}\left(2+y+\dfrac{2t(1+y)\Gamma\sigma^2}{1+t\sigma^2}\right)\end{array}\right)
$$

$$
q_A^{Y-R} = \frac{(1+y)(2-(1-k)t(2+y))}{8(1+y)-2(2+y)^2\theta} - \frac{(1+y)\theta}{8(1+y)-2(2+y)^2\theta}\left(2+y+\frac{ty\Gamma\sigma^2}{1+t\sigma^2}\right)
$$

$$
q_B^{Y-R} = \frac{2(1-k)t(1+y)}{2\theta\left(4(1+y)-(2+y)^2\theta\right)} + \frac{y\theta + t(y(1+\Gamma(2-\theta)) + 2\Gamma(1-\theta))\theta\sigma^2}{2\theta\left(4(1+y)-(2+y)^2\theta\right)(1+t\sigma^2)}
$$

$$
w_A^{Y-R} = \frac{(1+y)((1-k)ty)}{4(1+y)-(2+y)^2\theta} + \frac{(1+y)(1-\theta)}{4(1+y)-(2+y)^2\theta}\left(2+\frac{t(2+y)\Gamma\theta\sigma^2}{1+t\sigma^2}\right)
$$

$$w_B^{Y-R} = \frac{(1-k)t(2(1+y)-(2+y)\theta)}{4(1+y)-(2+y)^2\theta} + \frac{(1-\theta)\theta}{4(1+y)-(2+y)^2\theta}\left(2+y+\frac{2t(1+y)\Gamma\sigma^2}{1+t\sigma^2}\right)$$

$$\pi_R^{Y-R} = \frac{1}{4}\left(\begin{array}{c} -4t^2 + \frac{(1+y)(2-(1-k)t(2+y)-(2+y)\theta)((1+y)(2-(1-k)ty)-(2+y(2+y))\theta)}{\left(4(1+y)-(2+y)^2\theta\right)^2} + \\ \frac{(2(1-k)t(1+y)+y\theta)(((1-k)t(1+y)(2-(2+y)\theta))-\theta(2+3y-(1+y)(2+y)\theta))}{\theta\left(4(1+y)-(2+y)^2\theta\right)^2} \end{array}\right)$$

$$\pi_M^{Y-R} = \frac{(1+y)\left((1-k)^2t^2(1+y)+(1-k)ty\theta+(1-\theta)\theta\right)}{2\theta\left(4(1+y)-(2+y)^2\theta\right)}$$

$$\pi_{SC}^{Y-R} = \frac{1}{4}\left(\begin{array}{c} -4t^2 + \frac{(1+y)(2-(1-k)t(2+y)-(2+y)\theta)((1+y)(2-(1-k)ty)-(2+y(2+y))\theta)}{\left(4(1+y)-(2+y)^2\theta\right)^2} + \\ \frac{(2(1-k)t(1+y)+y\theta)(((1-k)t(1+y)(2-(2+y)\theta))-\theta(2+3y-(1+y)(2+y)\theta))}{\theta\left(4(1+y)-(2+y)^2\theta\right)^2} \end{array}\right) + \frac{(1+y)\left((1-k)^2t^2(1+y)+(1-k)ty\theta+(1-\theta)\theta\right)}{2\theta\left(4(1+y)-(2+y)^2\theta\right)}$$

7. The new utility function in the case of non-sharing in the market information (N′):

$$p_A^{N'} = \frac{5-(1-k)t+\theta}{2(3+\theta)} + \frac{2t\Gamma(1+\theta)\sigma^2}{2(3+\theta)(1+t\sigma^2)}$$

$$p_B^{N'} = \frac{1+\theta(4+\theta)}{2(3+\theta)} + \frac{t(4+\theta+k(2+\theta))}{2(3+\theta)} + \frac{4\Gamma\sigma^2 t}{2(3+\theta)(1+t\sigma^2)}$$

$$q_A^{N'} = \frac{1}{2(3+\theta)}\left(2+\theta+\frac{2t\Gamma\sigma^2}{1+t\sigma^2}\right) - \frac{(1-k)(1+\theta)t}{2(1-\theta)(3+\theta)}$$

$$q_B^{N'} = \frac{(1-k)t}{(1-\theta)(3+\theta)} - \frac{1}{2(3+\theta)}\left(1-\frac{2t\Gamma\sigma^2}{1+t\sigma^2}\right)$$

$$w_A^{N'} = \frac{1}{2}$$

$$w_B^{N'} = \frac{(1-k)t+\theta}{2}$$

$$\pi_R^{N'} = \frac{1-\theta-t^3(11+(2-k)k-4\theta(2+\theta))\sigma^2}{4(1-\theta)(3+\theta)(1+t\sigma^2)} - \frac{t^2(11+2k(1+2(2-\theta)\theta k)+(1-k)(1-\theta)\sigma^2)}{4(1-\theta)(3+\theta)(1+t\sigma^2)} - \frac{t\left(1-k-\sigma^2\left(1+4\sigma^2\right)\right)}{4(3+\theta)(1+t\sigma^2)}$$

$$\pi_M^{N'} = \frac{1-(1-k)(1-(1-k)t-\theta)t-\theta}{2(1-\theta)(3+\theta)}$$

$$\pi_{SC}^{N'} = \frac{1-\theta-t^3(11+(2-k)k-4\theta(2+\theta))\sigma^2}{4(1-\theta)(3+\theta)(1+t\sigma^2)} - \frac{t^2(11+2k(1+2(2-\theta)\theta k)+(1-k)(1-\theta)\sigma^2)}{4(1-\theta)(3+\theta)(1+t\sigma^2)}$$
$$-\frac{t(1-k-\sigma^2(1+4\sigma^2))}{4(3+\theta)(1+t\sigma^2)} + \frac{1-(1-k)(1-(1-k)t-\theta)t-\theta}{2(1-\theta)(3+\theta)}$$

8. The new utility function in the case of sharing in the market information (Y′):

$$p_A^{Y'} = \frac{5+\theta-(1-k)t}{2(3+\theta)} + \frac{t\Gamma(2+\theta)\sigma^2}{(3+\theta)(1+t\sigma^2)}$$

$$p_B^{Y'} = \frac{1+\theta(4+\theta)+t(4+\theta+k(2+\theta))}{2(3+\theta)} + \frac{t\Gamma(5+\theta)\sigma^2}{2(3+\theta)(1+t\sigma^2)}$$

$$q_A^{Y'} = \frac{1}{2(3+\theta)}\left(2+\Gamma+\theta-\frac{\Gamma}{1+t\sigma^2}\right) - \frac{(1-k)t(1+\theta)}{2(1-\theta)(3+\theta)}$$

$$q_B^{Y'} = \frac{2(1-k)t}{2(1-\theta)(3+\theta)} - \frac{1}{2(3+\theta)}\left(1-\Gamma+\frac{\Gamma}{1+t\sigma^2}\right)$$

$$w_A^{Y'} = \frac{1}{2}\left(1+\Gamma-\frac{\Gamma}{1+t\sigma^2}\right)$$

$$w_B^{Y'} = \frac{1}{2}\left((1-k)t+\Gamma+\theta-\frac{\Gamma}{1+t\sigma^2}\right)$$

$$\pi_R^{Y'} = \frac{1-\theta-t^3(11+(2-k)k-4\theta(2+\theta))\sigma^2}{4(1-\theta)(3+\theta)(1+t\sigma^2)} - \frac{t^2(11+k(2-k)-4\theta(2+\theta)+(1-k)(1-\theta)\sigma^2)}{4(1-\theta)(3+\theta)(1+t\sigma^2)}$$
$$-\frac{t(1-k-\sigma^2(1+\sigma^2))}{4(3+\theta)(1+t\sigma^2)}$$

$$\pi_M^{Y'} = \frac{1-\theta+(1-k)^2t^3\sigma^2+(1-k)t^2(1-k-\sigma^2(1-\theta))-t(1-\theta)(1-k+\sigma^2(1+\sigma^2))}{2(1-\theta)(3+\theta)(1+t\sigma^2)}$$

$$\pi_{SC}^{Y'} = \frac{1-\theta-t^3(11+(2-k)k-4\theta(2+\theta))\sigma^2}{4(1-\theta)(3+\theta)(1+t\sigma^2)} - \frac{t^2(11+k(2-k)-4\theta(2+\theta)+(1-k)(1-\theta)\sigma^2)}{4(1-\theta)(3+\theta)(1+t\sigma^2)}$$
$$-\frac{t(1-k-\sigma^2(1+\sigma^2))}{4(3+\theta)(1+t\sigma^2)} + \frac{1-\theta+(1-k)^2t^3\sigma^2+(1-k)t^2(1-k-\sigma^2(1-\theta))-t(1-\theta)(1-k+\sigma^2(1+\sigma^2))}{2(1-\theta)(3+\theta)(1+t\sigma^2)}$$

9. The supply chain uses the solution of information sharing in the case of uncovering all the market share (Y-N):

$$p_A^{Y-N} = \frac{3V}{4}$$

$$p_B^{Y-N} = \frac{1}{4}\left(kt+3\left(lt+V\theta+\Gamma-\frac{\Gamma}{1+t\sigma^2}\right)\right)$$

$$\lambda_A^{Y-N} = \frac{V}{2\rho}$$

$$\lambda_B^{Y-N} = \frac{1}{2\rho}\left(V\theta+\Gamma-\frac{\Gamma}{1+t\sigma^2}-(k-l)t\right)$$

$$w_A^{Y-N} = \frac{V}{2}$$

$$w_B^{Y-N} = \frac{1}{2}\left(V\theta + \Gamma - \frac{\Gamma}{1+t\sigma^2} - (k-l)t\right)$$

$$\pi_R^{Y-N} = \frac{\left((k-l)^2t^2 + V\left(V\left(1+\theta^2\right) - 2(k-l)t\theta\right)\right)M + t\left(M\left(V^2 + (-kt+lt+V\theta)^2\right) - 8t^2\rho\right)\sigma^2 + Mt\sigma^4 - 8t^2\rho}{8(\rho + t\rho\sigma^2)}$$

$$\pi_M^{Y-N} = \frac{M\left(V^2 + (kt-lt-V\theta)^2 + t\left(V^2 + (kt-lt-V\theta)^2\right)\sigma^2 + t\sigma^4\right)}{4(\rho + t\rho\sigma^2)}$$

$$\pi_{SC}^{Y-N} = \frac{\left((k-l)^2t^2 + V\left(V\left(1+\theta^2\right) - 2(k-l)t\theta\right)\right)M + t\left(M\left(V^2 + (-kt+lt+V\theta)^2\right) - 8t^2\rho\right)\sigma^2 + Mt\sigma^4 - 8t^2\rho}{8(\rho + t\rho\sigma^2)} + \frac{M\left(V^2 + (kt-lt-V\theta)^2 + t\left(V^2 + (kt-lt-V\theta)^2\right)\sigma^2 + t\sigma^4\right)}{4(\rho + t\rho\sigma^2)}$$

10. The supply chain uses the solution of information sharing in the case of uncovering all the market share and cost-sharing (Y-N-C):

$$\pi_R^{Y-N-C} = \frac{\left((k-l)^2t^2 + V\left(V\left(1+\theta^2\right) - 2(k-l)t\theta\right)\right)M + t\left(M\left(V^2 + (-kt+lt+V\theta)^2\right) - 8t^2\rho\right)\sigma^2 + Mt\sigma^4 - 8t^2\rho}{8(\rho + t\rho\sigma^2)} + F$$

$$\pi_M^{Y-N-C} = \frac{M\left(V^2 + (kt-lt-V\theta)^2 + t\left(V^2 + (kt-lt-V\theta)^2\right)\sigma^2 + t\sigma^4\right)}{4(\rho + t\rho\sigma^2)} - F$$

11. The supply chain uses the solution of information sharing in the case of uncovering all the market share and alliance (Y-N-CO):

$$p_A^{Y-N-CO} = \frac{V}{2}$$

$$p_B^{Y-N-CO} = \frac{1}{2}\left((k+l)t + \Gamma + V\theta - \frac{\Gamma}{1+t\sigma^2}\right)$$

$$\lambda_A^{Y-N-CO} = \frac{V}{\rho}$$

$$\lambda_B^{Y-N} = \frac{V\theta + \Gamma - (k-l)t}{\rho} - \frac{\Gamma}{(1+t\sigma^2)\rho}$$

$$w_A^{Y-N-CO} = \frac{V}{2}$$

$$w_B^{Y-N-CO} = \frac{1}{2}\left(V\theta + \Gamma - (k-l) - \frac{\Gamma}{1+t\sigma^2}\right)$$

$$\pi_{SC}^{Y-N-CO}=\frac{M\left(V^2+(V\theta-kt+lt)^2\right)-2t^2\rho+t\left(M\left(V^2+(V\theta-kt+lt)^2\right)-2t^2\rho\right)\sigma^2+Mt\sigma^4}{2(\rho+t\rho\sigma^2)}$$

12. The supply chain uses the solution of information sharing in the case of delay strategy (Y-DE):

(1) $U_A^{Y-DE}>U_B^{Y-DE}$

$$p_A^{Y-DE}=\frac{6\delta+t(1-\delta)(1-k\delta)-\theta+\frac{t\Gamma\theta\sigma^2}{1+t\sigma^2}-\frac{t\Gamma(1+\delta)(2+\delta)\theta^2\sigma^2}{1+t\sigma^2}-\delta\theta\left(4+\delta-5\Gamma\left(1-\frac{1}{1+t\sigma^2}\right)\right)}{8\delta-2(1+\delta)^2\theta}$$

$$p_B^{Y-DE}=\frac{2t\delta(3+k\delta)-t(1+\delta)(1+(2+k)\delta)\theta}{2\delta\left(4\delta-(1+\delta)^2\theta\right)}-\frac{1}{2\delta\left(4\delta-(1+\delta)^2\theta\right)}\left(\frac{\theta(\theta-\delta(5-3\theta+\delta(1-2\theta))+t(\theta-\delta(5+6\Gamma+\delta)+(\Gamma+\Gamma\delta(4+\delta)+\delta(3+2\delta))\theta)\sigma^2)}{1+t\sigma^2}\right)$$

$$q_A^{Y-DE}=\frac{\delta(2-\theta+\Gamma\theta)+\dfrac{\Gamma(1-\delta)\theta}{1+t\sigma^2}-t(1+\delta)(1-k\delta)-(1+\Gamma)\theta}{8\delta-2(1+\delta)^2\theta}$$

$$q_B^{Y-DE}=\frac{\delta\left(2t(1-k\delta)+\theta\left(1-\delta+\dfrac{t\Gamma(2-\theta-\delta\theta)\sigma^2}{1+t\sigma^2}\right)\right)}{2\theta\left(4\delta-(1+\delta)^2\theta\right)}$$

$$w_A^{Y-DE}=\frac{1}{2}\left(1+\frac{t\Gamma\theta\sigma^2}{1+t\sigma^2}\right)$$

$$w_B^{Y-DE}=\frac{t-kt\delta+\theta+\dfrac{t\Gamma\theta\sigma^2}{1+t\sigma^2}}{2\delta}$$

$$\pi_R^{Y-DE}=\frac{t(1-\delta)(1-k\delta)\theta+t^2\left((1-k\delta)^2-16\delta\theta+4(1+\delta)^2\theta^2\right)+\dfrac{(1-\theta)\theta(\delta+t\delta\sigma^2+t\theta\sigma^4)}{1+t\sigma^2}}{4\theta\left(4\delta-(1+\delta)^2\theta\right)}$$

$$\pi_M^{Y-DE}=\frac{t^2(1-k\delta)^2+t(1-\delta)(1-k\delta)\theta+\dfrac{(1-\theta)\theta(\delta+t\delta\sigma^2+t\theta\sigma^4)}{1+t\sigma^2}}{2\theta\left(4\delta-(1+\delta)^2\theta\right)}$$

$$\pi_{SC}^{Y-DE}=\frac{t(1-\delta)(1-k\delta)\theta+t^2\left((1-k\delta)^2-16\delta\theta+4(1+\delta)^2\theta^2\right)+\dfrac{(1-\theta)\theta(\delta+t\delta\sigma^2+t\theta\sigma^4)}{1+t\sigma^2}}{4\theta\left(4\delta-(1+\delta)^2\theta\right)}+\frac{t^2(1-k\delta)^2+t(1-\delta)(1-k\delta)\theta+\dfrac{(1-\theta)\theta(\delta+t\delta\sigma^2+t\theta\sigma^4)}{1+t\sigma^2}}{2\theta\left(4\delta-(1+\delta)^2\theta\right)}$$

(2) $U_A^{Y-DE}<U_B^{Y-DE}$

$$p_A^{Y-DE} = \frac{\begin{pmatrix} -\delta^2(1-kt+t(1-kt+\Gamma)\sigma^2)+\theta(t-\theta+t(t+\Gamma-\theta-2\Gamma\theta)\sigma^2) \\ +\delta(6-t-4\theta-kt\theta+t(6-t+5\Gamma-(4+kt+3\Gamma)\theta)\sigma^2) \end{pmatrix}}{2(2\delta(2-\theta)-\delta^2-\theta^2)(1+t\sigma^2)}$$

$$p_B^{Y-DE} = \frac{2t\delta(3+k\delta)-t(1+\delta)(1+(2+k)\delta)\theta}{2\delta\left(4\delta-(1+\delta)^2\theta\right)} - \frac{1}{2\delta\left(4\delta-(1+\delta)^2\theta\right)}\left(\frac{\theta(\theta-\delta(5-3\theta+\delta(1-2\theta))+t(\theta-\delta(5+6\Gamma+\delta)+(\Gamma+\Gamma\delta(4+\delta)+\delta(3+2\delta))\theta)\sigma^2)}{1+t\sigma^2}\right)$$

$$q_A^{Y-DE} = \frac{\delta(2-\theta+\Gamma\theta)+\frac{\Gamma(1-\delta)\theta}{1+t\sigma^2}-t(1+\delta)(1-k\delta)-(1+\Gamma)\theta}{8\delta-2(1+\delta)^2\theta}$$

$$q_B^{Y-DE} = \frac{\delta\left(2t(1-k\delta)+\theta\left(1-\delta+\frac{t\Gamma(2-\theta-\delta\theta)\sigma^2}{1+t\sigma^2}\right)\right)}{2\theta\left(4\delta-(1+\delta)^2\theta\right)}$$

$$w_A^{Y-DE} = \frac{1}{2}\left(1+\frac{t\Gamma\theta\sigma^2}{1+t\sigma^2}\right)$$

$$w_B^{Y-DE} = \frac{t-kt\delta+\theta+\frac{t\Gamma\theta\sigma^2}{1+t\sigma^2}}{2\delta}$$

$$\pi_R^{Y-DE} = \frac{t(1-\delta)(1-k\delta)\theta+t^2\left((1-k\delta)^2-16\delta\theta+4(1+\delta)^2\theta^2\right)+\frac{(1-\theta)\theta(\delta+t\delta\sigma^2+t\theta\sigma^4)}{1+t\sigma^2}}{4\theta\left(4\delta-(1+\delta)^2\theta\right)}$$

$$\pi_M^{Y-DE} = \frac{t^2(1-k\delta)^2+t(1-\delta)(1-k\delta)\theta+\frac{(1-\theta)\theta(\delta+t\delta\sigma^2+t\theta\sigma^4)}{1+t\sigma^2}}{2\theta\left(4\delta-(1+\delta)^2\theta\right)}$$

$$\pi_{SC}^{Y-DE} = \frac{t(1-\delta)(1-k\delta)\theta+t^2\left((1-k\delta)^2-16\delta\theta+4(1+\delta)^2\theta^2\right)+\frac{(1-\theta)\theta(\delta+t\delta\sigma^2+t\theta\sigma^4)}{1+t\sigma^2}}{4\theta\left(4\delta-(1+\delta)^2\theta\right)} + \frac{t^2(1-k\delta)^2+t(1-\delta)(1-k\delta)\theta+\frac{(1-\theta)\theta(\delta+t\delta\sigma^2+t\theta\sigma^4)}{1+t\sigma^2}}{2\theta\left(4\delta-(1+\delta)^2\theta\right)}$$

13. The supply chain uses the solution of information sharing in the case of new technology of big data (Y-NEW):

$$BCB = b/c:$$

$$p_A^{Y-NEW} = \frac{3}{4}\left(1+\frac{t\Gamma\theta\sigma^2}{1+t\sigma^2}\right)$$

$$p_B^{Y-NEW} = \frac{1}{4}\left(3b+c+(3+k)t+3\theta\left(1+\Gamma-\frac{\Gamma}{1+t\sigma^2}\right)\right)$$

$$q_A^{Y-NEW} = \frac{1-b+c-t+kt-\theta}{4(1-\theta)}$$

$$q_B^{Y-NEW} = \frac{b-c+(1-k)t+\dfrac{t\Gamma(1-\theta)\theta\sigma^2}{1+t\sigma^2}}{4(1-\theta)\theta}$$

$$w_A^{Y-NEW} = \frac{1}{2}\left(1+\frac{t\Gamma\theta\sigma^2}{1+t\sigma^2}\right)$$

$$w_B^{Y-NEW} = \frac{1}{2}\left(b-c+t-kt+\theta+\frac{t\Gamma\theta\sigma^2}{1+t\sigma^2}\right)$$

$$\pi_R^{Y-NEW} = \frac{1}{16}\left(1-16t^2+\frac{(b-c+t-kt)^2}{1-\theta}+\frac{(b-c+t-kt)^2}{\theta}+\frac{t\theta\sigma^4}{1+t\sigma^2}\right)+F$$

$$\pi_M^{Y-NEW} = \frac{(b-c+t-kt)^2+(1-\theta)\theta+t\left((b-c+t-kt)^2+\theta-\theta^2\right)\sigma^2+t(1-\theta)\theta^2\sigma^4}{8(1-\theta)\theta(1+t\sigma^2)}-F$$

$$\pi_{SC}^{Y-NEW} = \tfrac{1}{16}\left(1-16t^2+\frac{(b-c+t-kt)^2}{1-\theta}+\frac{(b-c+t-kt)^2}{\theta}+\tfrac{t\theta\sigma^4}{1+t\sigma^2}\right)+ \frac{(b-c+t-kt)^2+(1-\theta)\theta+t\left((b-c+t-kt)^2+\theta-\theta^2\right)\sigma^2+t(1-\theta)\theta^2\sigma^4}{8(1-\theta)\theta(1+t\sigma^2)}$$

Appendix B. Proof of Propositions

Appendix B.1. Proposition 1

$$c_B^N - c_B^Y = \frac{4t(1-\theta)\theta(1+t\sigma^2)}{1-k+((1-k)t+2\Gamma(1-\theta)\theta)\sigma^2} - \frac{4t(1-\theta)\theta(1+t\sigma^2)}{1-k-((1-k)t+\Gamma(1-\theta)\theta)\sigma^2}$$
$$\Rightarrow c_B^N < c_B^Y$$

Appendix B.2. Proposition 2

$$\pi_R^N - \pi_R^Y = \tfrac{1}{16}\left(1+\frac{t^2(1-(2-k)k-16(1-\theta)\theta)}{(1-\theta)\theta}+\frac{4t\theta\sigma^4}{1+t\sigma^2}\right)-\tfrac{1}{16}\left(1+\frac{t^2(1-(2-k)k-16(1-\theta)\theta)}{(1-\theta)\theta}+\frac{t\theta\sigma^4}{1+t\sigma^2}\right)$$
$$= \frac{3t\theta\sigma^4}{16(1+t\sigma^2)}$$
$$\Rightarrow \pi_R^N > \pi_R^Y$$

$$\pi_M^N - \pi_M^Y = \tfrac{1}{8}\left(1+\frac{(1-k)^2t^2}{(1-\theta)\theta}\right)-\tfrac{1}{8}\left(1+\frac{(1-k)^2t^2}{(1-\theta)\theta}+\frac{t\theta\sigma^4}{1+t\sigma^2}\right)$$
$$= -\frac{t\theta\sigma^4}{8(1+t\sigma^2)}$$
$$\Rightarrow \pi_M^N < \pi_M^Y$$

$$\pi_{SC}^{N}-\pi_{SC}^{Y}=\tfrac{1}{16}\left(3+\frac{t^{2}\left(3(1-k)^{2}-16(1-\theta)\theta\right)}{(1-\theta)\theta}+\tfrac{4t\theta\sigma^{4}}{1+t\sigma^{2}}\right)-\tfrac{1}{16}\left(3+\frac{t^{2}\left(3(1-k)^{2}-16(1-\theta)\theta\right)}{(1-\theta)\theta}+\frac{3t\theta\sigma^{4}}{1+t\sigma^{2}}\right)$$
$$=\frac{t\theta\sigma^{4}}{16+16t\sigma^{2}}$$
$$\Rightarrow\pi_{M}^{N}>\pi_{M}^{Y}$$

Appendix B.3. Proposition 3

The proof is same as Proposition 2, which omits this section.

Appendix B.4. Proposition 4

The proof is same as Proposition 2, which omits this section.

Appendix B.5. Proposition 5

The proof is same as Proposition 2, which omits this section.

Appendix B.6. Proposition 6

The proof is same as Proposition 2, which omits this section.

Appendix B.7. Proposition 7

The proof is same as Proposition 2, which omits this section.

Appendix B.8. Proposition 8

The proof is same as Proposition 2, which omits this section.

Appendix B.9. Proposition 9

The proof is same as Proposition 2, which omits this section.

Appendix B.10. Proposition 10

The proof is same as Proposition 2, which omits this section.

References

1. Yetis, H.; Karakose, M.; Baygin, N. Blockchain-based mass customization framework using optimized production management for industry 4.0 applications. *Eng. Sci. Technol. Int. J.* **2022**, *36*, 101151. [CrossRef]
2. Ganji, E.N.; Shah, S.; Coutroubis, A. Mass customization and technology functions: Enhancing market responsiveness. In Proceedings of the 2018 IEEE International Conference on Engineering, Technology and Innovation (ICE/ITMC), Stuttgart, Germany, 17–20 June 2018; 1–6. [CrossRef]
3. Iarovyi, S.; Lastra, J.L.M.; Haber, R.; del Toro, R. From artificial cognitive systems and open architectures to cognitive manufacturing systems. In Proceedings of the 2015 IEEE 13th International Conference on Industrial Informatics (INDIN), Cambridge, UK, 22–24 July 2015; 1225–1232. [CrossRef]
4. Karaköse, M.; Yetiş, H. A cyberphysical system based mass-customization approach with integration of Industry 4.0 and smart city. *Wirel. Commun. Mob. Comput.* **2017**, *2017*, 1058081. [CrossRef]
5. Investopedia. Mass customization. Investopedia. Available online: https://www.investopedia.com/terms/m/masscustomization.asp#:~:text=Mass%20customization%20allows%20a%20customer,a%20semi-custom%20final%20product (accessed on 5 September 2022).
6. Singh, S.; Sarin, S.C. Modeling and analysis of a new production methodology for achieving mass customization. *Int. J. Prod. Res.* **2024**, *62*, 183–203. [CrossRef]
7. Lee, G.; Kim, H.Y. Human vs. AI: The battle for authenticity in fashion design and consumer response. *J. Retail. Consum. Serv.* **2024**, *77*, 103690. [CrossRef]
8. Mindas, M.; Bednar, S. Mass customization in the context of Industry 4.0: Implications of variety-induced complexity. *Adv. Ind. Eng.* **2016**, *21*, 21–38.

9. Aheleroff, S.; Philip, R.; Zhong, R.Y.; Xu, X. The degree of mass personalisation under Industry 4.0. *Procedia CIRP* **2019**, *81*, 1394–1399. [CrossRef]
10. Ouyang, L.; Yuan, Y.; Wang, F.-Y. A blockchain-based framework for collaborative production in distributed and social manufacturing. In Proceedings of the 2019 IEEE International Conference on Service Operations and Logistics, and Informatics (SOLI), Zhengzhou, China, 6–8 November 2019; pp. 76–81. [CrossRef]
11. Song, Z.; Sun, Y.; Wan, J.; Huang, L.; Zhu, J. Smart e-commerce systems: Current status and research challenges. *Electron. Mark.* **2019**, *29*, 221–238. [CrossRef]
12. Dou, Z.; Sun, Y.; Wu, Z.; Wang, T.; Fan, S.; Zhang, Y. The architecture of mass customization–social Internet of Things system: Current research profile. *ISPRS Int. J. Geo-Inf.* **2021**, *10*, 653. [CrossRef]
13. Kouhpayeh, M.G.; Ziari, E.D.; Rousta, A. Designing and offering a customer trust model in the marketing ecosystem based on blockchain technology. *Int. J. Innov. Manag. Organ. Behav.* **2024**, *4*, 1–10. [CrossRef]
14. Qu, Y.; Gao, L.; Xiang, Y.; Shen, S.; Yu, S. FedTwin: Blockchain-enabled adaptive asynchronous federated learning for digital twin networks. *IEEE Netw.* **2022**, *36*, 183–190. [CrossRef]
15. Guo, S.; Choi, T.-M.; Chung, S.-H. Self-design fun: Should 3D printing be employed in mass customization operations? *Eur. J. Oper. Res.* **2022**, *299*, 883–897. [CrossRef]
16. Jain, P.; Garg, S.; Kansal, G. A TISM approach for the analysis of enablers in implementing mass customization in Indian manufacturing units. *Prod. Plan. Control* **2023**, *34*, 173–188. [CrossRef]
17. Longo, F.; Padovano, A.; Cimmino, B.; Pinto, P. Towards a mass customization in the fashion industry: An evolutionary decision aid model for apparel product platform design and optimization. *Comput. Ind. Eng.* **2021**, *162*, 107742. [CrossRef]
18. Qi, Y.; Mao, Z.; Zhang, M.; Guo, H. Manufacturing practices and servitization: The role of mass customization and product innovation capabilities. *Int. J. Prod. Econ.* **2020**, *228*, 107747. [CrossRef]
19. Shao, X.-F. What is the right production strategy for horizontally differentiated product: Standardization or mass customization? *Int. J. Prod. Econ.* **2020**, *223*, 107527. [CrossRef]
20. Wu, Q.; Liao, K.; Deng, X.; Marsillac, E. Achieving automotive suppliers' mass customization through modularity. *J. Manuf. Technol. Manag.* **2019**, *31*, 306–329. [CrossRef]
21. Qu, L.; Wen, F.; Huang, H.; Wang, Z. Aggregation-chain: A consortium blockchain based multi-chain data sharing framework with efficient query. *Cluster Comput.* **2025**, *28*, 46. [CrossRef]
22. Xu, X.; Chen, X.; Chen, J.; Cheng, T.C.E.; Yu, Y.; Liu, S.S. Adoption of blockchain considering platform's information sharing and service effort under the cap-and-trade scheme. *Int. J. Prod. Res.* **2024**, *62*, 6688–6712. [CrossRef]
23. Zheng, K.; Zhang, Z.; Chen, Y.; Wu, J. Blockchain adoption for information sharing: Risk decision-making in spacecraft supply chain. *Enterp. Inf. Syst.* **2021**, *15*, 8. [CrossRef]
24. Du, M.; Chen, Q.; Chen, J.; Ma, X. An optimized consortium blockchain for medical information sharing. *IEEE Trans. Eng. Manag.* **2021**, *68*, 1677–1689. [CrossRef]
25. Dwivedi, S.K.; Amin, R.; Vollala, S. Blockchain-based secured information sharing protocol in supply chain management system with key distribution mechanism. *J. Inf. Secur. Appl.* **2020**, *54*, 102554. [CrossRef]
26. Wang, Z.; Wang, T.; Hu, H.; Gong, J.; Ren, X.; Xiao, Q. Blockchain-based framework for improving supply chain traceability and information sharing in precast construction. *Autom. Constr.* **2020**, *111*, 103063. [CrossRef]
27. Si, H.; Sun, C.; Li, Y.; Qiao, H.; Shi, L. IoT information sharing security mechanism based on blockchain technology. *Future Gener. Comput. Syst.* **2019**, *101*, 1028–1040. [CrossRef]
28. Crettez, B.; Hayek, N.; Martín-Herrán, G. Existence of equilibrium in a dynamic supply chain game with vertical coordination, horizontal competition, and complementary goods. *Eur. J. Oper. Res.* **2025**, *321*, 302–313. [CrossRef]
29. Quadir, A.; Raj, A.; Foropon, C.R. Sharing demand information in competing supply chains with greening efforts. *J. Environ. Manag.* **2025**, *373*, 123626. [CrossRef] [PubMed]
30. Gago-Rodríguez, S.; Márquez-Illescas, G.; Núñez-Nickel, M. Bargaining power as moderator of the "delay costs effect" in supply chain negotiations. *Manag. Account. Res.* **2021**, *51*, 100737. [CrossRef]
31. Hosseini, S.M.; Paydar, M.M. Discount and advertisement in ecotourism supply chain. *Asia Pac. J. Tour. Res.* **2021**, *26*, 668–684. [CrossRef]
32. Zhang, C.; Wang, Y.; Ma, P. Optimal channel strategies in a supply chain under green manufacturer financial distress with advance payment discount. *Int. Trans. Oper. Res.* **2021**, *28*, 1347–1370. [CrossRef]
33. He, P.; He, Y.; Shi, C.V.; Xu, H.; Zhou, L. Cost-sharing contract design in a low-carbon service supply chain. *Comput. Ind. Eng.* **2020**, *139*, 106160. [CrossRef]
34. Wu, D.; Chen, J.; Li, P.; Zhang, R. Contract coordination of dual channel reverse supply chain considering service level. *J. Clean. Prod.* **2020**, *260*, 121071. [CrossRef]
35. Yu, B.; Wang, J.; Lu, X.; Yang, H. Collaboration in a low-carbon supply chain with reference emission and cost learning effects: Cost sharing versus revenue sharing strategies. *J. Clean. Prod.* **2020**, *250*, 119460. [CrossRef]

36. Wang, Z.; Brownlee, A.E.I.; Wu, Q. Production and joint emission reduction decisions based on two-way cost-sharing contract under cap-and-trade regulation. *Comput. Ind. Eng.* **2020**, *146*, 106549. [CrossRef]
37. Fan, J.; Ni, D.; Fang, X. Liability cost sharing, product quality choice, and coordination in two-echelon supply chains. *Eur. J. Oper. Res.* **2020**, *284*, 514–537. [CrossRef]
38. Gilotra, M.; Pareek, S.; Mittal, M.; Dhaka, V. Effect of Carbon Emission and Human Errors on a Two-Echelon Supply Chain under Permissible Delay in Payments. *Int. J. Math. Eng. Manag. Sci.* **2020**, *5*, 225–236. [CrossRef]
39. Raza, S.A. Supply chain coordination under a revenue-sharing contract with corporate social responsibility and partial demand information. *Int. J. Prod. Econ.* **2018**, *205*, 1–14. [CrossRef]
40. Jullien, B.; Reisinger, M.; Rey, P. Personalized Pricing and Distribution Strategies. *Manag. Sci.* **2023**, *69*, 1687–1702. [CrossRef]
41. Jiang, B.; Yang, B. Quality and Pricing Decisions in a Market with Consumer Information Sharing. *Manag. Sci.* **2019**, *65*, 272–285. [CrossRef]
42. Ha, A.Y.; Tian, Q.; Tong, S. Information Sharing in Competing Supply Chains with Production Cost Reduction. *Manuf. Serv. Oper. Manag.* **2017**, *19*, 246–262. [CrossRef]
43. Wu, X.; Zhang, F. Home or Overseas? An Analysis of Sourcing Strategies Under Competition. *Manag. Sci.* **2014**, *60*, 1223–1240. [CrossRef]
44. Khanjari, N.E.; Iravani, S.; Shin, H. The Impact of the Manufacturer-Hired Sales Agent on a Supply Chain with Information Asymmetry. *Manuf. Serv. Oper. Manag.* **2014**, *16*, 76–88. [CrossRef]
45. Yang, X.; Cai, G.; Ingene, C.A.; Zhang, J. Manufacturer Strategy on Service Provision in Competitive Channels. *Prod. Oper. Manag.* **2020**, *29*, 72–89. [CrossRef]
46. Wang, J.; Liu, J.; Wang, F.; Yue, X. Blockchain technology for port logistics capability: Exclusive or sharing. *Transp. Res. Part B: Methodol.* **2021**, *149*, 347–392. [CrossRef]
47. Luo, S.; Choi, T. E-commerce supply chains with considerations of cyber-security: Should governments play a role? *Prod. Oper. Manag.* **2022**, *31*, 2107–2126. [CrossRef]

mathematics

MDPI

Article

Manipulation Game Considering No-Regret Strategies

Julio B. Clempner

Instituto Politécnico Nacional, Mexico City 07320, Mexico; julio@clempner.name or jclempnerk@ipn.mx

Abstract: This paper examines manipulation games through the lens of Machiavellianism, a psychological theory. It analyzes manipulation dynamics using principles like hierarchical perspectives, exploitation tactics, and the absence of conventional morals to interpret interpersonal interactions. Manipulators intersperse unethical behavior within their typical conduct, deploying deceptive tactics before resuming a baseline demeanor. The proposed solution leverages Lyapunov theory to establish and maintain Stackelberg equilibria. A Lyapunov-like function supports each asymptotically stable equilibrium, ensuring convergence to a Nash/Lyapunov equilibrium if it exists, inherently favoring no-regret strategies. The existence of an optimal solution is demonstrated via the Weierstrass theorem. The game is modeled as a three-level Stackelberg framework based on Markov chains. At the highest level, manipulators devise strategies that may not sway middle-level manipulated players, who counter with best-reply strategies mirroring the manipulators' moves. Lower-level manipulators adjust their strategies in response to the manipulated players to sustain the manipulation process. This integration of stability analysis and strategic decision-making provides a robust framework for understanding and addressing manipulation in interpersonal contexts. A numerical example focusing on the oil market and its regulations highlights the findings of this work.

Keywords: Machiavellianism; manipulation; no regret; Markov chains; game theory

MSC: 91A10; 91A27; 91A50; 91A65; 91A80; 90C40

1. Introduction

1.1. Brief Review

The inclusion of believable behavioral assumptions into traditional economic models has sparked extensive debate. Manipulation emerges as a prominent concern, with game-theoretic studies revealing significant individual disparities in scenarios allowing such behavior. This work focuses on Machiavellianism—a personality trait associated with manipulation, callousness, and moral indifference—as studied in personality psychology. While enriching economic models with believable behavioral assumptions enhances their descriptive accuracy, it also introduces complexities and challenges in predicting human behavior and its economic outcomes. Thus, understanding the implications of incorporating these assumptions is crucial for advancing economic theory and policy [1].

Christie and Geis [2] utilized adapted Machiavellian remarks to investigate variations in human behavior, drawing from Machiavelli's dichotomy of individuals as either manipulators or the manipulated. This classification reflects a worldview steeped in power dynamics and strategic interactions. Machiavellian individuals are characterized by a propensity to exploit others, harbor a pessimistic view of human nature, and demonstrate a lack of moral accountability. The concept of Machiavellianism underscores the absence of coordination in self-serving, conflict-laden situations, particularly in games where interests diverge.

Christie and Geis further delineated Machiavellian traits into three key components:

- Worldview shaped by manipulators and manipulated (views component), lacking sympathy; manipulated individuals also engage in manipulation.
- Lack of regard for conventional morality (morality component), indifferent to actions like lying and cheating.
- Emphasis on pragmatic problem-solving over ideological commitments, utilizing power strategies (tactics component) for personal goals rather than ideological ones.

Understanding Machiavellianism offers valuable insights into the intricacies of social dynamics, decision-making, and interpersonal relationships. This psychological framework helps unravel how individuals navigate power structures, negotiate conflicting interests, and strategize to achieve personal goals. By dissecting Machiavellian traits—such as manipulation, pragmatic problem-solving, and moral flexibility—researchers can better understand the complexities of human behavior. It sheds light on how individuals employ strategic thinking in diverse contexts, from interpersonal interactions to organizational settings, revealing patterns of influence, persuasion, and control. Such analysis is crucial for comprehending motivations, fostering cooperation, and addressing conflicts in various social and professional environments [3,4].

1.2. Related Work

Allen and Gorton [5] delved into the concept of manipulation equilibrium, emphasizing the impact of stock price dynamics and their implications for welfare. Their work highlights how manipulation strategies can disrupt market equilibrium, leading to unintended consequences for various stakeholders. Building on this, Bagnoli and Lipman [6] demonstrated that efforts to manipulate stock prices often result in a decline in pre-bid stock prices, which can inflate takeover bids. This inflation can create barriers to successful acquisitions, particularly in times when the market experiences limited takeover activity, effectively creating an environment where genuine value is obscured by artificial price alterations.

Furthering the discussion on manipulation, Clempner [7] introduced a comprehensive game-theoretical framework designed to simulate such manipulation, integrating reinforcement learning techniques to explore moral considerations surrounding these strategies. This innovative approach sheds light on the ethical dimensions of decision-making in financial markets, illustrating how self-interest can clash with broader market integrity.

In a broader empirical study, Cumming et al. [8] examined a dataset comprising suspected instances of stock price manipulation across nine countries over an eight-year period. Their findings revealed significant adverse effects of market manipulation on innovation, suggesting that such practices stifle creativity and investment in new ideas. Interestingly, their research pointed out that the detrimental impacts are particularly severe in jurisdictions characterized by weak intellectual property rights coupled with strong shareholder protections, indicating that the regulatory environment plays a critical role in shaping market behavior. Clempner's contributions continued with the development of a manipulation model tailored for repeated Stackelberg security games [9], which explores strategic interactions over time among players with hierarchical roles. Additionally, Clempner [10] introduced a Machiavellian game model founded on ergodic Bayesian Markov processes, further enriching the discourse on strategic manipulation in competitive environments. This work addresses how players with Machiavellian traits might exploit informational asymmetries to gain an advantage. Finally, significant advancements in the study of Machiavellianism games were presented by [11], who explored the nuances of strategic deception and manipulation within competitive settings, emphasizing the psychological and behavioral aspects that drive such strategies. Collectively, these studies provide a comprehensive view of how manipulation in financial markets not only affects stock prices but also has far-reaching implications for overall market efficiency and ethical standards.

In this theoretical framework, the sender, armed with private information, formulates a signaling strategy designed to influence the decision-making process of the receiver. This strategic communication is crucial because it allows the sender to steer the receiver's actions in a desired direction. To explore extensions of this model that incorporate multiple senders, one can refer to the research conducted by Milgrom [12] and Krishna [13], who delve into the dynamics and interactions that arise when several senders attempt to convey information to a common receiver.

The idea of private signal-based persuasion has been investigated in diverse scenarios, ranging from two-player, two-action games [14] to unanimity voting systems [15]. These contexts provide valuable insights into how information asymmetries can shape competitive interactions and collective decision-making. Focusing specifically on the interaction between a single sender and a receiver, Kamenica and Gentzkow [16] developed a Bayesian persuasion framework. This model outlines the conditions under which the signals sent by the sender can be beneficial and delineates optimal strategies for signaling that enhance the sender's objectives. Bergemann and Morris [17] further explored the role of a mechanism designer within the Bayesian persuasion framework. Their research examines how such a designer selects an appropriate information structure to fulfill specific objectives, shedding light on the strategic considerations involved in information dissemination and persuasion. Expanding on their previous findings, Gentzkow and Kamenica [18] introduced a model with multiple senders, analyzing scenarios where a vast array of potential signals exists. They employed a lattice structure to facilitate intuitive comparisons and combinations of these signals, enriching the understanding of how senders can effectively convey their information in a competitive landscape.

Additionally, Brocas et al. [19] proposed a novel model where two competing agents allocate resources to acquire public information about an uncertain state of the world. Their analysis focuses on the equilibrium strategies for information acquisition, revealing how these strategies evolve when one agent faces greater costs in gathering information. This research highlights the complexities involved in information competition and the tactical decisions that arise from resource allocation. In the realm of political communication, Gul and Pesendorfer [20] developed a model in which two senders with opposing interests share asymmetric information regarding a binary state of the world. This study illustrates how differing objectives can influence the nature of information conveyed and the strategic behavior of the senders involved. Moreover, several authors, including [21–24], have explored frameworks in which the information revealed by the sender is exclusively accessible to the recipients. These works emphasize the significance of control over information flow and the implications it has for decision-making processes in various contexts.

Collectively, these studies contribute to a deeper understanding of the mechanisms of persuasion and information asymmetry, highlighting the strategic complexities involved in both competitive and cooperative environments.

1.3. Main Results

This paper concerns manipulation games based on the psychological theory of Machiavellianism, which analyzes the world from the Machiavellian perspective using three guiding principles: view, tactics, and immoral behavior. View is related to the existence of manipulating and manipulated individuals, corresponding to a hierarchical scheme represented by a Stackelberg game where all the participants are involved in some degree of manipulation. Tactics are concerned with employing interpersonal exploitation strategies as a method for control. Immoral refers to a behavior lacking conventional morals that would prohibit their actions, which is designed using a Bayesian–Markov game theory approach. A manipulator is someone who may temporarily stray from steady conduct by committing an unethical act and then reverting to their usual behavior.

We propose a solution to the manipulation game problem involving Lyapunov theory, which supports the existence and stability of Nash equilibria: each asymptotically stable equilibrium point admits the so-called Lyapunov-like function that converges to a Nash/Lyapunov equilibrium point if it exists. The asymptotic behavior of a Lyapunov-like function admits by definition no-regret strategies.

The idea behind the representation of the game is that it is a three-level Stackelberg game, where the manipulating players choose a manipulation strategy at the highest level that might not convince the manipulated players in the middle level. A best-reply strategy that equals the manipulating strategy can be selected by the player who is being manipulated. In order to ensure the manipulation process, the manipulating player at the lower level then chooses a strategy once more based on the strategy of the manipulated players.

2. Markov Games

In a discrete-time, finite-horizon setting, we define an environment characterized by private and independent valuations [25]. Machiavellian players, denoted as $\iota \in \mathcal{N}$, where $\mathcal{N} = \{1, 2, ..., n\}$, incur a cost at each period $t \in T$, with $T \subseteq \mathbb{N}$. This cost is influenced by the current physical allocation $y_t^\iota \in Y_t^\iota$ and the player's type $\theta_t^\iota \in \Theta_t^\iota$. The cost function φ^ι for Machiavellian player ι is represented as $\varphi^\iota(y_t^\iota, \theta_t^\iota) \geq 0$, where φ^ι maps $Y_t^\iota \times \Theta_t^\iota$ to $\mathbb{R}^+$.

At time period t, the type vector is represented as $\theta_t = (\theta_t^1, ..., \theta_t^n) \in \Theta$, where Θ_t is defined as the product $\times_{\iota \in \mathcal{N}} \Theta_t^\iota$. The set of permissible allocations in period t can depend on the vector of prior allocations $y_t = (y_t^1, ..., y_t^n) \in Y$, where $Y_t = \times_{\iota \in \mathcal{N}} Y_t^\iota$. We denote by Θ_t^{-l} the product $\times_{\iota \in \mathcal{N}, \iota \neq l} \Theta_t^\iota$, and by $\Delta(Y_t)$ the set of all probability distributions over Y_t. It is assumed that Y_t^ι, Θ_t^ι, and T are finite sets.

For each Machiavellian player ι, there exists a shared prior distribution $\mu^\iota(\theta_0^\iota)$. The player's current type, θ_t^ι, along with their current action, y_t^ι, determines the probability distribution for the next period's state variable θ_{t+1}^ι on Θ_{t+1}^ι. We assume that this probability distribution is described by a transition function (or stochastic kernel) $F^\iota(\theta_{t+1}^\iota | \theta_t^\iota, y_t^\iota)$.

We assume that the type θ_t^ι of Machiavellian player ι follows a Markov process over the state space Θ_t^ι, and the dynamics adhere to the Markov condition, expressed as follows:

$$F^\iota(\theta_{t+1}^\iota | \theta_t^\iota, y_t^\iota, \theta_{t-1}^\iota, y_{t-1}^\iota \dots, \theta_0^\iota, y_0^\iota) = F^\iota(\theta_{t+1}^\iota | \theta_t^\iota, y_t^\iota).$$

The sequence of states and actions for each player ι is represented by $\{(\Theta_t, Y_t) | t\}$. In this case, the process described by a Markov chain is fully characterized by the transition function $F^\iota(\theta_{t+1}^\iota | \theta_t^\iota, y_t^\iota)$ and the common prior distribution $\mu^\iota(\theta_0^\iota)$, where $\mu^\iota(\theta_t^\iota)$ belongs to $\Delta(\Theta_t^\iota)$, the set of probability distributions over Θ_t^ι. It is further assumed that each chain $(\mu^\iota, F^\iota(\theta_{t+1}^\iota | \theta_t^\iota, y_t^\iota))$ is ergodic [9,25].

The cost functions $\varphi^\iota(y_t^\iota, \theta_t^\iota)$ and the transition functions $F^\iota(\theta_{t+1}^\iota | \theta_t^\iota, y_t^\iota)$ are considered common knowledge among all agents ι. The shared prior distribution $\mu^\iota(\theta_0^\iota)$ and the transition function $F^\iota(\theta_{t+1}^\iota | \theta_t^\iota, y_t^\iota)$ are assumed to be independent for each agent. At the start of each period t, every agent ι privately observes their type θ_t^ι. By the end of the period, after choosing an action $y_t^\iota \in Y_t^\iota$, the cots for that period are realized. Asymmetric information arises from each player's private observation of θ_t^ι at every time t. Due to the independence of the prior distributions $\mu^\iota(\theta_0^\iota)$ and the transition functions $F^\iota(\theta_{t+1}^\iota | \theta_t^\iota, y_t^\iota)$ for each player ι, the information available to player ι, determined by θ_{t+1}^ι, does not depend on the type θt^l of any other player $l \neq \iota$. After this, each participant simultaneously sends messages $m_t^\iota \in M_t^\iota$, and the entire message profile is made public. Additionally, we assume that $M_t^\iota = \Theta_t^\iota$ from this point onward.

The relationship $\varrho^\iota(a_t^\iota|y_t^\iota)$ for the Machiavellian players is given by $\varrho^\iota : Y^\iota \to \Delta(Y^\iota)$. The set of actions that are admissible is defined by

$$\mathcal{Y}_{adm}^\iota = \left\{ \varrho^\iota(a_t^\iota|y_t^\iota) | \sum_{a_t^\iota \in Y^\iota} \varrho^\iota(a_t^\iota|y_t^\iota) = 1, y_t^\iota \in Y^\iota \right\},$$

The relationship $\varrho^\iota(a_t^\iota|y_t^\iota)$ represents the likelihood that a Machiavellian player ι believes that a_t^ι is an action y_t^ι.

A (behavioral) strategy $\sigma^\iota(m_t^\iota|\theta_t^\iota)$ for player ι is a function $\sigma^\iota : \Theta_t^\iota \to \Delta(M_t^\iota)$, which indicates the probability that player ι associates a message m_t^ι with type θ_t^ι. The set of (behavioral) strategies is defined as follows:

$$\Sigma_{adm}^\iota = \left\{ \sigma^\iota(m_t^\iota|\theta_t^\iota) \geq 0 \middle| \sum_{m_t^\iota \in M_t^\iota} \sigma^\iota(m_t^\iota|\theta_t^\iota) = 1, \theta_t^\iota \in \Theta_t^\iota \right\}.$$

A policy refers to a series $\{\pi^\iota(y_t^\iota|m_t^\iota)\}$ where, for each time period t, $\pi^\iota(y_t^\iota|m_t^\iota)$ represents a stochastic function on Y^ι. The collection of all possible policies is represented by the symbol Π. We have that

$$\Pi_{adm}^\iota = \left\{ \pi^\iota(y_t^\iota|m_t^\iota) \geq 0 \middle| \sum_{y_t^\iota \in Y_t^\iota} \pi^\iota(y_t^\iota|m_t^\iota) = 1, m_t^\iota \in M_t^\iota \right\}$$

Considering the Markovian nature of the type processes, the utility at time t starts at the type vector θ_t^ι, accompanied by the (behavioral) strategy $\sigma^\iota(\theta_t^\iota|m_t^\iota)$, the policy $\pi^\iota(y_t^\iota|\theta_t^\iota)$, and the ex ante probability $\mu^\iota(\theta_t^\iota)$. These elements can be expressed as

$$\begin{aligned} &\Phi^\iota(\pi,\sigma) = \\ &\sum_{t\in T}\sum_{\theta_t^\iota\in\Theta_t^\iota}\sum_{m_t^\iota\in M_t^\iota}\sum_{y_t^\iota\in Y_t^\iota}\sum_{\theta_{t+1}^\iota\in\Theta_{t+1}^\iota} \varphi^\iota(y_t^\iota,\theta_t^\iota)F^\iota(\theta_{t+1}^\iota|\theta_t^\iota,y_t^\iota)\prod_{\iota\in\mathcal{N}}\pi^\iota(y_t^\iota|m_t^\iota)\sigma^\iota(m_t^\iota|\theta_t^\iota)\mu^\iota(\theta_t^\iota) \\ &= \sum_{t\in T}\sum_{\theta_t^\iota\in\Theta_t^\iota}\sum_{m_t^\iota\in M_t^\iota}\sum_{y_t^\iota\in Y_t^\iota} r^\iota(y_t^\iota,\theta_t^\iota)\prod_{\iota\in\mathcal{N}}\pi^\iota(y_t^\iota|m_t^\iota)\sigma^\iota(m_t^\iota|\theta_t^\iota)\mu^\iota(\theta_t^\iota), \end{aligned}$$

where $r^\iota(y_t^\iota,\theta_t^\iota) = \sum_{\theta_{t+1}^\iota\in\Theta_{t+1}^\iota} \varphi^\iota(y_t^\iota,\theta_t^\iota)F^\iota(\theta_{t+1}^\iota|\theta_t^\iota,y_t^\iota)$.

We consider that players are aware of their respective costs. A policy π along with the strategy σ work together to minimize the cost function $\Phi^\iota(\pi,\sigma)$, which operates according to the following rule:

$$(\pi^*,\sigma^*) := Arg \min_{\pi\in\Pi_{adm}} \min_{\sigma\in\Sigma_{adm}} \sum_{\iota\in\mathcal{N}} \Phi(\pi,\sigma). \tag{1}$$

The policy π^* and strategy σ^* satisfy the conditions of a Bayesian–Nash equilibrium, holding true for all σ, such that

$$\Phi^\iota(\pi^*,\sigma^*) \leq \Phi^\iota(\pi^\iota,\pi^{-\iota*},\sigma^\iota,\sigma^{-\iota*}). \tag{2}$$

where the policy $\pi^* = (\pi^{1*},...,\pi^{n*})$ and the strategy $\sigma^* = (\sigma^{1*},...,\sigma^{n*})$ are called *Bayesian–Nash equilibrium*, where π and σ fulfill that $\pi^{-\iota*} = (\pi^{1*},...,\pi^{\iota-1*},\pi^{\iota+1,*},\ ...,\pi^{n*})$ and $\sigma^{-\iota*} = (\sigma^{1*},...,\sigma^{\iota-1*},\sigma^{\iota+1,*},...,\sigma^{n*})$ [26].

Here is an explanation in words. This mathematical expression describes a game-theoretic framework where a policy π and a strategy σ are optimized to minimize a cost function $\Phi^\iota(\pi,\sigma)$. The goal for cost minimization is to find the optimal policy π^* and strategy σ^* that minimize the overall cost function $\Phi(\pi,\sigma)$ across a set of entities $\mathcal{N}$. For the Bayesian–Nash equilibrium, the optimal policy π^* and strategy σ^* must satisfy the conditions of a Bayesian–Nash equilibrium. This means that no single entity ι can reduce its individual cost $\Phi^\iota(\pi,\sigma)$ by unilaterally deviating from the optimal policy or strategy, while others maintain their optimal choices. This framework ensures that each entity ι adopts its optimal policy $\pi^{\iota*}$ and strategy $\sigma^{\iota*}$, given that all others are also acting optimally $(\pi^{-\iota*},\sigma^{-\iota*})$. The result is a state where no entity has an incentive to change its actions unilaterally, which aligns with the concept of equilibrium in game theory. In summary, the expression defines a systematic approach to optimizing policies and strategies in a multi-agent system, ensuring equilibrium while minimizing the overall and individual costs.

Let us introduce the ξ-variable as follows

$$\xi^\iota(\theta_t^\iota, m_t^\iota, y_t^\iota) := \pi^\iota(y_t^\iota|m_t^\iota)\sigma^\iota(m_t^\iota|\theta_t^\iota)\mu^\iota(\theta_t^\iota). \tag{3}$$

such that

$$\Xi^\iota_{adm} := \Bigg\{ \xi^\iota(\theta_t^\iota, m_t^\iota, y_t^\iota) \Bigg| \sum_{\theta_t^\iota\in\Theta_t^\iota}\sum_{m_t^\iota\in\Theta_t^\iota}\sum_{y_t^\iota\in Y_t^\iota} \xi^\iota(\theta_t^\iota, m_t^\iota, y_t^\iota) = 1,$$
$$\sum_{m_t^\iota\in\Theta_t^\iota}\sum_{y_t^\iota\in Y_t^\iota} \xi^\iota(\theta_t^\iota, m_t^\iota, y_t^\iota) = \mu^\iota(\theta_t^\iota) > 0,$$
$$\sum_{m_t^\iota\in\Theta_t^\iota}\sum_{y_t^\iota\in Y_t^\iota}\sum_{\theta_t^\iota\in\Theta_t^\iota} [\delta_{\theta_t^\iota\theta_{t+1}^\iota} - F^\iota(\theta_{t+1}^\iota|\theta_t^\iota, y_t^\iota)]\xi^\iota(\theta_t^\iota, m_t^\iota, y_t^\iota) = 0,$$
$$\sum_{m_t^\iota\in\Theta_t^\iota}\sum_{y_t^\iota\in Y_t^\iota}\sum_{\theta_t^\iota\in\Theta_t^\iota} P^\iota(m_t^\iota|\vartheta_t^\iota)\xi^\iota(\theta_t^\iota, m_t^\iota, y_t^\iota) \geq 0,\ \vartheta_t^\iota \in \Theta_t^\iota \Bigg\}$$

such that $P^\iota(m_t^\iota|\theta_t^\iota) = (\sigma^\iota(m_t^\iota|\theta_t^\iota))^{-1}$ is the inverse and it exists; then, the policy $\pi^\iota(y_t^\iota|m_t^\iota)$ and the strategy $\sigma^\iota(m_t^\iota|\theta_t^\iota)$ form a Bayesian–Nash equilibrium. In this equilibrium, each player minimizes their expected cost for every $\iota \in \mathcal{N}$

$$\Phi^\iota(\pi^{\iota*}(y_t^\iota|m_t^\iota), \sigma^*(m_t^\iota|\theta_t^\iota)) \leq \Phi^\iota(\pi^\iota(y_t^\iota|m_t^\iota), \sigma(m_t^\iota|\theta_t^\iota)).$$

The variable $\xi^\iota(\theta_t^\iota, m_t^\iota, y_t^\iota)$ defined in Equation (3) must satisfy the following individual nonlinear optimization problem if and only if

$$\Phi(\pi,\sigma) = \sum_{\iota\in\mathcal{N}} \tilde{\Phi}^\iota(\xi) \to \min_{\xi\in\Xi_{adm}} \tag{4}$$

where

$$\tilde{\Phi}^\iota(\xi) = \sum_{\theta_t^\iota\in\Theta_t^\iota}\sum_{m_t^\iota\in\Theta_t^\iota}\sum_{y_t^\iota\in Y_t^\iota} \varphi^\iota(y_t^\iota, \theta_t^\iota) \prod_{\iota\in\mathcal{N}} \xi^\iota(\theta_t^\iota, m_t^\iota, y_t^\iota)$$

and $\Xi_{adm} = \times^\iota_{\iota\in\mathcal{N}}\Xi_{adm}$.

Remark 1. *We have that*

$$\sum_{y_t^\iota\in Y_t^\iota} \pi^\iota(y_t^\iota|m_t^\iota) = 1, \quad \sum_{m_t^\iota\in\Theta_t} \sigma(m_t^\iota|\theta_t^\iota) = 1, \quad \sum_{\theta_t^\iota\in\Theta_t^\iota} \mu^\iota(\theta_t^\iota) = 1.$$

The solution of the problem (4) is defined as ξ^*.

The variables $\pi^{\iota*}(y_t^\iota|m_t^\iota)$ can be derived from $\xi^\iota(\theta_t^\iota, m_t^\iota, y_t^\iota)$ in the following way:

$$\pi^{\iota*}(y_t^\iota|m_t^\iota) = \frac{1}{|\Theta_t^\iota|} \sum_{\theta_t^\iota \in \Theta_t^\iota} \frac{\xi^\iota(\theta_t^\iota, m_t^\iota, y_t^\iota)}{\sum_{y_t^\iota \in Y_t^\iota} \xi^\iota(\theta_t^\iota, m_t^\iota, y_t^\iota)}. \tag{5}$$

To retrieve $\sigma^{\iota*}(m_t^\iota|\theta_t^\iota)$, for each player $\iota \in \mathcal{N}$, the relevant quantity is defined as follows:

$$\sigma^{\iota*}(m_t^\iota|\theta_t^\iota) = \frac{\sum_{y_t^\iota \in Y_t^\iota} \xi^\iota(\theta_t^\iota, m_t^\iota, y_t^\iota)}{\sum_{m_t^\iota \in M_t^\iota} \sum_{y_t^\iota \in Y_t^\iota} \xi^\iota(\theta_t^\iota, m_t^\iota, y_t^\iota)}. \tag{6}$$

As well as the distribution $\mu^{\iota*}(\theta_t^\iota)$, we have

$$\mu^{\iota*}(\theta_t^\iota) = \sum_{m_t^\iota \in M_t^\iota} \sum_{y_t^\iota \in Y_t^\iota} \xi^\iota(\theta_t^\iota, m_t^\iota, y_t^\iota) > 0. \tag{7}$$

The formulas that minimize Equation (4) with respect to the variables $\xi^\iota(\theta_t^\iota, m_t^\iota, y_t^\iota)$ have been determined, along with those needed to recover the strategy $\sigma^{\iota*}(m_t^\iota|\theta_t^\iota)$, the policy $\pi^{\iota*}(y_t^\iota|m_t^\iota)$, and the distribution $\mu^{\iota*}(\theta_t^\iota)$. When players use a reporting strategy $\sigma^\iota(m_t^\iota|\theta_t^\iota)$, they are minimizing the expected cost described by Equation (4). The calculated strategy profile $\sigma^{\iota*}(m_t^\iota|\theta_t^\iota)$ and policy $\pi^{\iota*}(y_t^\iota|m_t^\iota)$ constitute a Bayesian–Nash equilibrium.

3. No-Regret Strategy

To tackle this issue, we propose modeling the state-value function Φ^ι using a linear framework, defined in terms of $\pi^\iota(y_t^\iota|m_t^\iota)\sigma^\iota(m_t^\iota|\theta_t^\iota)$. The aim is to develop a strategy that strictly minimizes the trajectory value. Accordingly, Φ^ι is formulated in a recursive matrix structure, facilitating efficient computations and in-depth analysis across various states and control parameters. This formulation optimizes the policy by enhancing both performance and decision-making, providing a systematic approach for achieving optimal control in dynamic systems [25].

3.1. The Cost Function

The cost function Φ^ι for any fixed strategies $\pi^\iota(y_t^\iota|m_t^\iota)\sigma^\iota(m_t^\iota|\theta_t^\iota)$ is defined over all possible state–action combinations. It represents the expected value of taking action y_t^ι and sending a message m_t^ι considering state θ_t^ι and subsequently following $\pi^\iota(y_t^\iota|m_t^\iota)\sigma^\iota(m_t^\iota|\theta_t^\iota)$. This function enables the evaluation of the long-term cost associated with a particular strategy. The Φ^ι-values for all the states in open format can be expressed by

$$\begin{aligned}
&\Phi^\iota(\pi, \sigma) = \\
&\sum_{t \in T} \sum_{\theta_t^\iota \in \Theta_t^\iota} \sum_{m_t^\iota \in M_t^\iota} \sum_{y_t^\iota \in Y_t^\iota} \sum_{\theta_{t+1}^\iota \in \Theta_{t+1}^\iota} \varphi^\iota(y_t^\iota, \theta_t^\iota) F^\iota(\theta_{t+1}^\iota|\theta_t^\iota, y_t^\iota) \pi^\iota(y_t^\iota|m_t^\iota) \sigma^\iota(m_t^\iota|\theta_t^\iota) \mu^\iota(\theta_t^\iota) \\
&= \sum_{t \in T} \sum_{\theta_t^\iota \in \Theta_t^\iota} \sum_{m_t^\iota \in M_t^\iota} \sum_{y_t^\iota \in Y_t^\iota} r^\iota(y_t^\iota, \theta_t^\iota) \pi^\iota(y_t^\iota|m_t^\iota) \sigma^\iota(m_t^\iota|\theta_t^\iota) \mu^\iota(\theta_t^\iota),
\end{aligned}$$

where $r^\iota(y_t^\iota, \theta_t^\iota) = \sum_{\theta_{t+1}^\iota \in \Theta_{t+1}^\iota} \varphi^\iota(y_t^\iota, \theta_t^\iota) F^\iota(\theta_{t+1}^\iota|\theta_t^\iota, y_t^\iota)$.

$$\mathsf{E}(\Phi_{t+1}^\iota(\pi, \sigma)) = \sum_{t \in T} \sum_{\theta_t^\iota \in \Theta_t^\iota} \sum_{m_t^\iota \in M_t^\iota} \sum_{y_t^\iota \in Y_t^\iota} r^\iota(y_t^\iota, \theta_t^\iota) \pi^\iota(y_t^\iota|m_t^\iota) \sigma^\iota(m_t^\iota|\theta_t^\iota) \mu^\iota(\theta_t^\iota), \tag{8}$$

such that $\varphi^\iota(y_t^\iota, \theta_t^\iota)$ is a loss value at type θ_t^ι when the action y_t^ι is applied (without loss of generality it can be assumed to be positive) and $\mu^\iota(\theta_t^\iota)$ for any given $\mu^\iota(\theta_0^\iota)$ is defined as follows

$$\mu^\iota(\theta_{t+1}^\iota) = \sum_{t\in T}\sum_{\theta_t^\iota\in\Theta_t^\iota}\sum_{m_t^\iota\in M_t^\iota}\sum_{y_t^\iota\in Y_t^\iota} F^\iota(\theta_{t+1}^\iota|\theta_t^\iota, y_t^\iota)\pi^\iota(y_t^\iota|m_t^\iota)\sigma^\iota(m_t^\iota|\theta_t^\iota)\mu^\iota(\theta_t^\iota)$$

or, in matrix format,

$$\boldsymbol{\mu}_{t+1}^\iota = (\mathbf{F}_t^\iota)_t^\top \boldsymbol{\mu}_t^\iota,$$

$$\mathbf{F}_t^\iota := \sum_{t\in T}\sum_{\theta_t^\iota\in\Theta_t^\iota}\sum_{m_t^\iota\in M_t^\iota}\sum_{y_t^\iota\in Y_t^\iota} F^\iota(\theta_{t+1}^\iota|\theta_t^\iota, y_t^\iota)\pi^\iota(y_t^\iota|m_t^\iota)\sigma^\iota(m_t^\iota|\theta_t^\iota).$$

Remark 2. *We will suppose that*

$$\varphi^\iota(y_t^\iota, \theta_t^\iota) > 0, \textit{ for all } \iota \in \mathcal{N}.$$

Considering

$$\mathsf{E}(\Phi_{t+1}^\iota(\pi,\sigma)) =$$

$$\sum_{t\in T}\sum_{\theta_t^\iota\in\Theta_t^\iota}\sum_{m_t^\iota\in M_t^\iota}\sum_{y_t^\iota\in Y_t^\iota}\sum_{\theta_{t+1}^\iota\in\Theta_{t+1}^\iota} r^\iota(y_t^\iota, \theta_t^\iota)\pi^\iota(y_t^\iota|m_t^\iota)\sigma^\iota(m_t^\iota|\theta_t^\iota)\mu^\iota(\theta_t^\iota) =$$

$$\sum_{t\in T}\sum_{\theta_t^\iota\in\Theta_t^\iota}\sum_{m_t^\iota\in M_t^\iota}\sum_{y_t^\iota\in Y_t^\iota}\sum_{\theta_{t+1}^\iota\in\Theta_{t+1}^\iota} [\varphi^\iota(y_t^\iota, \theta_t^\iota) + c] \cdot \left[F^\iota(\theta_{t+1}^\iota|\theta_t^\iota, y_t^\iota)\pi^\iota(y_t^\iota|m_t^\iota)\sigma^\iota(m_t^\iota|\theta_t^\iota)\mu^\iota(\theta_t^\iota) - c\right]$$

the minimization of the state-value function $\mathsf{E}(\Phi_{t+1}^\iota(\pi,\sigma))$ *is equivalent to the minimization of the function* $\mathsf{E}(\tilde{\Phi}_{t+1}^\iota(\pi,\sigma))$ *where*

$$\tilde{\Phi}_{t+1}^\iota(\pi,\sigma) = \tilde{\Phi}_{t+1}^\iota(\pi,\sigma) + c,$$

which is strictly positive if we take

$$c > \max_{y_t^\iota\in Y_t^\iota, \theta_t^\iota\in\Theta_t^\iota} \varphi^\iota(y_t^\iota, \theta_t^\iota). \tag{9}$$

In a vector format, the Formula (8) can be expressed as

$$\mathsf{E}(\Phi_{t+1}^\iota(\pi,\sigma)) =$$

$$\sum_{t\in T}\sum_{\theta_t^\iota\in\Theta_t^\iota}\left[\sum_{m_t^\iota\in M_t^\iota}\sum_{y_t^\iota\in Y_t^\iota}\sum_{\theta_{t+1}^\iota\in\Theta_{t+1}^\iota} r^\iota(y_t^\iota, \theta_t^\iota)\pi^\iota(y_t^\iota|m_t^\iota)\sigma^\iota(m_t^\iota|\theta_t^\iota)\right]\mu^\iota(\theta_t^\iota) = \langle \mathbf{R}_t^\iota, \boldsymbol{\mu}_t^\iota\rangle,$$

where

$$\mathbf{R}_t^\iota := \sum_{m_t^\iota\in M_t^\iota}\sum_{y_t^\iota\in Y_t^\iota}\sum_{\theta_{t+1}^\iota\in\Theta_{t+1}^\iota} r^\iota(y_t^\iota, \theta_t^\iota)\pi^\iota(y_t^\iota|m_t^\iota)\sigma^\iota(m_t^\iota|\theta_t^\iota),$$

$$r^\iota(y_t^\iota, \theta_t^\iota) := \sum_{\theta_{t+1}^\iota\in\Theta_{t+1}^\iota} \varphi^\iota(y_t^\iota, \theta_t^\iota)F^\iota(\theta_{t+1}^\iota|\theta_t^\iota, y_t^\iota),$$

$$\mathbf{U}_t^\iota := \sum_{t\in T}\sum_{\theta_t^\iota\in\Theta_t^\iota}\sum_{m_t^\iota\in M_t^\iota}\sum_{y_t^\iota\in Y_t^\iota} F^\iota(\theta_{t+1}^\iota|\theta_t^\iota, y_t^\iota)\pi^\iota(y_t^\iota|m_t^\iota)\sigma^\iota(m_t^\iota|\theta_t^\iota).$$

Remark 3. *Lyapunov theory plays a critical role in ensuring the stability of dynamic systems by providing conditions under which a system remains in equilibrium. Beyond stability, Lyapunov theory also offers a pathway for these systems to converge toward a Nash equilibrium, where all participants optimize their strategies. This dual function—guaranteeing both stability and convergence to optimal outcomes—enhances the practical applicability of the framework. It ensures that not only will the system remain stable over time, but it will also naturally evolve toward a* *state of optimal strategic interactions. This makes the framework robust and suitable for real-world applications requiring both stability and optimality.*

3.2. Optimal Stationary Strategy

Let us first introduce the following Lemma.

Lemma 1. *Any nonstationary strategy $\xi^{\iota}(\theta_t^{\iota}, m_t^{\iota}, y_t^{\iota})$ cannot improve the average cost obtained by employing an optimal stationary strategy $\xi^{\iota*}(\theta^{\iota*}, m^{\iota*}, y^{\iota*})$ defined on a compact set. An optimal solution exists.*

Proof. According to the Weierstrass theorem, any nonstationary bounded strategy $\xi^{\iota}(\theta_t^{\iota}, m_t^{\iota}, y_t^{\iota})$, defined on a compact set, necessarily contains a convergent subsequence that satisfies the following relationships:

$$\liminf_{n\to\infty} \tfrac{1}{t} \sum_{n=1}^{t} \tilde{\Phi}^{\iota}(\xi^{\iota}(\theta_t^{\iota}, m_t^{\iota}, y_t^{\iota})) \geq \liminf_{t\to\infty} \tfrac{1}{t} \sum_{n=1}^{t} \min_{k=1,\ldots,n} \tilde{\Phi}^{\iota}(\xi^{\iota}(\theta_k^{\iota}, m_k^{\iota}, y_k^{\iota})) \geq$$
$$\liminf_{n\to\infty} \tfrac{1}{t} \sum_{n=1}^{t} \liminf_{k\to\infty} \tilde{\Phi}^{\iota}(\xi^{\iota}(\theta_{t_k}^{\iota}, m_{t_k}^{\iota}, y_{t_k}^{\iota})) = \tilde{\Phi}^{\iota}(\xi^{\iota**}(\theta^{\iota**}, m^{\iota**}, y^{\iota**})),$$

where $\tilde{\Phi}^{\iota}(\xi^{\iota}(\theta^{\iota}, m^{\iota}, y^{\iota}))$ is assumed to be a monotonically decreasing functional with respect to each component $\xi^{\iota'}(\theta^{\iota'}, m^{\iota'}, y^{\iota'})$, when the others are held constant. Additionally,

$$\liminf_{k\to\infty} \tilde{\Phi}^{\iota}(\xi^{\iota}(\theta_{t_k}^{\iota}, m_{t_k}^{\iota}, y_{t_k}^{\iota})) := \tilde{\Phi}^{\iota}(\xi^{\iota**}(\theta^{\iota**}, m^{\iota**}, y^{\iota**})$$

This lower bound is achieved by setting

$$\xi^{\iota}(\theta_t^{\iota}, m_t^{\iota}, y_t^{\iota}) = \xi^{\iota*}(\theta^{\iota*}, m^{\iota*}, y^{\iota*}) = \xi^{\iota**}(\theta^{\iota**}, m^{\iota**}, y^{\iota**})$$

and this is because

$$\liminf_{n\to\infty} \frac{1}{n} \sum_{h=1}^{n} \tilde{\Phi}^{\iota}(\xi^{\iota*}(\theta^{\iota*}, m^{\iota*}, y^{\iota*})) = \tilde{\Phi}^{\iota}(\xi^{\iota*}(\theta^{\iota*}, m^{\iota*}, y^{\iota*})) = \tilde{\Phi}^{\iota}(\xi^{\iota**}(\theta^{\iota**}, m^{\iota**}, y^{\iota**})).$$

□

Remark 4. *It is essential to highlight that this property holds exclusively for average cost functionals defined on compact sets of strategies.*

Proposition 1. *Let S be the unit simplex in $\mathbf{R}^M$, that is,*

$$S = \{\kappa \in \mathbf{R}^{|Y|} | \sum_{y_t^{\iota} \in Y_t^{\iota}} \kappa(y_t^{\iota}) = 1, \kappa(y_t^{\iota}) \geq 0\}.$$

Then,

$$\min_{\kappa \in S} \sum_{y_t^{\iota} \in Y_t^{\iota}} g(y_t^{\iota})\kappa(y_t^{\iota}) = \min_{y_t^{\iota} \in Y_t^{\iota}} g(y_t^{\iota}) = g(e),$$

such that the minimum is achieved at least for $\kappa = \left(\underbrace{0, 0, \ldots, 0, 1}_{e}, 0, \ldots, 0\right)$.

Proof. We have that

$$\sum_{y_t^{\iota} \in Y_t^{\iota}} g(y_t^{\iota})\kappa(y_t^{\iota}) \geq \sum_{y_t^{\iota} \in Y_t^{\iota}} (\min g(y_t^{\iota})\kappa(y_t^{\iota})) = \min g(y_t^{\iota}) \sum_{y_t^{\iota} \in Y_t^{\iota}} \kappa(y_t^{\iota}) = \min g(y_t^{\iota}) = g(e),$$

and the equality is achieved at least for $\kappa = \left(\underbrace{0,0,\ldots,0,1}_{e},0,\ldots,0\right)$. □

As a result, it holds that

$$\boldsymbol{\Psi}^{\iota}_{t+1} = \langle \mathbf{R}^{\iota}_{t}, \boldsymbol{\mu}^{\iota}_{t} \rangle = \sum_{\theta^{\iota}_{t} \in \Theta^{\iota}_{t}} \mathbf{R}^{\iota}_{t} \boldsymbol{\mu}^{\iota}_{t} \leq$$

$$\sum_{\theta^{\iota}_{t} \in \Theta^{\iota}_{t}} \min_{\pi^{\iota}(y^{\iota}_{t}|m^{\iota}_{t})\sigma^{\iota}(m^{\iota}_{t}|\theta^{\iota}_{t})} \mathbf{R}^{\iota}_{t} \boldsymbol{\mu}^{\iota}_{t} = \sum_{\theta^{\iota}_{t} \in \Theta^{\iota}_{t}} \boldsymbol{\mu}^{\iota}_{t} \min_{y^{\iota}_{t} \in Y^{\iota}_{t}} \left[\sum_{\theta^{\iota}_{t+1} \in \Theta^{\iota}_{t+1}} \varphi^{\iota}(y^{\iota}_{t}, \theta^{\iota}_{t}) F^{\iota}(\theta^{\iota}_{t+1} | \theta^{\iota}_{t}, y^{\iota}_{t}) \right].$$

At this point, let us introduce the following general definition of a Lyapunov-like function Given fixed history of the process $(\boldsymbol{\mu}^{\iota}_{0}, \pi^{\iota}(y^{\iota}_{0}|m^{\iota}_{0}), \sigma^{\iota}(m^{\iota}_{0}|\theta^{\iota}_{0}), \ldots, \pi^{\iota}(y^{\iota}_{t}|m^{\iota}_{t}), \sigma^{\iota}(m^{\iota}_{t}|\theta^{\iota}_{t}))$

$$\min_{\pi^{\iota}(y^{\iota}_{t}|m^{\iota}_{t})\sigma^{\iota}(m^{\iota}_{t}|\theta^{\iota}_{t})} \mathbb{V}^{\iota}_{t+1} = \sum_{\theta^{\iota}_{t} \in \Theta^{\iota}_{t}} (\boldsymbol{\mu}^{\iota}_{0})_{\theta^{\iota}_{t}} \left[\sum_{\theta^{\iota}_{t+1} \in \Theta^{\iota}_{t+1}} \varphi^{\iota}(y^{\iota *}_{t}, \theta^{\iota}_{t}) F^{\iota}(\theta^{\iota}_{t+1} | \theta^{\iota}_{t}, y^{\iota *}_{t}) \right]. \tag{10}$$

the identity in (10) is achieved for the pure and stationary local-optimal policy.

$$\pi^{\iota *}(y^{\iota *}_{t} | m^{\iota *}_{t}) \sigma^{\iota *}(m^{\iota}_{t} | \theta^{\iota}_{t}) = \kappa_{y^{\iota *}_{t}(\theta^{\iota *}_{t}), \theta^{\iota *}_{t}} \quad t = 0, 1, \ldots \tag{11}$$

where $\kappa_{y^{\iota *}_{t}(m^{\iota *}_{t}), m^{\iota *}_{t}}$ is the Kronecker symbol and $y^{\iota *}_{t}(m^{\iota *}_{t})$ is an index for which

$$\sum_{\theta^{\iota}_{t+1} \in \Theta^{\iota}_{t+1}} \varphi^{\iota}(y^{\iota *}_{t}, \theta^{\iota}_{t}) F^{\iota}(\theta^{\iota}_{t+1} | \theta^{\iota}_{t}, y^{\iota *}_{t}) \leq \sum_{\theta^{\iota}_{t+1} \in \Theta^{\iota}_{t+1}} \varphi^{\iota}(y^{\iota}_{t}, \theta^{\iota}_{t}) F^{\iota}(\theta^{\iota}_{t+1} | \theta^{\iota}_{t}, y^{\iota}_{t}) := r^{\iota}(y^{\iota}_{t}, \theta^{\iota}_{t}), \ \forall y^{\iota}_{t} \in Y^{\iota}_{t} \tag{12}$$

As a result, we can state the following lemma.

Lemma 2. *Given a fixed local-optimal policy, the* $\mathbb{V}$*-values for all state–action pairs from (8) in the recursive matrix format become*

$$\mathbb{V}^{l}_{t+1} = \langle \mathbf{R}^{\iota *}_{t}, \boldsymbol{\mu}^{\iota *}_{t} \rangle \tag{13}$$

where

$$\mathbf{R}^{\iota *}_{t} := \sum_{\theta^{\iota}_{t+1} \in \Theta^{\iota}_{t+1}} \varphi^{\iota}(y^{\iota *}_{t}, \theta^{\iota}_{t}) F^{\iota}(\theta^{\iota}_{t+1} | \theta^{\iota}_{t}, y^{\iota *}_{t}) = \min_{y^{\iota}_{t} \in Y^{\iota}_{t}} r^{\iota}(y^{\iota}_{t}, \theta^{\iota}_{t}),$$

$$r^{\iota}(y^{\iota}_{t}, \theta^{\iota}_{t}) = \sum_{\theta^{\iota}_{t+1} \in \Theta^{\iota}_{t+1}} \varphi^{\iota}(y^{\iota}_{t}, \theta^{\iota}_{t}) F^{\iota}(\theta^{\iota}_{t+1} | \theta^{\iota}_{t}, y^{\iota}_{t}).$$

Remark 5. *Under the local-optimal strategy (11), the probability state vector* $\boldsymbol{\mu}^{\iota}_{t}$ *satisfies the following relation*

$$\boldsymbol{\mu}^{\iota}_{t+1} = (\mathbf{F}^{\iota}_{t})^{\top}_{t} \boldsymbol{\mu}^{\iota}_{t} = \left(\left(\mathbf{F}^{\iota^{*}} \right)^{\top} \right)^{t+1} \boldsymbol{\mu}^{\iota}_{0},$$

where

$$(\mathbf{F}^{\iota *}) := \sum_{y^{\iota}_{t} \in Y^{\iota}_{t}} F^{\iota}(\theta^{\iota}_{t+1} | \theta^{\iota}_{t}, y^{\iota}_{t}) \kappa_{y^{\iota *}_{t}(\theta^{\iota}_{t}), \theta^{\iota}_{t}} = F^{\iota}(\theta^{\iota}_{t+1} | \theta^{\iota}_{t}, y^{\iota *}_{t}).$$

Lemma 3. *The function* $\mathbb{V}^{\iota}_{t}$ *is a Lyapunov-like function (monotonically decreasing in time) that reaches an equilibrium point.*

Proof. The function $\mathbb{V}^{\iota}_{t}$ satisfies the following properties:

1. According to Remark 2, $\mathbb{V}^{\iota}_{t} > 0$.
2. According to Equations (10) and (11), $\Delta \mathbb{V}^{\iota}_{t} = \mathbb{V}^{\iota}_{t+1} - \mathbb{V}^{\iota}_{t} < 0$.

3. $\mathbb{V}_t^\iota = 0$ at the Lyapunov equilibrium point. It results straightforwardly from the following:

$$\sum_{\iota\in\mathcal{N}} \min_{\pi^\iota(y_t^\iota|m_t^\iota)\sigma^\iota(m_t^\iota|\theta_t^\iota)} \mathbb{V}_{t+1}^\iota = \sum_{\iota\in\mathcal{N}} \sum_{\theta_t^\iota\in\Theta_t^\iota} (\boldsymbol{\mu}_0^\iota)_{\theta_t^\iota} \left[\sum_{\theta_{t+1}^\iota\in\Theta_{t+1}^\iota} \varphi^\iota(y_t^{\iota*},\theta_t^\iota) F^\iota(\theta_{t+1}^\iota|\theta_t^\iota, y_t^{\iota*}) \right] =$$
$$\sum_{\iota\in\mathcal{N}} \sum_{t\in T} \sum_{\theta_t^\iota\in\Theta_t^\iota} \sum_{m_t^\iota\in M_t^\iota} \sum_{y_t^\iota\in Y_t^\iota} r^\iota(y_t^\iota,\theta_t^\iota)\pi^{\iota*}(y_t^\iota|m_t^\iota)\sigma^{\iota*}(m_t^\iota|\theta_t^\iota)\mu^{\iota*}(\theta_t^\iota)\times$$
$$\prod_{i\neq\iota\in\mathcal{N}} \pi^{i*}(x_t^i|m_t^i)\sigma^{i*}(m_t^{i*}|\theta_t^{i*})\mu^{i*}(\theta_t^{i*}) =$$
$$\sum_{\iota\in\mathcal{N}} \left(\sum_{t\in T} \min_{\pi^\iota(y_t^\iota|m_t^\iota)\sigma^\iota(m_t^\iota|\theta_t^\iota)} \sum_{m_t^\iota\in M_t^\iota} \sum_{y_t^\iota\in Y_t^\iota} \pi^\iota(y_t^\iota|m_t^\iota) \sum_{\theta_t^\iota\in\Theta_t^\iota} r^\iota(y_t^\iota,\theta_t^\iota)\sigma^\iota(m_t^\iota|\theta_t^\iota)\mu^\iota(\theta_t^\iota)\times \right.$$
$$\left. \prod_{i\neq\iota\in\mathcal{N}} \pi^{i*}(x_t^i|m_t^i)\sigma^{i*}(m_t^{i*}|\theta_t^{i*})\mu^{i*}(\theta_t^{i*}) \right) <$$
$$\sum_{\iota\in\mathcal{N}} \left(\sum_{t\in T} \sum_{m_t^\iota\in M_t^\iota} \sum_{y_t^\iota\in Y_t^\iota} \pi^\iota(y_t^\iota|m_t^\iota) \sum_{\theta_t^\iota\in\Theta_t^\iota} r^\iota(y_t^\iota,\theta_t^\iota)\sigma^\iota(m_t^\iota|\theta_t^\iota)\mu^\iota(\theta_t^\iota)\times \right.$$
$$\left. \prod_{i\neq\iota\in\mathcal{N}} \pi^{i*}(x_t^i|m_t^i)\sigma^{i*}(m_t^{i*}|\theta_t^{i*})\mu^{i*}(\theta_t^{i*}) \right) =$$
$$\sum_{\iota\in\mathcal{N}} \Phi_t^\iota(\pi^\iota,\pi^{-\iota*},\sigma^\iota,\sigma^{-\iota*})$$

Hence,

$$\sum_{\iota\in\mathcal{N}} (\Phi_t^\iota(\pi^*,\sigma^*) - \Phi_t^\iota(\pi^\iota,\pi^{-\iota*},\sigma^\iota,\sigma^{-\iota*})) < 0. \tag{14}$$

Given that the inequality in Equation (14) is valid for all admissible strategies π and σ, it is valid when $\pi^i = \pi^{i*}$ and $\sigma^i = \sigma^{i*}$ for $i \neq \iota$, considering

$$\Phi_t^\iota(\pi^*,\sigma_\pi^*) - \Phi_t^\iota(\pi^*,\sigma^\iota,\sigma^{-\iota*}) < 0. \tag{15}$$

Then, Equation (1) holds at (π^*,σ_π^*), and then $\mathbb{V}_t^\iota = 0$.

□

4. The Three-Level Manipulation Model

The manipulation model is a game in which the manipulating players move first and then the manipulated players move sequentially [27].

4.1. The Stackelberg Game

To simplify the descriptions below, let us introduce the new variables

$$\left.\begin{aligned} \alpha &:= col\left[\xi^l\right],\ A_{adm} := \Xi^l, (l=\overline{1,n}). \\ \beta &:= col\left[\xi^f\right],\ B_{adm} := \Xi^f\ (f=\overline{1,m}). \end{aligned}\right\} \tag{16}$$

where $col[\cdot]$ transforms a matrix in a vector format.

Consider a game involving manipulating players, where each manipulator's strategy is denoted by $\alpha^l \in A^l$ and $\alpha^{-l} \in A^{-l}$ represents the strategies of the other manipulating players, $\alpha = (\alpha^l,\alpha^{-l})$. Similarly, manipulated players have strategies $\beta^f \in B_{adm}^f$ and $\beta^{-f} \in B_{adm}^{-f}$ denotes the strategies of the other manipulated players, $\beta = (\beta^f,\beta^{-f})$. The manipulating players anticipate the reactions of the manipulated players, choosing their strategies $\alpha \in A_{adm}$ before the manipulated players respond with their best-reply strategies $\beta \in B_{adm}$.

- The individual aim for the manipulating players is to find $\alpha^{l*} \in A^l_{adm}$ for fixed $\beta(\alpha^{l*}, \alpha^{-l}) \in B^f_{adm}$ such that
$$v^l(\alpha^{l*}, \alpha^{-l}|\beta(\alpha^{l*}, \alpha^{-l})) := \tilde{\Phi}^l(\alpha^{l*}, \alpha^{-l}|\beta(\alpha^{l*}, \alpha^{-l})) - \tilde{\Phi}^l(\alpha^l, \alpha^{-l}|\beta(\alpha^{l*}, \alpha^{-l})) \leq \varepsilon,\ \varepsilon \geq 0, \quad (17)$$
valid for all $\alpha^l \in A^l_{adm}$, $\alpha^{-l} \in A^{-l}_{adm}$ and all $l = \overline{1,n}$. Below α^{l*} will be associated with the ε-Nash equilibrium in a Stackelberg–Nash game.

The individual objectives of the manipulating players can be expressed in a joint format by utilizing the regularized Tanaka function, as demonstrated below:

$$v_\delta(\alpha^{l*}, \alpha^{-l}|\lambda, \beta(\alpha^{l*}, \alpha^{-l})) := \sum_{l=1}^{n} v^l_\delta\left(\alpha^{l*}, \alpha^{-l}|\lambda^{l,i}, \lambda^{f,j}, \beta(\alpha^{l*}, \alpha^{-l})\right) \leq \varepsilon \quad (18)$$

for all $\alpha \in A$, $\lambda = [\lambda^{j,i}, \lambda^{f,j}]_{l=\overline{1,n}, f=\overline{1,m}, i=\overline{1,N}, j=\overline{1,N}}$ and

$$v^l_\delta\left(\alpha^{l*}, \alpha^{-l}|\lambda^{l,i}, \lambda^{f,j}, \beta(\alpha^{l*}, \alpha^{-l})\right) = \mathbb{L}^l_\delta(\alpha^*|\lambda^{l,i}, \lambda^{f,j}, \beta(\alpha^{l*}, \alpha^{-l})) - \mathbb{L}^l_\delta(\alpha|\lambda^{l,i}, \lambda^{f,j}, \beta(\alpha^l, \alpha^{-l})) \quad (19)$$

where $\mathbb{L}^l_\delta(\alpha|\lambda^{l,i}, \lambda^{f,j}, \beta(\alpha^l, \alpha^{-l}))$ corresponds with the Lagrange function.

The function $v_\delta(\alpha^{l*}, \alpha^{-l}|\lambda, \beta(\alpha^{l*}, \alpha^{-l}))$ is known as the regularized Tanaka function within the context of Stackelberg–Nash games. When both $\delta = 0$ and $\lambda = 0$, it corresponds to the original Tanaka function as outlined in [28].

- The individual aim for the manipulated players is to find $\beta^{f*} \in B^f_{adm}$ for a fixed $\alpha(\beta^{f*}, \beta^{-f}) \in A^l_{adm}$ such that
$$u^f(\beta^{f*}, \beta^{-f}|\alpha(\beta^{f*}, \beta^{-f})) := U^f(\beta^{f*}, \beta^{-f}|\alpha(\beta^{f*}, \beta^{-f})) - U^f(\beta^f, \beta^{-f}|\alpha(\beta^{f*}, \beta^{-f})) \leq \varepsilon,\ \varepsilon \geq 0, \quad (20)$$
valid for all $\beta^f \in B^f_{adm}$, $\beta^{-f} \in B^{-f}_{adm}$ and all $f = \overline{1,m}$. Below β^{f*} will be associated with the ε-Nash equilibrium in a Stackelberg–Nash game.

The individual objectives of the manipulated players can be expressed jointly through the regularized Tanaka function, as outlined below:

$$u_\delta(\beta^{f*}, \beta^{-f}|\lambda, \alpha(\beta^{f*}, \beta^{-f})) := \sum_{f=1}^{m} u^f_\delta\left(\beta^{f*}, \beta^{-f}|\lambda^{l,i}, \lambda^{f,j}, \alpha(\beta^{f*}, \beta^{-f})\right) \leq \varepsilon \quad (21)$$

for all $\beta \in B^f$, $\lambda = [\lambda^{j,i}, \lambda^{f,j}]_{l=\overline{1,n}, f=\overline{1,m}, i\in|\Theta_t|, j\in|\Theta_t|}$ and

$$u^f_\delta\left(\beta^{f*}, \beta^{-f}|\lambda^{l,i}, \lambda^{f,j}, \alpha(\beta^{f*}, \beta^{-f})\right) = \mathbb{L}^f_\delta(\beta^{f*}|\lambda^{l,i}, \lambda^{f,j}, \alpha(\beta^{f*}, \beta^{-f})) - \mathbb{L}^f_\delta(\beta^f|\lambda^{l,i}, \lambda^{f,j}, \alpha(\beta^f, \beta^{-f})) \quad (22)$$

In this context, the equilibrium concept for games is defined as Nash equilibrium for simultaneous play scenarios and Stackelberg equilibrium for games with a hierarchical structure.

Definition 1. *(Stackelberg Equilibrium) In a game featuring l-leaders, a strategy $\alpha^* \in A_{adm}$ is considered a Stackelberg–Nash equilibrium strategy for the leaders if*

$$\max_{\beta\in\rho(\alpha^*)} v_\delta(\alpha^*|\beta) \leq \max_{\beta\in\rho(\alpha)} v_\delta(\alpha\xi|\beta)$$

such that

$$\zeta(\alpha) = \{\beta \in B_{adm} \mid u_\delta(\beta|\alpha) \leq u_\delta(\beta'|\xi) \forall \beta' \in B_{adm}\}$$

is the best-reply strategy set of the followers.

The previously defined Stackelberg equilibrium can be reformulated for the followers by replacing the set $\zeta(\alpha)$ with the set of Nash equilibria. In this context, if the leaders adopt the strategy α; the optimal response of the followers will be a Nash equilibrium. In this scenario, the Nash equilibrium represents a state where, given the leaders' strategy α, the followers have optimized their own strategies. They cannot improve their outcomes by unilaterally changing their decisions, as doing so would lead to a less favorable result. This adjustment transforms the dynamics of the game, as the followers must consider not only their responses but also the broader implications of their strategies within the context of Nash equilibrium. To expand upon this, we can envision a multi-level decision-making environment where the leaders, acting as primary decision-makers, set the stage for the followers' reactions. By defining their strategy α, the leaders create a framework within which the followers must operate. The followers, recognizing the leaders' strategic choices, analyze their own positions and responses. In essence, their responses are not isolated actions but part of a larger strategic equilibrium that includes the leaders' moves.

Definition 2. *A strategy $\alpha^* \in A_{adm}$ of the manipulating player together with the collection $\beta^* \in B_{adm}$ of the manipulated players is said to be a Stackelberg–Nash equilibrium $(\varepsilon = 0)$ if*

$$(\alpha^*, \beta^*) \in Arg \max_{\lambda \geq 0} \min_{\beta^f, \beta^{-f} \in B_{adm}} \min_{\alpha' \in A_{adm}} \left\{ v_\delta(\alpha'|\beta) | u_\delta(\beta^f, \beta^{-f}|\lambda, \alpha(\beta^f, \beta^{-f})) \leq 0 \right\}. \tag{23}$$

Using the Lagrange multiplier approach, we can represent (23) as follows:

$$\mathbb{L}_\delta(\alpha', \beta^f, \beta^{-f}, \mu, \lambda) \to \max_{\mu, \lambda \geq 0} \min_{\beta^f, \beta^{-f} \in B_{adm}} \min_{\alpha' \in A_{adm}}, \tag{24}$$

where

$$\mathbb{L}_\delta(\alpha', \beta^f, \beta^{-f}, \mu, \lambda) = v_\delta(\alpha'|\beta') + \mu u_\delta(\beta^f, \beta^{-f}|\lambda, \alpha(\beta^f, \beta^{-f})). \tag{25}$$

With $\delta > 0$, the considered functions become strictly convex, providing the uniqueness of the considered conditional optimization problem (24). Notice also that the Lagrange function in (25) satisfies the saddle-point condition, namely, for all $\alpha' \in A_{adm}$ and $\lambda \geq 0$ we have

$$\mathbb{L}_\delta(\alpha'^*, \beta^{f*}, \beta^{-f*}, \mu, \lambda) \leq \mathbb{L}_\delta(\alpha'^*, \beta^{f*}, \beta^{-f*}, \mu^*, \lambda^*) \leq \mathbb{L}_\delta(\alpha', \beta^f, \beta^{-f}, \mu^*, \lambda^*). \tag{26}$$

4.2. Tri-Level Manipulation Game

The costs associated with different hierarchical roles are assumed to be distributed as follows: the outermost manipulators incur a cost v_δ, the players being manipulated bear a cost u_δ, and the innermost manipulators are assigned a cost w_δ. These costs are thought to resemble Lyapunov-like functions.

Tri-level game problems have the following general form

$$\begin{aligned}
&\min_{\alpha \in A_{adm}} v_\delta(\alpha|\beta, \gamma) \\
&s.t \\
&\quad \phi(\alpha, \beta, \gamma) \leq 0,\ \beta, \gamma \in \zeta(\alpha) \\
&\quad \min_{\beta \in B_{adm}} u_\delta(\beta|\alpha, \gamma) \\
&\quad s.t \\
&\quad \phi(\alpha, \beta, \gamma) \leq 0,\ \alpha, \gamma \in \zeta(\beta) \\
&\quad\quad \min_{\gamma \in A_{adm}} w_\delta(\gamma|\alpha, \beta) \\
&\quad\quad s.t \\
&\quad\quad \phi(\alpha, \beta, \gamma) \leq 0,\ \alpha, \beta \in \zeta(\gamma)
\end{aligned}$$

where $\zeta(\alpha)$, $\zeta(\beta)$, and $\zeta(\gamma)$ are the solution set of the corresponding problem. The complement variables are fixed; for instance, for computing $\min_{\alpha \in A_{adm}} v_\delta(\alpha|\beta,\gamma)$ the variables β and γ are fixed.

At the initial stage of this framework, the manipulating players choose a manipulation strategy $\alpha \in A_{adm}$. In response, at the second stage, the manipulated players react by selecting a manipulation strategy $\beta \in B_{adm}$. In the final stage, the manipulating players, after considering both the manipulation strategy β and the manipulation strategy α, steer the system by selecting a strategy $\gamma \in A_{adm}$ that persuades the manipulated players.

At the highest level, the strategy $\alpha \in A_{adm}$ represents the approach taken by the manipulating players who initiate the game by choosing a manipulation tactic α. This strategy may involve actively influencing the manipulated players. By controlling $\alpha \in A_{adm}$, the manipulating players aim to allocate resources in a way that builds an infrastructure system as resilient to manipulation as possible. The set A_{adm} often encompasses one or more critical resource limitations that govern these actions.

The strategy $\beta \in B_{adm}$ represents the actions of the manipulated players at the middle level of the hierarchical structure. These players observe the manipulative tactics $\alpha \in A_{adm}$ and $\gamma \in A_{adm}$. Their goal is to resist the manipulation from the manipulating players to the greatest extent possible. Best-reply strategies at the middle level function as a defensive mechanism, effectively countering manipulation attempts from higher or lower levels. This counterbalancing dynamic enhances the system's resilience by neutralizing external influences. As a result, it improves strategic robustness, ensuring stability and adaptability in complex, multi-level decision-making environments.

The strategy of the manipulating players, acting as operators, is reflected at the deepest level through the minimization of $w_\delta(\gamma|\alpha,\beta)$, where they select $\gamma \in A_{adm}$ to reduce the system's operational costs. It is important to note that the strategy γ can be influenced by both the manipulation strategy α and the response strategy β, which are determined according to $u_\delta(\beta|\alpha,\gamma)$. While the manipulating players may represent different real entities, they all pursue the same manipulation objectives.

Continuous refinement by lower-level players ensures their strategies remain aligned with those of higher-level players. This adaptability is vital for maintaining effectiveness, as it allows the system to respond to changing conditions and evolving strategies within the hierarchy, ensuring that decisions remain optimal and responsive to dynamic environments.

Decision-making at each level generates feedback loops that significantly influence the entire game's dynamics. These loops create a continual exchange of information and adjustments between levels, ensuring that the system evolves toward stable outcomes. This interconnected process enhances the system's predictability, allowing for more reliable forecasts and better strategic alignment over time.

The main point of this solution lies in its asymptotic stability, which means that even if players start with different strategies, as the game evolves, they will naturally gravitate toward a stable equilibrium, without needing to regret any past decisions. No matter how the manipulative players attempt to influence the middle level, or how the lower level responds, the structure of the game ensures that, in the long run, stability will prevail.

This concept of stability also ensures that players can adopt no-regret strategies, meaning they will not have to second-guess their moves, because over time, their actions will converge to a point that best suits the entire dynamic.

5. Numerical Example: Oil Market and Regulatory Influence

This example considers controllable Bayesian–Markov chains with Lyapunov-like cost functions to model manipulation in a hierarchical oil market system vs. the regulatory influence. The decision-making framework involves players that manipulate each other strategically in a three-level Stackelberg framework.

Consider a oil market system as follows. There are three type of players: (a) outermost manipulating players: a lobbying group representing oil companies, aiming to influence regulatory priorities; (b) middle-level manipulated players: national regulators, balancing energy production, environmental sustainability, and economic growth; (c) innermost manipulating players: oil producers, who adjust production and investment strategies based on regulatory policies that persuade the manipulated players.

Consider a market that evolves over discrete types $\{\theta_1, \theta_2, \theta_3, \theta_4\} \in \Theta_{adm}$, representing oil and environmental conditions: θ_1—high demand, low environmental degradation; θ_2—moderate demand, moderate degradation; θ_3—steady demand, unaltered degradation; and θ_4—low demand, high degradation.

Each player can take three possible actions $\{y_1, y_2, y_3\} \in \mathcal{Y}_{adm}$. The transition probabilities $F^l(\theta_{t+1}|\theta_t, y_t)$ depend on the current type θ_t and actions y_t by players at all levels: y_1—lobbying actions (e.g., media campaigns and policy framing); y_2—regulatory decisions (e.g., subsidies and emissions limits); and y_3—oil producers' production and investment strategies.

The outermost manipulators that represent the lobbying group incur a cost v_δ, which is the cost of lobbying activities, the players being manipulated, regulators, bear a cost u_δ, which optimize policies based on their perception, penalizing deviations from environmental and energy demand equilibrium, and the innermost manipulators, energy producers, are assigned a cost w_δ, which adjust their strategies to minimize profit fluctuations and operational costs. These costs are thought to resemble Lyapunov-like functions

$$\min_{\pi^l(y_t^l|m_t^l)\sigma^l(m_t^l|\theta_t^l)} \mathbb{V}_{t+1}^l = \sum_{\theta_t^l \in \Theta_t^l} (\mu_0^l)_{\theta_t^l} \left[\sum_{\theta_{t+1}^l \in \Theta_{t+1}^l} \varphi^l(y_t^{l*}, \theta_t^l) F^l(\theta_{t+1}^l|\theta_t^l, y_t^{l*}) \right].$$

Each level minimizes a cost function that encourages stability.

The strategic manipulation will be as follows:

- The outermost manipulating players manipulate transition probabilities $F^l(\theta_{t+1}^l|\theta_t^l, y_t^l)$ into $F^l(\theta_{t+1}^l|\theta_t^l, y_t^{l*})$, i) sponsoring biased reports exaggerating oil demand and ii) framing oil fuels as critical for economic growth. Their goal is to influence manipulated players (regulators') cost function u_δ.
- Manipulated players in the middle-level best reply to manipulated transition probabilities $F^l(\theta_{t+1}^l|\theta_t^l, y_t^{l*})$ by implementing policies by applying $min\ \pi^f(y_t^{f*}|m_t^f)\sigma^f(m_t^f|\theta_t^f)$ to $F^f(\theta_{t+1}^f|\theta_t^f, y_t^{f*})$, such as i) subsidies for oil fuel production and ii) relaxed penalties for environmental violations. Their objective is to influence manipulating players' cost functions v_δ w_δ.
- The innermost manipulating players again improve their strategies $F^l(\theta_{t+1}^l|\theta_t^l, y_t^{l*})$ based on $F^f(\theta_{t+1}^f|\theta_t^f, y_t^{f*})$, leading to the following: i) increased production exploiting subsidies and ii) reduced investment in renewable energy.

Considering

$$F^l(:|:,1) = \begin{bmatrix} 0.8181 & 0.6596 & 0.8003 & 0.0835 \\ 0.8175 & 0.5186 & 0.4538 & 0.1332 \\ 0.7224 & 0.9730 & 0.4324 & 0.1734 \\ 0.1499 & 0.6490 & 0.8253 & 0.3909 \end{bmatrix} \quad F^l(:|:,2) = \begin{bmatrix} 0.8314 & 0.5269 & 0.2920 & 0.1672 \\ 0.8034 & 0.4168 & 0.4317 & 0.1062 \\ 0.0605 & 0.6569 & 0.0155 & 0.3724 \\ 0.3993 & 0.6280 & 0.9841 & 0.1981 \end{bmatrix}$$

$$F^{\iota}(:\mid :,3)=\begin{bmatrix}0.4897 & 0.0527 & 0.5479 & 0.3015\\ 0.3395 & 0.7379 & 0.9427 & 0.7011\\ 0.9516 & 0.2691 & 0.4177 & 0.6663\\ 0.9203 & 0.4228 & 0.9831 & 0.5391\end{bmatrix}$$

and random cost matrices, the resulting optimal strategies are shown in Figure 1 of the oil market Stackelberg game.

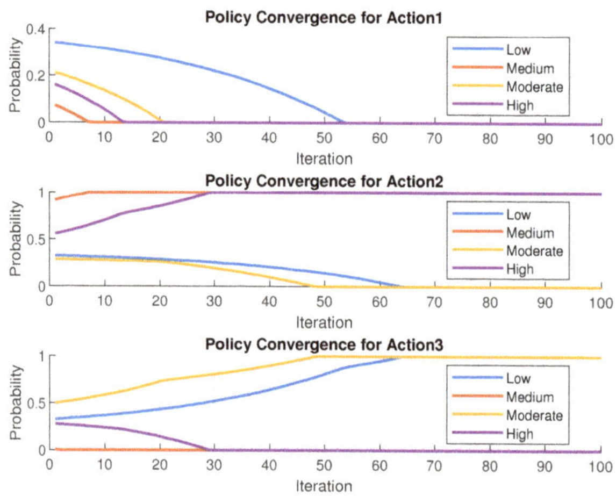

Figure 1. Convergence of the policy.

$$\pi^{\iota}(y_t^{\iota *}|m_t^{\iota}) = \begin{bmatrix}0 & 0. & 1\\ 0 & 1 & 0\\ 0 & 0 & 1\\ 0 & 1 & 0\end{bmatrix}$$

The manipulation occurs via transforming $F^{\iota}(\theta_{t+1}^{\iota}|\theta_t^{\iota}, y_t^{\iota})$ into $F^{\iota}(\theta_{t+1}^{\iota}|\theta_t^{\iota}, y_t^{\iota *})$, misleading regulators and energy producers into suboptimal but predictable decisions:

- Outermost manipulating players: $F^{l}(\theta_{t+1}^{l}|\theta_t^{l}, y_t^{l*})$ maximizes influence and profit by ensuring system inertia favors oil fuels.
- Middle-level manipulated players: $F^{f}(\theta_{t+1}^{f}|\theta_t^{f}, y_t^{f*})$ implements suboptimal policies based on manipulated dynamics.
- Innermost manipulating players: $F^{l}(\theta_{t+1}^{l}|\theta_t^{l}, y_t^{l*})$ exploits short-term gains, worsening long-term environmental and economic sustainability.

6. Conclusions

We proposed a solution to the manipulation game problem by applying Lyapunov theory, demonstrating both the existence and stability of Nash equilibria within the system. Each equilibrium that is asymptotically stable is associated with a Lyapunov-like function, which converges towards a Nash/Lyapunov equilibrium when such an equilibrium is present. This convergence guarantees the system's robustness, as the asymptotic properties of the function promote no-regret strategies, ensuring players have little incentive to deviate from their chosen course of action in the long run.

Adaptation is crucial for maintaining a strategic advantage in a dynamic environment. The ability to adjust based on observed responses allows the system to stay aligned with its evolving objectives. This ongoing adjustment process ensures that strategies remain relevant, effective, and responsive to changing conditions, fostering long-term success and competitiveness.

The manipulation game is modeled as a hierarchical, three-level Stackelberg framework. At the top level, the manipulating players select a strategy aimed at influencing the middle-level participants. However, the strategy chosen by the manipulators may not always succeed in persuading those on the middle tier. In response, the middle-level players "who are being manipulated" adopt a best-reply strategy that can effectively neutralize or match the manipulation, creating a counterbalancing dynamic. This interaction embodies the core of strategic manipulation, where each player's decisions are interdependent and reactions must be carefully calibrated.

To ensure the success of the manipulation, the player operating at the lowest level of the hierarchy must refine and adjust their strategy in response to the behavior of the manipulated players above them. This bottom-up adjustment is crucial to aligning with the overall manipulation scheme, as it allows the lower-level manipulator to account for and anticipate the reactions of the middle-tier participants. In essence, the success of the manipulation depends not just on the initial strategy but also on continuous adaptation based on observed responses at higher levels of the hierarchy.

This structured approach underscores the complexity of manipulation in multi-tiered strategic games. Each layer of decision-making "whether it is at the top, middle, or lower level" creates feedback loops that influence the broader dynamics of the game. By using Lyapunov theory to map out stable equilibria, we provide a framework that ensures the system gravitates towards a stable and predictable outcome, even in the face of manipulation. The incorporation of no-regret strategies further reinforces the stability and long-term viability of these equilibria, suggesting that the players' strategic choices will converge to an optimal solution over time. This comprehensive solution offers valuable insights into both the theoretical and practical applications of manipulation in game theory, with implications for economics, cybersecurity, and political strategy.

Funding: This research received no external funding.

Data Availability Statement: All data required for this article are included within this article, further inquiries can be directed to the corresponding author.

Conflicts of Interest: The authors declare no conflicts of interest.

References

1. Jones, D.N.; Paulhus, D.L. *Handbook of Individual Differences in Social Behavior*; Chapter Machiavellianism; The Guilford Press: New York, NY, USA, 2009; pp. 93–108.
2. Christie, R.; Geis, F. *Studies in Machiavellianism*; Academic Press: Cambridge, MA, USA, 2013.
3. Spielberger, C.D.; Butcher, J. *Advances in Personality Assessment*; Routledge: London, UK, 2013.
4. Geis, F. *Dimensions of Personality*; Wiley: New York, NY, USA, 1978; Chapter Machiavellianism, pp. 305–363.
5. Allen, F.; Gorton, G. Stock price manipulation, market microstructure and asymmetric information. *Eur. Econ. Rev.* **1992**, *36*, 624–630. [CrossRef]
6. Bagnoli, M.; Lipman, B.L. Stock Price Manipulationthrough Takeover Bids. *Rand J. Econ.* **1996**, *27*, 124–147. [CrossRef]
7. Clempner, J.B. A game theory model for manipulation based on Machiavellianism: Moral and ethical behavior. *J. Artif. Soc. Soc. Simul.* **2017**, *20*, 12. [CrossRef]
8. Cumming, D.; Ji, S.; Peter, R.; Tarsalewska, M. Market manipulation and innovation. *J. Bank. Financ.* **2020**, *120*, 105957. [CrossRef]
9. Clempner, J.B. Learning machiavellian strategies for manipulation in Stackelberg security games. *Ann. Math. Artif. Intell.* **2022**, *90*, 373–395. [CrossRef]
10. Clempner, J.B. A Manipulation Game Based on Machiavellian Strategies. *Int. Game Theory Rev.* **2022**, *24*, 2150015. [CrossRef]

11. Sanchez-Rabaza, J.; Rocha-Martinez, J.M.; Clempner, J.B. Characterizing Manipulation via Machiavellianism. *Mathematics* **2023**, *11*, 4143. [CrossRef]
12. Milgrom, P.; Roberts, J. Relying on the information of interested parties. *RAND J. Econ.* **1986**, *17*, 18–32. [CrossRef]
13. Krishna, V.; Morgan, J. A model of expertise. *Q. J. Econ.* **2001**, *116*, 747–775. [CrossRef]
14. Taneva, I. Information Design. *Am. Econ. J. Microecon.* **2019**, *11*, 151–185. [CrossRef]
15. Bardhi, A.; Guo, Y. Modes of persuasion toward unanimous consent. *Theor. Econ.* **2018**, *13*, 1111–1149. [CrossRef]
16. Kamenica, E.; Gentzkow, M. Bayesian Persuasion. *Am. Econ. Rev.* **2011**, *101*, 2590–2615. [CrossRef]
17. Bergemann, D.; Morris, S. Information design, Bayesian persuasion, and Bayes correlated equilibrium. *Am. Econ. Rev.* **2016**. [CrossRef]
18. Gentzkow, M.; Kamenica, E. Bayesian persuasion with multiple senders and rich signal spaces. *Games Econ. Behav.* **2017**, *104*, 411–429. [CrossRef]
19. Brocas, I.; Carrillo, J.D.; Palfrey, T.R. Information gatekeepers: Theoryand experimental evidence. *Econ. Theory* **2012**, *51*, 649–676. [CrossRef]
20. Gul, F.; Pesendorfer, W. The war of information. *Rev. Econ. Stud.* **2012**, *79*, 707–734. [CrossRef]
21. Eső, P.; Szentes, B. Optimal Information Disclosure in Auctionsand the Handicap Auction. *Rev. Econ. Stud.* **2007**, *74*, 705–731. [CrossRef]
22. Bergemann, D.; Pesendorfer, M. Information structuresin optimal auctions. *J. Econ. Theory* **2007**, *137*, 580–609. [CrossRef]
23. Rayo, L.; Segal, I. Optimal information disclosure. *J. Political Econ.* **2010**, *118*, 949–987. [CrossRef]
24. Li, H.; Shi, X. Discriminatory Information Disclosure. *Discrim. Inf. Discl.* **2017**, *107*, 3363–3385. [CrossRef]
25. Clempner, J.B.; Poznyak, A.S. *Optimization and Games for Controllable Markov Chains: Numerical Methods with Application to Finance and Engineering*; Springer: Berlin/Heidelberg, Germany, 2023.
26. Harsanyi, J.C.; Selten, R. A generalized Nash solutionfor two-person bargaining games with incomplete information. *Manag. Sci.* **1972**, *18*, 80–106. [CrossRef]
27. Clempner, J.B. A tri-level approach for computing Stackelberg Markov game equilibrium: Computational analysis. *J. Comput. Sci.* **2023**, *68*, 101995. [CrossRef]
28. Tanaka, K.; Yokoyama, K. On ϵ-equilibrium point in a noncooperative n-person game. *J. Math. Anal. Appl.* **1991**, *160*, 413–423. [CrossRef]

mathematics

MDPI

Article

Innovation Prioritization Decisions in the Product–Service Supply Chain: The Impact of Data Mining and Information Sharing Strategies

Jinfa Shi [1], Wei Liu [1] and Yongqiang Su [2,*]

[1] School of Management and Economics, North China University of Water Resources and Electric Power, Zhengzhou 450046, China
[2] School of Management Engineering, Henan University of Engineering, Zhengzhou 451191, China
* Correspondence: suyongqiang0104@163.com

Abstract: Driven by the servicing and digital transformation of manufacturing enterprises, product and service innovation for manufacturers and service providers to promote integrated solutions collaboratively has become an important way for enterprises to maintain market competitiveness. Building on this foundation, this paper develops an innovation priority decision model for the product–service supply chain, which comprises manufacturers and service providers, considering the data mining and information sharing strategies of service providers. It analyzes the optimal decisions and profits of the members when product innovation is prioritized as well as when service innovation is prioritized, and subsequently explores the selection of innovation strategies for the product–service supply chain under varying conditions. The results of the study show that, firstly, service providers' data mining and information sharing strategies are not always favorable to the innovation decisions of both parties. Only when data resources can be transformed into real innovation value at a reasonable cost can data mining and information sharing play the role of 'external incentives' to promote collaborative innovation between the two parties. Secondly, when service providers do not adopt data mining and information sharing strategies, the efficiency of product and service innovation plays a decisive role in innovation prioritization. The party with high innovation efficiency adopts the sub-priority innovation strategy, which can lead to a larger market share for the innovation results. Finally, under the service provider's data mining and information sharing strategy, the innovation priority selection of the product–service supply chain depends on the information value transformation ability of the manufacturer and the service provider. Moreover, the profits of manufacturers and service providers under the same innovation priority do not always 'advance or retreat together', and there may be cases where one of them suffers a loss of profits. This study provides a theoretical basis for the choice of innovation strategies given to manufacturers and service providers, and promotes the development of collaborative innovation between them.

Keywords: product innovation; service innovation; data resources; priority; decision-making

MSC: 91-10; 91A80

1. Introduction

Driven by the wave of global digital transformation and smart manufacturing, services carried in the product entity have become a common means for manufacturing enterprises to realize added value during the product technology life cycle [1]. More and more manufacturing enterprises promote the digital transformation of their new product development by 'leveraging' digital service enterprises [2], and, through the integration and management of the latter's information technology, the Internet of Things, big data analysis [3], and other means, they provide customers with integrated product–service customized solutions [4,5], realizing a seamless transition from product manufacturing to service delivery. For example, Haier cooperates with Aliyun to provide a complete set of smart home solutions, through which smart home appliances are connected to the cloud, and users can use mobile phone apps to control and manage home appliances to achieve equipment interconnectivity remotely. Sany Heavy Industries and Baidu Cloud developed the 'Root Cloud Platform' to achieve real-time data collection of engineering machinery and equipment worldwide, through remote monitoring of the working status of the equipment, to pre-judge potential problems, and to achieve accurate maintenance and repair. Compared with the independent provision of products or services, hybrid products can achieve greater customer value creation through the optimal matching of products and services [6]. However, under the impetus of servicing and digitizing manufacturing enterprises, customers' demand for innovation in products and value-added services is increasing, and manufacturing enterprises and service enterprises must maintain their competitiveness in the market through continuous innovation. As a result, hybrid product offerings, driven by both product and service innovation, have become the specific focus of many manufacturing companies' servicing efforts [7].

The wide application of digital scenarios provides more possibilities to achieve synergy in product and service innovation. The empowerment of digital intelligence technology enables the various experiences, advice, and information that traditional market insights rely on to be stored in the form of massive numbers of data [8]. The collected data can be organized to form an information flow so that scientific and precise actions can be taken under the guidance of data support [9,10]. For example, the Root Cloud Platform not only helps Sany Heavy Industry optimize equipment operation and maintenance, but also analyzes the efficiency of excavators under different working conditions through the data collected from users' use and feedback. By understanding the different requirements of customers for the performance of equipment under different geological conditions, Sany Heavy Industry can improve product design and launch machinery and equipment that are more suitable for specific working conditions. In the delivery of integrated solutions for innovative products and services, service providers provide last-mile digital value-added services, which also means that service providers are able to achieve access to product operation and customer feedback data over a relatively long period of time through technologies such as the Internet of Things (IoT) and big data [11]. Digital service companies use advanced analytics to mine and analyze these data to gain insights into customer needs [12] as well as correlative information that can improve the quality of their products and services [13]. As a result, manufacturers often require service providers to report the necessary data during the contracting process. The value of these 'non-monetary transaction' data becomes the intermediary that drives product innovation. At the same time, the continuous emergence of back-office product innovation triggers the iteration of front-office service innovation [14], and the complementary nature of these two types of innovations drives overall innovation and competitiveness enhancement.

In the traditional product-led logic, enterprises usually regard products as the core carriers of value, and product innovation becomes the main means of enterprise competition, while service innovation enhances the use of products and strengthens customer experience by optimizing services. Under this innovation logic, service innovation is more often used as a support for and an extension of product innovation [7]. As the digitalization process of manufacturing enterprises advances, the higher level of product intelligence supports the development of higher-order services [15]. It creates an interaction and transaction environment where multiple subjects participate and diverse scenarios coexist, and a massive number of data/amount of information is generated by customers in the process of service experience [16]. Deep mining and analysis based on massive data breaks the unidirectional pattern of relying on product innovation to drive service innovation. From only being able to access explicit knowledge, such as knowledge of product usage, functional defects, and improvement opinions, it has evolved to draw on the tacit knowledge hidden in product operation and customer feedback data [1]. This helps decision-makers to identify unmet innovation needs and generate new innovation insights [17,18]. Service innovation is no longer just a passive response to product innovation, but may become the source driving overall innovation. In this new innovation ecosystem, the boundaries between product and service innovation become more blurred, and this change requires us to rethink the roles of manufacturers and service providers in the innovation process. Through rational decision-making with regard to the priority of product and service innovation, we can ensure the internal matching of innovative products and services with the external demand market, and realize greater value creation for both supply and demand.

2. Literature Review

The research area of this paper primarily encompasses two aspects: data resource mining and information sharing among supply chain members and the innovation of the product–service supply chain. Consequently, the related literature review was conducted from these two perspectives.

2.1. Supply Chain Innovation Decisions

Currently, scholars are examining innovation decision-making in product–service supply chains from various perspectives. Several scholars have explored the factors influencing innovation: Du et al. [19] investigated the effects of supply chain information asymmetry and retailer overconfidence on members' decisions regarding innovation inputs and revenues utilizing a signaling game approach. Liu et al. [20] explores the impact of remanufacturing process innovation on closed-loop supply chains from the perspective of government subsidies. Qu et al. [21] examines joint innovation investment and pricing decisions for revenue-sharing contracts and customer value. Xing et al. [22] constructed a green innovation supply chain consisting of suppliers, manufacturers, and retailers, examining the effects of various government subsidies on this supply chain. Yi et al. [23] employed a differential game approach to analyze the impact of disappointment aversion on supply chain members' innovation decisions and proposed a coordination mechanism that includes subsidies and revenue matching. Another group of scholars has examined the issue of innovation choice within product–service supply chains: Liu et al. [24] found that retailers provide extended warranty services only when the cost advantage is significant. Furthermore, manufacturers can enhance retailers' incentives to affiliate through innovation investment, fostering cooperation between both parties. Yu et al. [25] compared and analyzed optimal decision-making under two innovation modes, product innovation alone and retailer equity-financed product innovation, considering suppliers' varying dominant positions and highlighting the influence of multiple factors on the choice of innovation mode. Li et al. [26] proposed two innovation strategies—closed innovation and leading user innovation—and analyzed the influence of consumer heterogeneity on each strategy's innovation decisions. Lin et al. [27] investigated the equilibrium results under three innovation strategies: manufacturer's non-innovation, independent innovation, and technology purchasing in the context of uncertain technical efficiency. Jiang et al. [28] examine the conditions applicable to new energy vehicle enterprises for implementing process or model innovation, discussing the impact of government subsidies on innovation decisions. Current research on innovation primarily emphasizes product innovation; however, the strong correlation between product and service innovation, which jointly influences demand, remains unexamined. Furthermore, there is a paucity of research addressing decision-making issues related to innovation prioritization stemming from the blurred boundaries between product and service innovations. Moreover, the decision-making problem regarding innovation prioritization, resulting from the blurred boundaries between product and service innovation, has received limited scholarly attention.

2.2. Supply Chain Information Sharing

A data resource is a kind of resource that contains great value. Through their data mining and sharing strategy, an enterprise can obtain effective data and analyze, store, and use this data resource, which can increase the revenue of the enterprise [17]. Brinch [29] emphasizes that the analysis of data resource value is emerging as a significant research hotspot in the field of supply chain management. A related area of inquiry is the sharing of demand information within the supply chain. Research on demand information sharing in supply chains primarily investigates the effects of various factors on information sharing and its impact on decision-making among stakeholders. For instance, Zhang et al. [30] found that, when both manufacturers and retailers forecast market demand and retailers possess asymmetric information, they will voluntarily share this information when the manufacturer's innovation efficiency is high. Shi et al. [31] examined product network externalities and found that demand information sharing between retailers and manufacturers correlates with the network externality coefficient, making information sharing profitable when this coefficient is high. Wang et al. [32] investigated dynamic strategies for knowledge sharing and emission reduction benefits in low-carbon technologies within green supply chains using differential game methods, identifying optimal trajectories for suppliers' and manufacturers' knowledge stocks and emission reduction benefits under various strategies. Tang et al. [33] analyzed the impact of information prediction and sharing strategies on manufacturing companies' supply chains under independent and joint innovation models. Nie et al. [34] discovered that incorporating remanufactured products into the product line can create a win–win–win scenario for retailers, manufacturers, and the environment through effective information sharing strategies. Wei et al. [35] revealed that, irrespective of the supplier's distribution model and the retailer's information sharing strategy, greater consumer sensitivity to the product's green level correlates with higher expected profits, product prices, and green levels among channel members. The existing literature primarily considers information sharing as a factor influencing supply chain decision-making, neglecting the value of data resources as economic assets. Data resources hold significant potential value that can be transformed into business profits through effective acquisition, analysis, storage, and utilization. This paper, therefore, discusses the impact of data resource sharing behavior on product and service innovation by incorporating service providers' data resource sharing into the supply chain and considering the economic value and uncertainty associated with data resources.

To address the issues in the existing research, this study considers two scenarios—information sharing and non-sharing by service providers—and constructs four decision-making models based on product innovation priority strategies and service innovation priority strategies. It analyzes the optimal decision-making, market demand, and profits of each member by solving the different models. This study addresses the following questions: What conditions influence product and service innovation priorities within the product–service innovation supply chain? What impact do service providers' demand information sharing strategies have on their product and service innovation levels and pricing decisions under the same priority level, and how does this affect profit levels? Further comparisons of market demand for innovative products and services, along with manufacturers' and service providers' profits under varying prioritization levels, reveal how different factors influence innovation prioritization choices in product–service supply chains. The research findings aim to provide managerial insights into the collaborative innovation decisions made by manufacturers and service providers.

3. Model Construction and Analysis

3.1. Problem Description and Assumptions

Consider a product–service supply chain consisting of a digital service provider (hereinafter referred to as 'service provider', S) and a product manufacturer (hereinafter referred to as 'manufacturer', M) which, together, provide an integrated solution, a 'product + digital service' (hereinafter referred to as 'integrated solution'), to a customer. Among them, the service provider provides digital services such as product system modification and additional function expansion of the integrated solution. The manufacturer produces products according to customer and market demand, combines them with the digital services provided by the service provider, and delivers a comprehensive solution integrating products and services to the customer. In order to meet the increasingly individualized needs of customers, both parties must continuously improve and innovate their products and services. Manufacturers, for example, upgrade technology through modular product design, intelligent product upgrades, and ecosystem product design. Service providers, on the other hand, offer innovative services through remote monitoring, troubleshooting for data-driven service optimization, and customized solutions. In the process of implementing product and service innovations, the manufacturer and the service provider first decide on an innovation prioritization strategy and the service provider decides whether to share information. With the service provider adopting an information sharing strategy, if both parties decide to prioritize product innovation, the manufacturer initiates product modification and innovation activities and sets the final sales price of the integrated solution. The service provider then innovates the service and also decides on the unit service price. After both parties have completed their innovations, the manufacturer delivers to the market a comprehensive solution that integrates the innovative products and services. At the end of the sales period, the service provider collects product operation and customer feedback data during the service process, converts them into demand information, and shares them with the manufacturer to guide further innovation. When both parties decide to prioritize service innovation, the service provider takes the lead in initiating service innovation activities and determines the unit service price. At the same time, the service provider shares service innovation information and customer demand information with the manufacturer so that the manufacturer can adapt its product innovation program to the new service characteristics. The manufacturer then innovates the product, sets the selling price, and brings the integrated solution to the market. In cases where the service provider does not share information, the two parties set innovation levels and prices according to different innovation priorities. The product service innovation supply chain structure is shown in Figure 1.

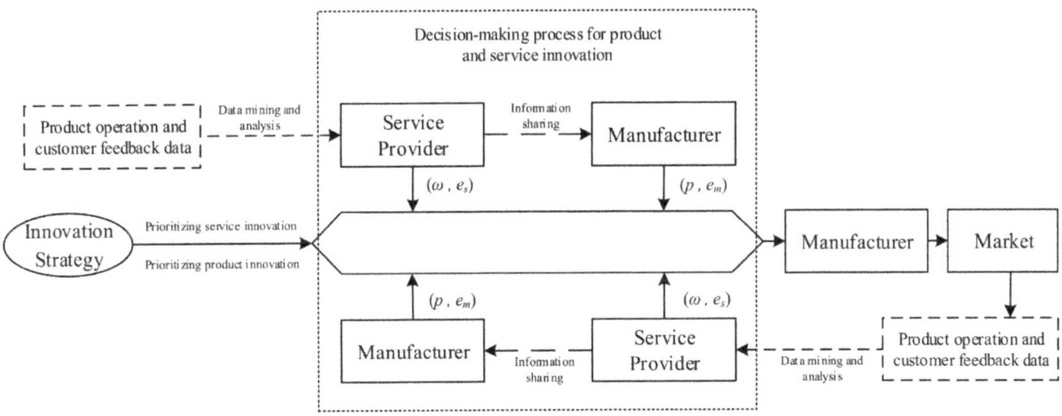

Figure 1. The structure of the product–service innovation supply chain.

Demand Assumption: Product innovation enhances the performance, stability, and functionality of hardware, software, or systems, making integrated solutions more relevant and flexible, and increasing the attractiveness of the solutions. Service innovation, on the other hand, enhances customer satisfaction and loyalty by providing personalized, continuously optimized services and improving service quality and responsiveness [36]. This has led customers to focus more and more on the overall value provided by integrated solutions, rather than just on single products or services. Therefore, it is further assumed that both the level of product innovation and the level of service innovation have a positive impact on the demand for integrated solutions. With reference to similar studies [37,38], the demand for innovative integrated solutions is portrayed as $D = a - bp + \lambda e_m + \beta e_s$, where a is the potential market demand for integrated solutions; p is the selling price of integrated solutions; b is the price elasticity coefficient of integrated solutions, which measures the sensitivity of the market demand to the selling price; and λ is the product innovation utility and β is the service innovation utility, which represent the degree of customer demand for product and service innovations, respectively.

Innovation Cost Assumptions: Consistent with most innovation cost function assumptions [39–41], the innovation input cost functions of manufacturers and service providers are convex functions with respect to the level of innovation. Assuming that the level of service innovation by the service provider is e_s and the level of product innovation by the manufacturer is e_m, the cost function of service innovation input by the service provider is $c(e_s) = \frac{1}{2}\mu_s e_s^2$ and the cost function of product innovation input by the manufacturer is $c(e_m) = \frac{1}{2}\mu_m e_m^2$, where μ_s represents the cost coefficient of service innovation by the service provider and μ_m is the cost coefficient of product innovation by the manufacturer. For the convenience of the study, the initial production cost of products and services in this study is set to 0, which does not affect the conclusions of the study.

Information Sharing Costs and Benefits Assumptions: According to Ma [42] and Correani [43], the ability of an enterprise to 'process' massive data resources into valuable knowledge or information depends largely on the enterprise's ability to analyze and process massive data and big data. This includes the ability to make data-driven decisions, data productization, and value realization, which is expressed in terms of shorter decision-making cycles, cost reductions as a result of data optimization, and the number of innovative products supported by data. Manufacturers and service providers have the ability to convert data resources into information values, k_m and k_s. In this study, integrated solutions are sold to customers as integrated innovative products and services. Therefore, the number of data resources acquired by the service provider is equal to the number of its services under the product innovation priority model [44]. The value of the shared information received by the manufacturer is then quantified as $k_s D$. In the service innovation priority model, the service provider transforms the collected market demand information into service innovation, and the level of its service innovation represents the data value of the data resources that the service provider converts into information. The manufacturer converts the shared data value into product innovation knowledge, which increases the value of the manufacturer's product innovation [45] by $k_m e_s$. Additionally, the cost of information sharing by service providers in both models is categorized as $\eta_s D$ and $\eta_s e_s$, including, for example, technical operation and maintenance, data conversion, and communication costs [46]. Among them, η_s is the cost coefficient of information sharing for service providers.

The primary parameters utilized in this paper are presented in Table 1 below.

Table 1. Description of the primary parameters.

Notation	Description
a	The potential market demand
b	The price elasticity coefficient
λ	The product innovation utility
β	The service innovation utility
μ_m	The cost coefficient of product innovation
μ_s	The cost coefficient of service innovation
η_s	The cost coefficient of information sharing
k_m	Information value transformation coefficient of manufacturer
k_s	Information value transformation coefficient of service provider
e_m	The level of product innovation
e_s	The level of service innovation
ω	The unit service price
p	The selling price
π_m	The profits of manufacturer
π_s	The profits of service provider

For the sake of clarity, superscripts SP, DP, SPI, and DPI represent service innovation priority without information sharing, product innovation priority without information sharing, service innovation priority with information sharing, and product innovation priority with information sharing, respectively. The subscripts m and s denote manufacturers and service providers, respectively. The symbol '*' indicates the optimal solution.

3.2. Model Construction

This study considers two scenarios—information sharing and non-sharing by service providers—and constructs four decision-making models based on product innovation priority strategies and service innovation priority strategies. Among them, the DP model and SP model do not take into account the information-sharing behavior of service providers, while the DPI model and SPI model do. The utility functions for manufacturers and service providers in the four models are as follows:

(DP Model) In the product innovation priority model, which does not take into account information value conversion and sharing by service providers, the manufacturer takes the lead in completing the product innovation and setting the final sales price for the integrated solution to be released in the market. The service provider determines the level of service innovation and the price per unit of service based on the manufacturer's decision program.

$$\pi_m^{DP} = (p - \omega)(a - bp + \lambda e_m + \beta e_s) - \frac{1}{2}\mu_m e_m^2 \tag{1}$$

$$\pi_s^{DP} = \omega(a - bp + \lambda e_m + \beta e_s) - \frac{1}{2}\mu_s e_s^2 \tag{2}$$

(SP Model) In the service innovation priority model, which does not take into account information value conversion and sharing by service providers, the service provider takes the lead in service innovation and sets the unit price of the innovative service. The manufacturer determines the level of product innovation and the final selling price of the integrated solution released in the market based on the service provider's decision scenario.

$$\pi_m^{SP} = (p - \omega)(a - bp + \lambda e_m + \beta e_s) - \frac{1}{2}\mu_m e_m^2 \tag{3}$$

$$\pi_s^{SP} = \omega(a - bp + \lambda e_m + \beta e_s) - \frac{1}{2}\mu_s e_s^2 \tag{4}$$

In the DP and SP models, $(a - bp + \lambda e_m + \beta e_s)$ is the overall market demand created by the manufacturer and the service provider. Equation (3) is the manufacturer's utility function, where $(p - \omega)$ and $\frac{1}{2}\mu_m e_m^2$ represent the manufacturer's profit gained from the unit of the integrated solution and the total cost of product innovation, respectively. Equation (4) is the service provider's utility function, where ω and $\frac{1}{2}\mu_s e_s^2$ represent the service provider's profit gained from the unit of the integrated solution and the total cost of service innovation, respectively.

(DPI Model) Similar to the DP model's decision-making process, the service provider commits to converting the product operation and customer feedback data resources obtained during the delivery of the integrated solution into valuable information and sharing it with the manufacturer. Considering the value of the information shared by the service provider, the manufacturer decides the level of product innovation and sets the selling price of the integrated solution. The service provider decides on the level of service innovation and the price per unit of service based on the manufacturer's decision scheme.

$$\pi_m^{DPI} = (p - \omega)(a - bp + \lambda e_m + \beta e_s) + k_s(a - bp + \lambda e_m + \beta e_s) - \frac{1}{2}\mu_m e_m^2 \tag{5}$$

$$\pi_s^{DPI} = \omega(a - bp + \lambda e_m + \beta e_s) - \eta_s(a - bp + \lambda e_m + \beta e_s) - \frac{1}{2}\mu_s e_s^2 \tag{6}$$

In the DPI model, the value of the information that the manufacturer receives from the service provider's data sharing is $k_s(a - bp + \lambda e_m + \beta e_s)$. At the same time, the service provider's cost of data resource sharing is $\eta_s(a - bp + \lambda e_m + \beta e_s)$.

(SPI Model) Similar to the SP model's decision process, the service provider takes the lead in completing the service innovation and determining the unit price of the innovative service. At the same time, market demand and service innovation information is shared with the manufacturer. Based on the service provider's innovation decisions and the value of the shared information, the manufacturer determines the level of product innovation and the final selling price of the innovative integrated solution released to the market.

$$\pi_m^{SPI} = (p - \omega + k_m e_s)(a - bp + \lambda e_m + \beta e_s) - \frac{1}{2}\mu_m e_m^2 \tag{7}$$

$$\pi_s^{SPI} = \omega(a - bp + \lambda e_m + \beta e_s) - \eta_s e_s - \frac{1}{2}\mu_s e_s^2 \tag{8}$$

In the SPI model, the preferred innovation service enables the manufacturer to obtain an additional value of $k_m e_s$ from the unit integrated solution. Meanwhile, the service provider's cost of sharing data resources is $\eta_s e_s$.

3.3. Model Analysis

By employing backward induction, the equilibrium solutions of the four models can be derived. The derivation of Propositions 1 and 2 is presented below, and the procedure for proving the remaining propositions is analogous. The proofs of the corollaries are included in Appendix A.

Proposition 1 (DP Model). *Manufacturers prioritize product innovation without considering service provider information sharing.*

(i) The optimal levels of product innovation (e_m^{DP*}), service innovation (e_s^{DP*}), price per unit of service (ω^{DP*}), and selling price (p^{DP*}) for both manufacturers and service providers are defined as follows:

$$e_m^{DP*} = \frac{a\lambda\mu_s}{4b\mu_m\mu_s - 2\beta^2\mu_m - \lambda^2\mu_s} \tag{9}$$

$$e_s^{DP*} = \frac{a\beta\mu_m}{4b\mu_m\mu_s - 2\beta^2\mu_m - \lambda^2\mu_s} \tag{10}$$

$$\omega^{DP*} = \frac{a\mu_m\mu_s}{4b\mu_m\mu_s - 2\beta^2\mu_m - \lambda^2\mu_s} \tag{11}$$

$$p^{DP*} = \frac{a\mu_m(3b\mu_s - \beta^2)}{b(4b\mu_m\mu_s - 2\beta^2\mu_m - \lambda^2\mu_s)} \tag{12}$$

(ii) The optimal profits for manufacturers (π_m^{DP*}) and service providers (π_s^{DP*}) are defined as follows:

$$\pi_m^{DP*} = \frac{a^2\mu_m\mu_s}{2(4b\mu_m\mu_s - 2\beta^2\mu_m - \lambda^2\mu_s)} \tag{13}$$

$$\pi_s^{DP*} = \frac{a^2\mu_s\mu_m^2(2b\mu_m - \lambda^2)}{2(4b\mu_m\mu_s - 2\beta^2\mu_m - \lambda^2\mu_s)^2} \tag{14}$$

Proof. The manufacturer first sets the sales premium r for the innovative integrated solution, and the sales price determined by the manufacturer is the price per unit of service provided by the service provider ω plus its sales premium r, which can be expressed as $p = \omega + r$. Substituting into Equations (1) and (2) and applying the inverse induction method of solving for Equation (2) to find the first-order derivatives of π_s^{DP} with respect to ω and e_s and making the first-order derivatives equal to 0 yields $\frac{\partial \pi_s^{DP}}{\partial \omega} = a + \lambda e_m +$

$\beta e_s - b\omega - b(r+\omega) = 0$, $\frac{\partial \pi_s^{DP}}{\partial e_s} = \beta\omega - \mu_s e_s = 0$. Therefore, the Hessian matrix of π_s^{DP} is $H_1 = \begin{bmatrix} -2b & \beta \\ \beta & -\mu_s \end{bmatrix}$. It can be shown that the Hessian matrix $|H_1|$ is negative definite when $2b\mu_s - \beta^2 > 0$. The service provider profit π_s^{DP} is a joint concave function of ω and e_s, and there exists an optimal solution with great value. Substituting the result of $\frac{\partial \pi_s^{DP}}{\partial \omega}$ and $\frac{\partial \pi_s^{DP}}{\partial e_s}$ being equal to 0 into Equation (1) and taking the first-order derivatives of r and e_m, respectively, the Hessian matrix of π_m^{DP} is $H_2 = \begin{bmatrix} \frac{2b^2 u_s}{\beta^2 - 2b\mu_s} & \frac{b\lambda\mu_s}{2b\mu_s - \beta^2} \\ \frac{b\lambda\mu_s}{2b\mu_s - \beta^2} & -\mu_m \end{bmatrix}$. When $4b\mu_m\mu_s - 2\beta^2\mu_m - \lambda^2\mu_s > 0$, the Hessian matrix $|H_2|$ is negative definite and the profit function π_m^{DP} is the joint concave function of r and e_m. Seeking its first-order derivatives equal to 0, the joint solution to the manufacturer's optimal level of innovation and the optimal sales premium, which will be brought into the $\frac{\partial \pi_s^{DP}}{\partial \omega}$, $\frac{\partial \pi_s^{DP}}{\partial e_s}$, are solved to obtain the optimal level of innovation of the service provider and the optimal unit price of service results for Formulas (10) and (11). The manufacturer's optimal sales price p is equal to the optimal unit service price ω plus r. Substituting the above optimal decision variables into the profit function of the manufacturer, the service provider can be obtained. In the product innovation priority, the manufacturer and the service provider profit results in Formulas (13) and (14), and Proposition 1 is proved. □

Proposition 2 (SP Model). *Service providers prioritize service innovation without considering service provider information sharing.*

(i) The optimal levels of product innovation (e_m^{SP*}), service innovation (e_s^{SP*}), price per unit of service (ω^{SP*}), and selling price (p^{SP*}) for both manufacturers and service providers are defined as follows:

$$e_m^{SP*} = \frac{a\lambda\mu_s}{4b\mu_m\mu_s - \beta^2\mu_m - 2\lambda^2\mu_s} \tag{15}$$

$$e_s^{SP*} = \frac{a\beta\mu_m}{4b\mu_m\mu_s - \beta^2\mu_m - 2\lambda^2\mu_s} \tag{16}$$

$$\omega^{SP*} = \frac{a\mu_s(2b\mu_m - \lambda^2)}{b(4b\mu_m\mu_s - \beta^2\mu_m - 2\lambda^2\mu_s)} \tag{17}$$

$$p^{SP*} = \frac{a\mu_s(3b\mu_m - \lambda^2)}{b(4b\mu_m\mu_s - \beta^2\mu_m - 2\lambda^2\mu_s)} \tag{18}$$

(ii) The optimal profits for manufacturers (π_m^{SP*}) and service providers (π_s^{SP*}) are defined as follows:

$$\pi_m^{SP*} = \frac{a^2\mu_m\mu_s^2(2b\mu_m - \lambda^2)}{2(4b\mu_m\mu_s - \beta^2\mu_m - 2\lambda^2\mu_s)^2} \tag{19}$$

$$\pi_s^{SP*} = \frac{a^2\mu_m\mu_s}{2(4b\mu_m\mu_s - \beta^2\mu_m - 2\lambda^2\mu_s)} \tag{20}$$

Proof. The inverse induction method is used to solve Equation (3) to find the first-order derivative of π_m^{SP} with respect to p, e_m and make it equal to 0 to obtain $\frac{\partial \pi_m^{SP}}{\partial p} = a + \lambda e_m + \beta e_s - bp - b(p-\omega) = 0$, $\frac{\partial \pi_m^{SP}}{\partial e_m} = \lambda(p-\omega) - \mu_m e_m = 0$. Then, the Hessian matrix of π_m^{SP} is $H_1 = \begin{bmatrix} -2b & \lambda \\ \lambda & -\mu_m \end{bmatrix}$. It can be seen that the Hessian matrix $|H_1|$ is negatively definite when $2b\mu_m - \lambda^2 > 0$. The manufacturer profit π_m^{SP} is a joint concave function of p and e_m, and there exists an optimal solution with a great value. Substituting the result of $\frac{\partial \pi_m^{SP}}{\partial p}$ and $\frac{\partial \pi_m^{SP}}{\partial e_m}$ equal to 0 into Equation (4) and taking the first-order derivatives of ω and e_s, respectively, it can be seen that the Hessian matrix of π_s^{SP} is $H_2 = \begin{bmatrix} \frac{2b^2 u_m}{\lambda^2 - 2b\mu_m} & \frac{b\beta\mu_m}{2b\mu_m - \lambda^2} \\ \frac{b\beta\mu_m}{2b\mu_m - \lambda^2} & -\mu_s \end{bmatrix}$. When $4b\mu_m\mu_s - \beta^2\mu_m - 2\lambda^2\mu_s > 0$, the Hessian matrix $|H_2|$ is negative definite and the profit function π_s^{SP} is the joint concave function of ω and e_s. Seeking its first-order derivatives equal to 0, the joint solution to the optimal level of innovation of the service provider and the optimal price of services, respectively, can be found for the results of Equations (16) and (17), with $\frac{\partial \pi_m^{SP}}{\partial p}$ and $\frac{\partial \pi_m^{SP}}{\partial e_m}$ being used to solve for the optimal level of innovation of the manufacturer and the optimal price of the results of the sales from Equations (15)–(18). Substituting the above optimal decision variables into the profit function of the manufacturer, the service provider can be obtained. When the service innovation is prioritized, the manufacturer and the service provider profits result in Equations (19) and (20), respectively, and Proposition 2 is proved. □

Corollary 1. *The optimal decision, when product innovation is prioritized, increases with the product innovation utility* λ *and decreases with the product innovation cost coefficient* μ_m*, without considering service provider information sharing.*

$$\frac{\partial e_m^{DP*}}{\partial \lambda} > 0,\ \frac{\partial e_s^{DP*}}{\partial \lambda} > 0,\ \frac{\partial \omega^{DP*}}{\partial \lambda} > 0,\ \frac{\partial p^{DP*}}{\partial \lambda} > 0,\ \frac{\partial \pi_m^{DP*}}{\partial \lambda} > 0,\ \frac{\partial \pi_s^{DP*}}{\partial \lambda} > 0;$$
$$\frac{\partial e_m^{DP*}}{\partial \mu_m} < 0,\ \frac{\partial e_s^{DP*}}{\partial \mu_m} < 0,\ \frac{\partial \omega^{DP*}}{\partial \mu_m} < 0,\ \frac{\partial p^{DP*}}{\partial \mu_m} < 0,\ \frac{\partial \pi_m^{DP*}}{\partial \mu_m} < 0,\ \frac{\partial \pi_s^{DP*}}{\partial \mu_m} < 0;$$

Corollary 2. *The optimal decision, when service innovation is prioritized, increases with the service innovation utility* β *and decreases with the service innovation cost coefficient* μ_s*, without considering service provider information sharing.*

$$\frac{\partial e_m^{SP*}}{\partial \beta} > 0,\ \frac{\partial e_s^{SP*}}{\partial \beta} > 0,\ \frac{\partial \omega^{SP*}}{\partial \beta} > 0,\ \frac{\partial p^{SP*}}{\partial \beta} > 0,\ \frac{\partial \pi_m^{SP*}}{\partial \beta} > 0,\ \frac{\partial \pi_s^{SP*}}{\partial \beta} > 0;$$
$$\frac{\partial e_m^{SP*}}{\partial \mu_s} < 0,\ \frac{\partial e_s^{SP*}}{\partial \mu_s} < 0,\ \frac{\partial \omega^{SP*}}{\partial \mu_s} < 0,\ \frac{\partial p^{SP*}}{\partial \mu_s} < 0,\ \frac{\partial \pi_m^{SP*}}{\partial \mu_s} < 0,\ \frac{\partial \pi_s^{SP*}}{\partial \mu_s} < 0;$$

Corollaries 1 and 2 suggest that both manufacturers and service providers benefit from increased service and product innovation effects. Greater product and service innovation utility implies greater customer acceptance of the innovative content and model. When service providers introduce more flexible and personalized services to give new functions and usage scenarios to their products, this expands the scope of application of the products, brings new market opportunities for manufacturers, and encourages manufacturers to innovate their products to meet different customer needs. At the same time, product innovation by manufacturers usually introduces new technologies, which change the operation mode and service process of existing service providers, prompting service providers to adjust and innovate their existing services to adapt to new products. Both parties cooperate in technology development and the production process, and the innovation effects promote each other to form a spiral development mode. On the contrary, the increase in the innovation cost coefficient implies that manufacturers and service providers have insufficient innovation capability and cannot meet the market demand despite investing more in innovation costs. The uncertainty of innovation leads both parties to be more cautious when investing in innovation, or even to reduce their investment in innovation, which leads to a decrease in the level of innovation.

Proposition 3 (DPI model). *Considering the scenario of information sharing by service providers, product innovation is prioritized.*

(i) The optimal levels of product innovation (e_m^{DPI*}), service innovation (e_s^{DPI*}), price per unit of service (ω^{DPI*}), and selling price (p^{DPI*}) for both manufacturers and service providers are defined as follows:

$$e_m^{DPI*} = \frac{(a + b(k_s - \eta_s))\lambda\mu_s}{4b\mu_m\mu_s - 2\beta^2\mu_m - \lambda^2\mu_s} \tag{21}$$

$$e_s^{DPI*} = \frac{(a + b(k_s - \eta_s))\beta\mu_m}{4b\mu_m\mu_s - 2\beta^2\mu_m - \lambda^2\mu_s} \tag{22}$$

$$\omega^{DPI*} = \frac{(a + b(3\eta_s + k_s))\mu_m\mu_s - \eta_s(2\beta^2\mu_m + \lambda^2\mu_s)}{4b\mu_m\mu_s - 2\beta^2\mu_m - \lambda^2\mu_s} \tag{23}$$

$$p^{DPI*} = \frac{a\mu_m(3b\mu_s - \beta^2) + b(\eta_s - k_s)(-\beta^2\mu_m - \lambda^2\mu_s + b\mu_m\mu_s)}{b(4b\mu_m\mu_s - 2\beta^2\mu_m - \lambda^2\mu_s)} \tag{24}$$

(ii) The optimal profits for manufacturers (π_m^{DPI*}) and service providers (π_s^{DPI*}) are defined as follows:

$$\pi_m^{DPI*} = \frac{(a + b(k_s - \eta_s))^2\mu_m\mu_s}{8b\mu_m\mu_s - 4\beta^2\mu_m - 2\lambda^2\mu_s} \tag{25}$$

$$\pi_s^{DPI*} = \frac{(a + b(k_s - \eta_s))^2\mu_m^2\mu_s(2b\mu_s - \beta^2)}{2(4b\mu_m\mu_s - 2\beta^2\mu_m - \lambda^2\mu_s)^2} \tag{26}$$

To ensure the existence and practical significance of the equilibrium solution, the following conditions must be satisfied: $2b\mu_s - \beta^2 > 0$, $4b\mu_m\mu_s - 2\beta^2\mu_m - \lambda^2\mu_s > 0$, $0 < \eta_s < \frac{(-a+bk_s)\beta^2\mu_m + b\mu_s(2a\mu_m + k_s(\lambda^2 - 2b\mu_m))}{b\mu_m(2b\mu_s - \beta^2)}$.

Proposition 4 (SPI model). *Considering the scenario of information sharing by service providers, service innovation is prioritized.*

(i) The optimal levels of product innovation (e_m^{SPI*}), service innovation (e_s^{SPI*}), price per unit of service (ω^{SPI*}), and selling price (p^{SPI*}) for both manufacturers and service providers are defined as follows:

$$e_m^{SPI*} = \frac{\lambda(a\mu_s - \eta_s(bk_m + \beta))}{4b\mu_m\mu_s - (bk_m + \beta)^2\mu_m - 2\lambda^2\mu_s} \tag{27}$$

$$e_s^{SPI*} = \frac{a(bk_m + \beta)\mu_m - 2\eta_s(2b\mu_m - \lambda^2)}{4b\mu_m\mu_s - (bk_m + \beta)^2\mu_m - 2\lambda^2\mu_s} \tag{28}$$

$$\omega^{SPI*} = \frac{(2b\mu_m - \lambda^2)(a\mu_s - \eta_s(bk_m + \beta))}{b(4b\mu_m\mu_s - (bk_m + \beta)^2\mu_m - 2\lambda^2\mu_s)} \tag{29}$$

$$p^{SPI*} = \frac{b^2k_m\mu_m(\eta_s - ak_m) + \lambda^2(\eta_s\beta - a\mu_s) - b(\eta_s k_m\lambda^2 + 3\eta_s\beta\mu_m + ak_m\beta\mu_m - 3a\mu_m\mu_s)}{b(4b\mu_m\mu_s - (bk_m + \beta)^2\mu_m - 2\lambda^2\mu_s)} \tag{30}$$

(ii) The optimal profits for manufacturers (π_m^{SPI*}) and service providers (π_s^{SPI*}) are defined as follows:

$$\pi_m^{SPI*} = \frac{\mu_m(2b\mu_m - \lambda^2)(b\eta_s k_m + \eta_s\beta - a\mu_s)^2}{2(4b\mu_m\mu_s - (bk_m + \beta)^2\mu_m - 2\lambda^2\mu_s)^2} \tag{31}$$

$$\pi_s^{SPI*} = \frac{a^2\mu_m\mu_s - 2\eta_s(a\mu_m(bk_m + \beta) - \eta_s(2b\mu_m - \lambda^2))}{2(4b\mu_m\mu_s - (bk_m + \beta)^2\mu_m - 2\lambda^2\mu_s)} \tag{32}$$

To ensure the existence and practical significance of the equilibrium solution, the following conditions must be satisfied: $2b\mu_m - \lambda^2 > 0$, $4b\mu_m\mu_s - (bk_m + \beta)^2\mu_m - 2\lambda^2\mu_s > 0$, $0 < \eta_s < \frac{a\mu_m(bk_m + \beta)}{2(2b\mu_m - \lambda^2)}$.

Corollary 3. *(i) In the service provider information sharing scenario, the optimal decision* e_m^{DPI*}, e_s^{DPI*}, π_m^{DPI*}, π_s^{DPI*} *for the product innovation prioritization decision increases with the increase in* *the service provider information value transformation coefficient* k_s *and decreases with the increase in the information sharing cost coefficient* η_s.

$$\frac{\partial e_m^{DPI*}}{\partial k_s} > 0, \frac{\partial e_s^{DPI*}}{\partial k_s} > 0, \frac{\partial \pi_m^{DPI*}}{\partial k_s} > 0, \frac{\partial \pi_s^{DPI*}}{\partial k_s} > 0;$$
$$\frac{\partial e_m^{DPI*}}{\partial \eta_s} < 0, \frac{\partial e_s^{DPI*}}{\partial \eta_s} < 0, \frac{\partial \pi_m^{DPI*}}{\partial \eta_s} < 0, \frac{\partial \pi_s^{DPI*}}{\partial \eta_s} < 0;$$

(ii) When $\beta^2\mu_m + \lambda^2\mu_s - b\mu_m\mu_s > 0$, *the optimal decision* p^{DPI*} *increases with the increase in the service provider information value transformation coefficient* k_s *and decreases with the increase in the information sharing cost coefficient* η_s. *When* $\beta^2\mu_m + \lambda^2\mu_s - b\mu_m\mu_s < 0$, *the optimal decision* p^{DPI*} *decreases with the increase in the service provider's information value transformation coefficient* k_s *and increases with the increase in the information sharing cost coefficient* η_s.

(iii) The optimal decision ω^{DPI*} *increases with the increase in the service provider's information value transformation coefficient* k_s. *When* $3b\mu_m\mu_s - 2\beta^2\mu_m - \lambda^2\mu_s > 0$, *the optimal decision* ω^{DPI*} *increases with the increase in the information sharing cost coefficient* η_s. *When* $3b\mu_m\mu_s - 2\beta^2\mu_m - \lambda^2\mu_s < 0$, *the optimal decision* ω^{DPI*} *decreases with the increase in the information sharing cost coefficient* η_s.

Corollary 4. *In the service provider information sharing scenario, the optimal decision to prioritize service innovation* $e_m^{SPI*}, e_s^{SPI*}, \omega^{SPI*}, p^{SPI*}, \pi_m^{SPI*}, \pi_s^{SPI*}$ *increases with the increase in the value transformation coefficient of the manufacturer's information* k_m *and decreases with the increase in the information sharing cost coefficient* η_s.

$$\frac{\partial e_m^{SPI*}}{\partial k_m} > 0, \frac{\partial e_s^{SPI*}}{\partial k_m} > 0, \frac{\partial \omega^{SPI*}}{\partial k_m} > 0, \frac{\partial p^{SPI*}}{\partial k_m} > 0, \frac{\partial \pi_m^{SPI*}}{\partial k_m} > 0, \frac{\partial \pi_s^{SPI*}}{\partial k_m} > 0$$
$$\frac{\partial e_m^{SPI*}}{\partial \eta_s} < 0, \frac{\partial e_s^{SPI*}}{\partial \eta_s} < 0, \frac{\partial \omega^{SPI*}}{\partial \eta_s} < 0, \frac{\partial p^{SPI*}}{\partial \eta_s} < 0, \frac{\partial \pi_m^{SPI*}}{\partial \eta_s} < 0, \frac{\partial \pi_s^{SPI*}}{\partial \eta_s} < 0$$

It follows from Corollaries 3 and 4 that, on the one hand, information sharing by service providers allows manufacturers to capture economic value beyond the benefits of the innovation itself. For example, by obtaining more innovation information based on product operation data and customer feedback data, the cost of innovation for the manufacturer can be reduced through information sharing. On the other hand, enterprises process and analyze product operation data, customer feedback, and market demand information to identify potential market opportunities and changes in consumer demand, improve R&D efficiency through optimal allocation of resources, and reduce the risk of innovation due to market uncertainty. In this case, manufacturers and service providers will tend to increase the level of innovation of their products and services to meet specific market demands. Customers are willing to pay higher prices to try and accept innovative products when the manufacturer's and service provider's innovations greatly satisfy their needs. In this case, both parties can set higher prices in order to make higher profits. On the contrary, higher information sharing costs mean that the service provider needs to invest more resources in information collection, processing, and delivery. The increase in information sharing costs means service providers have to undermine their investment in service innovation, leading to a decrease in the level of service innovation. The decrease in the level of service innovation directly affects the value of innovation information and indirectly increases the cost of product innovation for manufacturers, which leads to a decrease in the level of product innovation. In the case of a decline in the level of product and service innovation, service providers and manufacturers maintain market competitiveness and market share by lowering wholesale and selling prices.

Corollary 5 (DPI Model vs. DP Model).

(i) There exists $\overline{k_{s1}}$. *When* $0 < k_s < \overline{k_{s1}}$, *if* $0 < \eta_s < k_s$, *the level of product innovation, the level of service innovation, the selling price, the manufacturer's profit, and the service provider's profit under the product innovation priority strategy satisfy* $e_m^{DPI*} > e_m^{DP*} > 0$, $e_s^{DPI*} > e_s^{DP*} > 0$, $\pi_m^{DPI*} > \pi_m^{DP*}$, $\pi_s^{DPI*} > \pi_s^{DP*}$; *if* $k_s < \eta_s < \overline{\eta_{sd}}$, *the level of product innovation, the level of*

service innovation, the selling price, the manufacturer's profit, and the service provider's profit satisfy $0 < e_m^{DPI*} < e_m^{DP*}$, $0 < e_s^{DPI*} < e_s^{DP*}$, $\pi_m^{DPI*} < \pi_m^{DP*}$, $\pi_s^{DPI*} < \pi_s^{DP*}$. *When* $k_s > \overline{k_{s1}}$, *if* $0 < \eta_s < \overline{\eta_{sd}}$, *the level of product innovation, the level of service innovation, the selling price, the manufacturer's profit, and the service provider's profit satisfy* $e_m^{DPI*} > e_m^{DP*} > 0$, $e_s^{DPI*} > e_s^{DP*} > 0$, $\pi_m^{DPI*} > \pi_m^{DP*}$, $\pi_s^{DPI*} > \pi_s^{DP*}$. *The comparison between* p^{DPI*} *and* p^{DP*} *has the same threshold, but the magnitude of both depends on* $\beta^2\mu_m + \lambda^2\mu_s - b\mu_m\mu_s$.

Where $\overline{k_{s1}} = \frac{a\mu_m(\beta^2-2b\mu_s)}{b(2\beta^2\mu_m+(\lambda^2-4b\mu_m)\mu_s)}$, $\overline{\eta_{sd}} = \frac{(-a+bk_s)\beta^2\mu_m+b\mu_s(2a\mu_m+k_s(\lambda^2-2b\mu_m))}{b\mu_m(2b\mu_s-\beta^2)}$.

(ii) There exists $\overline{k_{s2}}$. *When* $0 < k_s < \overline{k_{s2}}$, *if* $0 < \eta_s < \frac{bk_s\mu_m\mu_s}{2\beta^2\mu_m+(\lambda^2-3b\mu_m)\mu_s}$, *the price of the service satisfies* $\omega^{DPI*} > \omega^{DP*}$; *if* $\frac{bk_s\mu_m\mu_s}{2\beta^2\mu_m+(\lambda^2-3b\mu_m)\mu_s} < \eta_s < \overline{\eta_{sd}}$, *the price of the service satisfies* $\omega^{DPI*} < \omega^{DP*}$. *When* $k_s > \overline{k_{s2}}$, *if* $0 < \eta_s < \overline{\eta_{sd}}$, *the price of the service satisfies* $\omega^{DPI*} > \omega^{DP*}$, *where* $\overline{k_{s2}} = \frac{a\mu_m(\beta^2-2b\mu_s)(2\beta^2\mu_m+\lambda^2\mu_s-3b\mu_m\mu_s)}{b(b\mu_m\mu_s-\beta^2\mu_m-\lambda^2\mu_s)(4b\mu_m\mu_s-2\beta^2\mu_m-\lambda^2\mu_s)}$.

Corollary 5 shows that the choice of information sharing strategy under the priority of product innovation is closely related to the information value conversion coefficient and information sharing cost coefficient of the service provider. Only when the service provider's information value conversion coefficient and information sharing cost coefficient satisfy certain conditions will the information sharing strategy increase the innovation level of both parties as well as bring excess profits. The market demand information acquired by the service provider is not always directly transformed into product innovation knowledge. Only if the service provider transforms it into knowledge that is highly compatible with the manufacturer's product innovation needs will information sharing be of sufficient economic value to the manufacturer. In addition, service providers incurring additional information sharing costs will demand higher service prices, which will directly affect manufacturers' product pricing and profit allocation. In conjunction with Corollary 3, the information sharing strategy benefits both parties more when the service provider's ability to transform the value of information is higher and the cost of information sharing is lower.

Corollary 6 (SPI model vs. SP model).

(i) There exists $\overline{\eta_{s1}}$. *When* $0 < \eta_s < \overline{\eta_{s1}}$, *the level of product innovation, the price per unit of service, and the manufacturer's profit under the service innovation priority decision satisfy* $e_m^{SPI*} > e_m^{SP*}$, $\omega^{SPI*} > \omega^{SP*}$, $\pi_m^{SPI*} > \pi_m^{SP*}$. *When* $\overline{\eta_{s1}} < \eta_s < \overline{\eta_{ss}}$, *the level of product innovation, the price per unit of service, and the manufacturer's profit satisfy* $e_m^{SPI*} < e_m^{SP*}$, $\omega^{SPI*} < \omega^{SP*}$, $\pi_m^{SPI*} < \pi_m^{SP*}$.

(ii) There exists $\overline{\eta_{s2}}$. *When* $0 < \eta_s < \overline{\eta_{s2}}$, *the level of service innovation satisfies* $e_s^{SPI*} > e_s^{SP*}$. *When* $\overline{\eta_{s2}} < \eta_s < \overline{\eta_{ss}}$, *the level of service innovation satisfies* $e_s^{SPI*} < e_s^{SP*}$.

(iii) There exists $\overline{\eta_{s3}}$. *When* $0 < \eta_s < \overline{\eta_{s3}}$, *the sales price satisfies* $p^{SPI*} > p^{SP*}$. *When* $\overline{\eta_{s3}} < \eta_s < \overline{\eta_{ss}}$, *the sales price satisfies* $p^{SPI*} < p^{SP*}$.

(iv) There exists $\overline{\eta_{s4}}$. *When* $0 < \eta_s < \overline{\eta_{s4}}$, *service provider profits satisfy* $\pi_s^{SPI*} > \pi_s^{SP*}$. *When* $\overline{\eta_{s4}} < \eta_s < \overline{\eta_{ss}}$, *service provider profits satisfy* $\pi_s^{SPI*} < \pi_s^{SP*}$.

Where $\overline{\eta_{ss}} = \frac{a\mu_m(bk_m+\beta)}{2(2b\mu_m-\lambda^2)}$, $\overline{\eta_{s1}} = \frac{abk_m(bk_m+2\beta)\mu_m\mu_s}{(bk_m+\beta)(-\beta^2\mu_m-2\lambda^2\mu_s+4b\mu_m\mu_s)}$; $\overline{\eta_{s2}} = \frac{abk_m\mu_m(-2\lambda^2\mu_s+\mu_m(bk_m\beta+\beta^2+4b\mu_s))}{2(\lambda^2-2b\mu_m)(\beta^2\mu_m+2\lambda^2\mu_s-4b\mu_m\mu_s)}$; $\overline{\eta_{s3}} = \frac{abk_m\mu_m(\beta^2\mu_m(bk_m+\beta)+b\mu_s(2\beta\mu_m+k_m(\lambda^2-b\mu_m)))}{(\lambda^2\beta+b^2k_m\mu_m-b(k_m\lambda^2+3\beta\mu_m))(\beta^2\mu_m+2\lambda^2\mu_s-4b\mu_m\mu_s)}$; $\overline{\eta_{s4}} = \frac{a\mu_m(bk_m+\beta)(-\beta^2\mu_m-2\lambda^2\mu_s+4b\mu_m\mu_s)-\sqrt{a^2\beta^2\mu_m^2(\beta^2\mu_m+2\lambda^2\mu_s-4b\mu_m\mu_s)((bk_m+\beta)^2\mu_m+2\lambda^2\mu_s-4b\mu_m\mu_s)}}{2(\lambda^2-2b\mu_m)(\beta^2\mu_m+2\lambda^2\mu_s-4b\mu_m\mu_s)}$.

Corollary 6 shows that the choice of information sharing strategy under service innovation prioritization is not only closely related to the information sharing cost coefficient of the service provider, but also significantly associated with the information value conversion coefficient of the manufacturer. If the manufacturer can effectively use the service innovation information and demand information shared by the service provider to accurately adjust the product design, improve the production process, and quickly launch innovative products in response to the market demand, this efficient information conversion can significantly increase the added value of service innovation to product innovation. Innovative products and services occupy a larger market share, which, in turn, will lead to the growth of service providers' profits and make up for the economic losses caused by service providers in the process of information sharing. However, the value of information sharing cannot be maximized if the manufacturer's ability to convert the value of information is weak, even if the service provider provides detailed information on market demand. In this case, there is limited improvement in the service provider's level of innovation and profitability, and the high cost of information sharing may even cause the service provider's profit to be lower than it would have been without information sharing.

The previous section analyzed the correlation between the optimal decision and each parameter as well as the impact of the service provider's information sharing strategy on the optimal decision and profit by solving different models. Comprehensively comparing the four models constructed, it is found that the key factors affecting the product–service innovation supply chain are the information value conversion coefficient of the manufacturer and the service provider and the information sharing cost coefficient of the service provider. Next, we further compare the market demand under different innovation priorities and analyze the impact of these parameters on the choice of supply chain innovation priorities.

Proposition 7. *(i) Market demand under product innovation prioritization and service innovation prioritization decisions in the scenario of no information sharing by the service provider:*

$$D^{DP*} = \frac{ab\mu_m\mu_s}{4b\mu_m\mu_s - 2\beta^2\mu_m - \lambda^2\mu_s} \tag{33}$$

$$D^{SP*} = \frac{ab\mu_m\mu_s}{4b\mu_m\mu_s - \beta^2\mu_m - 2\lambda^2\mu_s} \tag{34}$$

(ii) Market demand under product innovation prioritization and service innovation prioritization decisions in the scenario of information sharing by the service provider:

$$D^{DPI*} = \frac{b(a + b(k_s - \eta_s))\mu_m\mu_s}{4b\mu_m\mu_s - 2\beta^2\mu_m - \lambda^2\mu_s} \tag{35}$$

$$D^{SPI*} = \frac{b\mu_m(a\mu_s - \eta_s(bk_m + \beta))}{4b\mu_m\mu_s - (bk_m + \beta)^2\mu_m - 2\lambda^2\mu_s} \tag{36}$$

Corollary 7 (SP Model vs. DP Model). *Introduces two parameters, product innovation efficiency and service innovation efficiency, φ_m, φ_s. When the party with high innovation efficiency carries out innovation activities as a sub-priority, higher market demand can be obtained.*

A party with high innovation efficiency usually has better resource optimization and cost control capabilities, and can further reduce innovation costs on the basis of previous innovations, which helps it to be more competitive in the market. At the same time, by drawing on the technological foundation and market experience of the prior innovation process, it avoids innovation risks, makes use of existing market data and technological advances to carry out more optimized innovation, integrates the latest technology with market demand, and provides more mature and efficient products or services. This precise alignment enables them to quickly meet demand and gain market share when entering the market.

Corollary 8 (SPI Model vs. DPI Model).

Case 1: If $b^2\mu_m\mu_s - \sqrt{b^2\mu_m(2\beta^2\mu_m - \lambda^2\mu_s)} < 0$, *then* $\overline{k_{m1}} < \overline{k_{m2}}$.

There exist $\overline{k_{m1}}$, $\overline{k_{m2}}$, $\overline{k_{m3}}$. *When* $0 < k_m < \overline{k_{m1}}$, *if* $0 < \eta_s < \overline{\eta_{s5}}$, *the market demand under the product innovation priority and service innovation model satisfies* $D^{SPI*} < D^{DPI*}$;

if $\eta_s > \overline{\eta_{s5}}$, *the market demand satisfies* $D^{SPI*} > D^{DPI*}$. *When* $\overline{k_{m1}} < k_m < \overline{k_{m2}}$, *the market demand satisfies* $D^{SPI*} < D^{DPI*}$. *When* $\overline{k_{m2}} < k_m < \overline{k_{m3}}$, *if* $0 < k_s < \overline{k_{s3}}$, $0 < \eta_s < \overline{\eta_{s5}}$, *the market demand satisfies* $D^{SPI*} > D^{DPI*}$; *if* $0 < k_s < \overline{k_{s3}}$, $\eta_s > \overline{\eta_{s5}}$, *the market demand satisfies* $D^{SPI*} < D^{DPI*}$; *if* $k_s > \overline{k_{s3}}$, *market demand satisfies* $D^{SPI*} < D^{DPI*}$.

Case 2: If $b^2\mu_m\mu_s - \sqrt{b^2\mu_m(2\beta^2\mu_m - \lambda^2\mu_s)} > 0$, *then* $\overline{k_{m1}} > \overline{k_{m2}}$.

There exist $\overline{k_{m1}}$, $\overline{k_{m2}}$, $\overline{k_{m3}}$. *When* $0 < k_m < \overline{k_{m2}}$, *if* $0 < \eta_s < \overline{\eta_{s5}}$, *the market demand under the product innovation priority and service innovation model satisfies* $D^{SPI*} < D^{DPI*}$; *if* $\eta_s > \overline{\eta_{s5}}$, *the market demand satisfies* $D^{SPI*} > D^{DPI*}$. *When* $\overline{k_{m2}} < k_m < \overline{k_{m1}}$, *if* $0 < k_s < \overline{k_{s3}}$, *the market demand satisfies* $D^{SPI*} > D^{DPI*}$. *If* $k_s > \overline{k_{s3}}$, $0 < \eta_s < \overline{\eta_{s5}}$, *the market demand satisfies* $D^{SPI*} < D^{DPI*}$; *if* $k_s > \overline{k_{s3}}$, $\eta_s > \overline{\eta_{s5}}$, *the market demand satisfies* $D^{SPI*} > D^{DPI*}$. *When* $\overline{k_{m1}} < k_m < \overline{k_{m3}}$, *if* $0 < k_s < \overline{k_{s3}}$, $0 < \eta_s < \overline{\eta_{s5}}$, *market demand satisfies* $D^{SPI*} > D^{DPI*}$. *If* $k_s > \overline{k_{s3}}$, *the market demand satisfies* $D^{SPI*} < D^{DPI*}$.

Where $\overline{k_{s3}} = \frac{a(b^2k_m^2\mu_m+2bk_m\beta\mu_m-\beta^2\mu_m+\lambda^2\mu_s)}{b(4b\mu_m\mu_s-(bk_m+\beta)^2\mu_m-2\lambda^2\mu_s)}$,

$$\overline{\eta_{s5}} = -\frac{\mu_s(a(b^2k_m^2\mu_m+2bk_m\beta\mu_m-\beta^2\mu_m+\lambda^2\mu_s)+bk_s((bk_m+\beta)^2\mu_m+2(\lambda^2-2b\mu_m)\mu_s))}{2\beta^2(bk_m+\beta)\mu_m-(bk_m+\beta)(-\lambda^2+b(4+bk_m+\beta)\mu_m)\mu_s+2b(2b\mu_m-\lambda^2)\mu_s^2},$$

$$\overline{k_{m1}} = \frac{\lambda^2\mu_s+2\mu_s(\beta^2-b(2+\beta)\mu_s)+\sqrt{4\beta^4\mu_m^4+4\beta^2\mu_m\mu_s(\lambda^2-4b\mu_m)+\mu_s^2((\lambda^2-4b\mu_m)^2-8b^2\mu_m\mu_s(\lambda^2-2b\mu_m))}}{2b^2\mu_m\mu_s},$$

$$\overline{k_{m2}} = \frac{-b\beta\mu_m+\sqrt{2b^2\beta^2\mu_m^2-b^2\lambda^2\mu_m\mu_s}}{b^2\mu_m},\ \overline{k_{m3}} = \frac{-b\beta\mu_m+\sqrt{2}\sqrt{b^2\mu_m(2b\mu_m-\lambda^2)\mu_s}}{b^2\mu_m}.$$

Corollary 8 shows that, in the case of service provider information sharing, the choice of innovation priority for integrated solutions is closely related to the information value conversion coefficients of the manufacturer and the service provider. When the cost coefficient of data resource sharing and the information conversion coefficients of both parties satisfy a certain magnitude relationship, it is possible to decide the priority of product innovation according to the market demand. This means that the delivery and use of integrated solutions are not only expressed in the delivery of physical products and value-added services, but also include regular services, optimization services, and upgrading services in the process of product use and operation. The value of data resources such as customer needs and services for product innovation provides 'external incentives' for manufacturers and service providers to innovate.

4. Numerical Simulation

4.1. Numerical Calculation of the Model

Building upon the theoretical analysis, we establish the model parameters and compute the decision-making outcomes for various models. Taking into account the constraints necessary for the existence of equilibrium solutions, and drawing from the existing literature [11,47–49], the parameters are set as follows: $a = 5$, $b = 1.2$, $\lambda = 0.5$, $\beta = 0.6$, $k_m = 0.7$, $k_s = 1$, $\mu_m = 0.8$, $\mu_s = 1.5$, $\eta_s = 0.8$. By substituting the values of each parameter into the analytical solutions of the decision variables across the four innovation strategies, the results are presented in Table 2.

Table 2. Decision-making results under different decision-making models.

Decision Models	e_m	e_s	ω	p	D	π_m	π_s
DP	0.7798	0.4991	1.2477	3.4934	1.4972	3.1192	1.6812
SP	0.7942	0.5083	2.2104	3.4810	1.5248	1.6852	3.1766
DPI	0.8172	0.5230	2.1075	3.4611	1.5691	3.4258	1.8465
SPI	0.9471	0.9215	2.6362	3.5066	1.8185	2.3970	3.4200

As can be seen from Table 2, when service providers do not share information, the innovation level and market demand of both parties do not change significantly under different innovation priorities, and the party that prioritizes innovation is able to obtain higher profits. When service providers share information, the innovation level, market demand, and profits of both manufacturers and service providers increase. This also reflects that the service provider's data mining and information sharing strategy is conducive to improving the innovation performance of the supply chain.

4.2. Impact Analysis of Information Sharing Cost Coefficients

This section analyzes the impact of variations in information sharing costs on the levels of innovation, unit service pricing, sales prices, and profits of both parties, as illustrated in Figure 2. To satisfy the constraints derived from Propositions 3 and 4, the range of values for the information sharing cost coefficients is set to $\eta_s \in (0, 1.6]$. The remaining parameters are maintained using the values specified in Section 4.1 of this paper.

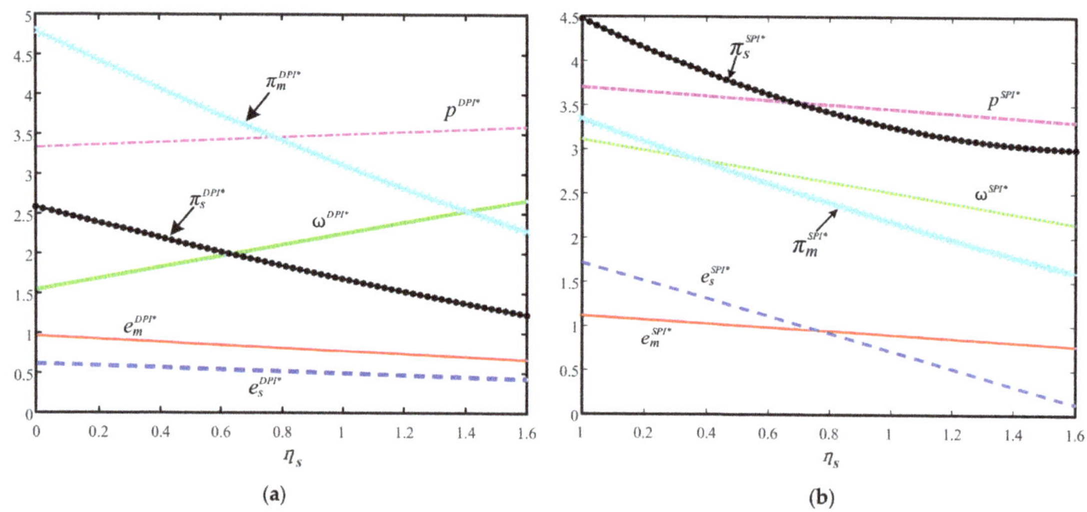

Figure 2. The impact of information sharing cost coefficients on optimal decision-making. (**a**) The effect of information sharing cost coefficients on optimal decision-making under the product innovation prioritization strategy. (**b**) The effect of information sharing cost coefficients on optimal decision-making under the service innovation prioritization strategy.

It can be seen that the optimal level of innovation and profits for both manufacturers and service providers decreases with increasing information sharing cost coefficients, irrespective of the innovation priority adopted. When service innovation is prioritized, the unit service price and selling price decrease with an increasing information sharing cost coefficient. When product innovation is prioritized, the price per unit of service and selling price increase with the information sharing cost coefficient. Combined with the analyses of Corollaries 3 and 7, the changes in unit service prices and selling prices with the information sharing cost coefficient depend on the innovation efficiency of products and services. Low innovation efficiency implies that the competitiveness enhancement of products and services brought about by innovation has relatively little impact on market demand, and firms are unable to strive for higher premiums through differentiated innovation. It also reflects the passive position of firms in the market, with their pricing strategies relying more on cost orientation. In the face of rising costs, firms often choose to shift cost pressures by raising prices in order to maintain their profitability.

4.3. Impact Analysis of the Information Value Transformation Coefficient

The impact of the information value transformation coefficients of both the manufacturer and the service provider on the levels of innovation, unit service pricing, selling price, and profits for both parties is analyzed in Figure 3. In this analysis, based on the constraints derived from Propositions 3 and 4, the ranges for the information value transformation coefficients of the manufacturer and the service provider are set to $k_m \in (0, 0.9]$ and $k_s \in (0, 2]$, respectively. The remaining parameters continue to be selected according to the values specified in Section 4.1 of this paper.

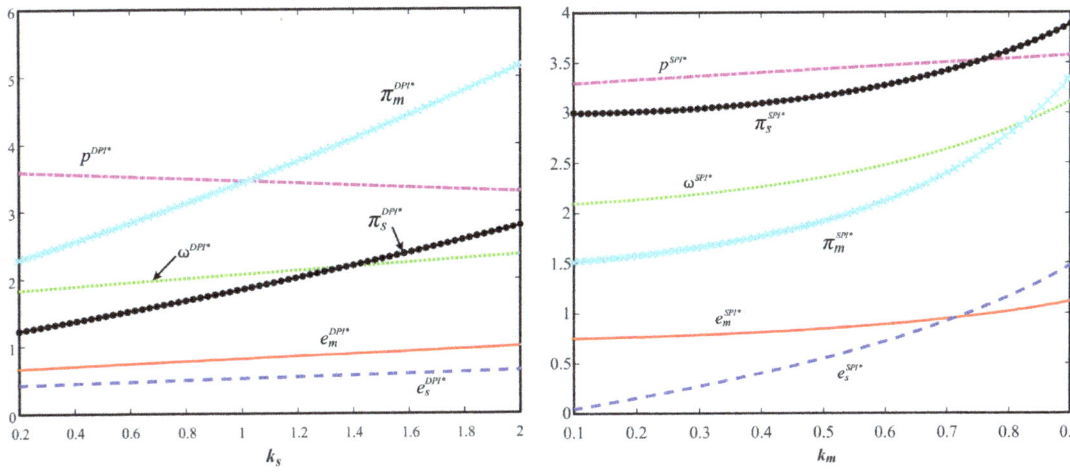

Figure 3. The impact of information value transformation coefficients on optimal decision-making. (**a**) Impact of service providers' information value transformation coefficient on optimal decision-making under product innovation prioritization strategies. (**b**) Impact of manufacturer's information value transformation coefficient on optimal decision-making under service innovation prioritization strategies.

It can be seen that the optimal level of innovation and profits for both manufacturers and service providers increase with the value of the information transformation coefficient, irrespective of the innovation priority adopted. When service innovation is prioritized, the selling price increases with the information value transformation coefficient. However, when product innovation is prioritized, the selling price decreases with the increase in the information value transformation coefficient. The main reason for this phenomenon is the different ways in which manufacturers obtain the value of information under the two innovation priorities. Under the product innovation priority, the manufacturer directly captures the economic value of the information shared by the service provider, and the value of the information increases with the increase in the service provider's information value transformation coefficient, which enlarges the manufacturer's profit margin. With low innovation efficiency, manufacturers attempt to capture a larger market share by lowering the selling price. In contrast, the value of information acquired by the manufacturer under the service innovation priority is reflected in the creation of differentiated products to meet market demand based on service innovation, and this differentiation allows for greater freedom in pricing, enabling the manufacturer to earn higher profits by raising prices.

4.4. Impact Analysis of Information Sharing and Non-Sharing

This section compares the levels of innovation, unit service pricing, sales prices, and profits for both parties when service providers adopt information sharing and non-sharing strategies under the same innovation priority. Furthermore, it identifies the relationship governed by the service provider's information sharing cost coefficient and the conversion coefficients of both the manufacturer and service provider's information value when the optimal decisions differ between the two strategies.

As illustrated in Figure 4, under the product innovation priority strategy, when the service provider's information value transformation coefficient exceeds the information sharing cost coefficient, the levels of innovation and profit for both parties under the service provider's information sharing strategy surpass those when they do not share. Under the service innovation priority strategy, when the service provider's information sharing cost coefficient and the manufacturer's information value transformation coefficient fulfill a specific relationship, the innovation decisions of both parties under the service provider's information sharing strategy exceed those made when not sharing.

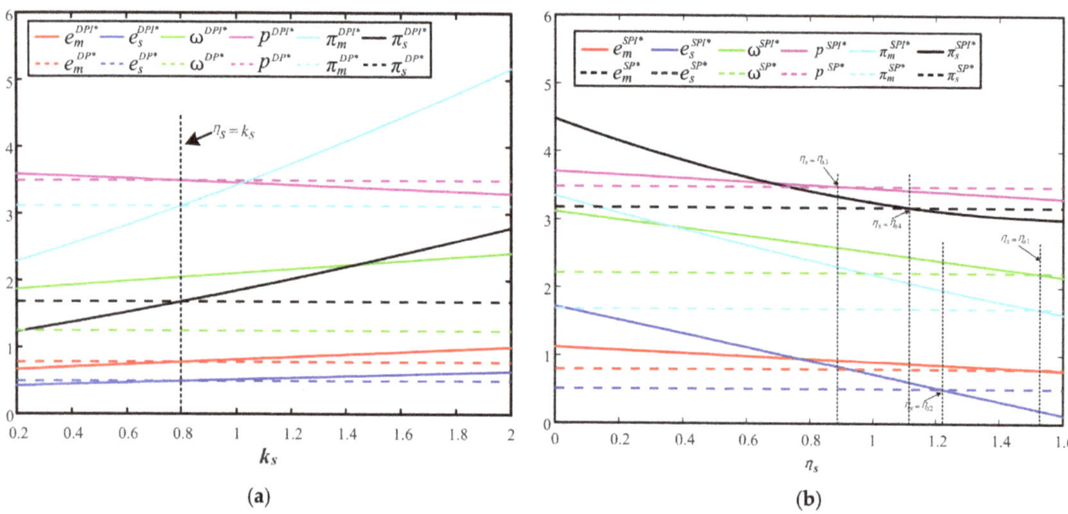

Figure 4. Differences in optimal decision-making between sharing and not sharing information at the same priorities. (**a**) Differences in optimal decision-making between information sharing and no information sharing under product innovation prioritization. (**b**) Differences in optimal decision-making between information sharing and no information sharing under service innovation prioritization.

4.5. Impact Analysis of Innovation Priorities

In order to show more intuitively the difference between market demand and profit when the information value conversion coefficients of the two parties are at different levels under different innovation priorities, we make a difference between the market demand and the profit values of the two parties under the priority of product innovation and the priority of service innovation, and compare the size of the market demand and profit under the two scenarios by the positive and negative values of the difference.

Figure 5a shows how the difference in market demand under the product innovation priority decision and the service innovation priority decision varies with the information value transformation coefficient k_m, k_s. From Figure 5a, it can be seen that there exists a critical range for the relationship between the size of the information value transformation coefficient k_m and the information value transformation coefficient k_s. When k_s is greater than this critical range, the market demand under the service innovation priority is smaller than that under the product innovation priority, and the opposite is true if it is less than this critical value. At the same time, it can also be found from the scope covered by the difference of market demand greater than zero that the enhanced information value transformation capability of the manufacturer under the service innovation priority strategy is more likely to increase the market demand. For this reason, measuring the information value transformation capability of manufacturers is a key indicator when deciding on innovation prioritization based on market demand.

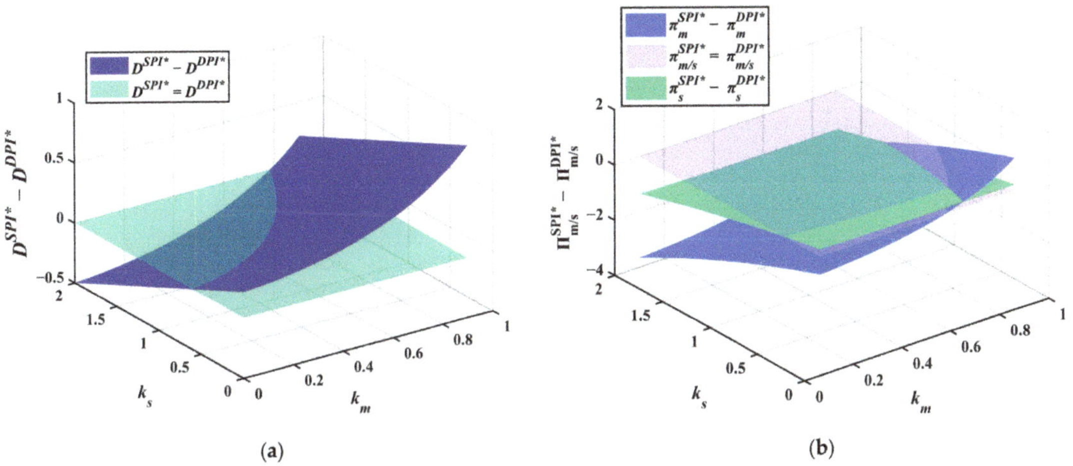

Figure 5. Impact of information value transformation coefficients on market demand and profits under different innovation priorities. (**a**) Differences in market demand under different innovation priorities. (**b**) Differences in profits between manufacturers and service providers under different innovation priorities.

Figure 5b shows how the difference between the manufacturer's and the service provider's profits under different innovation prioritization decisions varies with the information value transformation coefficient k_m and the information value transformation coefficient k_s. From Figure 5b, it can be seen that, when the manufacturer's information value transformation coefficient k_m and the service provider's information value transformation coefficient k_s satisfy a certain size relationship, either product innovation priority or service innovation priority leads to an increase in the profits of both parties compared to the other strategy. It can also be found that, when both information transformation coefficients are low, the service provider's profit under the service innovation priority strategy is higher than that of the product innovation priority strategy, but the product innovation priority leads to a higher profit for the manufacturer. This suggests that the profits of manufacturers and service providers under the same innovation prioritization do not always 'advance or retreat together', and there are cases where one of them loses profits. In this case, the development of an appropriate profit-sharing mechanism between the manufacturer and the service provider plays an important role in fostering collaborative innovation between the two parties.

5. Conclusions and Insights

In the context of the digital economy, data, recognized as a new type of production factor, have become a key resource for enterprises aiming to achieve economic benefits and sustainable competitive advantages. For manufacturers and service providers collaborating to deliver integrated solutions comprising 'products + digital services', data mining and analysis empower them to swiftly identify market trends and customer demands, thereby facilitating innovative decision-making. Within the framework of collaborative innovation between manufacturers and service providers, it is essential to examine the impact of service providers' data mining and information sharing strategies on the innovation decisions and priorities of both parties. Framing this as the research problem, this paper constructs and analyzes the optimal decision-making results of the DP, SP, DPI, and SPI models, examines the influence of various factors on decision-making, and subsequently explores the issue of innovation priority selection in the product–service innovation supply chain. The main findings are summarized below. Firstly, service providers' data resource mining and information sharing strategies are not always favorable to both parties' innovative decisions. It is certain that the optimal innovation decisions of manufacturers and service providers increase with the level of information value transformation of both parties and decrease with the increase in the information sharing cost coefficient. Therefore, only when data resources can be transformed into real innovation value at a reasonable cost can data mining and sharing really play the role of 'external incentives' to promote collaborative innovation between the two parties. Secondly, product and service innovation efficiency plays a decisive role in innovation prioritization choices when information is not shared, and the party with high innovation efficiency adopts sub-priority innovations, which can lead to a larger market share for the innovative integrated solution. Under the information sharing strategy, product and service innovation efficiency affects the pricing decision under the product innovation prioritization strategy. This dominates whether the supply chain gains higher profits by increasing pricing or captures a greater market share by lowering prices. Finally, the choice of innovation priority under the information sharing strategy depends on the ability of manufacturers and service providers to transform information value. When selecting innovation priorities based on market demand, service innovation is prioritized when the manufacturer's ability to transform information value is high. Product innovation is prioritized when the service provider's information value transformation capability is high. When selecting innovation priorities based on the profits of both parties, it is possible that one party's profits may be compromised.

Based on the research conclusions, the following managerial recommendations are derived: Firstly, enhance data resource sharing and optimize sharing costs. Data resource sharing can improve levels of service and product innovation. Manufacturers should actively establish cooperative relationships with service providers for data mining and information sharing, creating efficient mechanisms for information exchange. Service providers should optimize existing sharing processes through technological improvements and cost reductions, thereby enhancing the overall benefits of information sharing. Secondly, emphasize value transformation of information to improve innovation decision quality. Manufacturers and service providers should introduce advanced data analysis tools and techniques to strengthen their data analysis and information management capabilities, enhancing the value derived from transforming data resources into actionable information. Managers should accurately assess their capabilities to set reasonable innovation priority decisions and promote trust and support among collaborative innovation partners through transparent decision-making processes. Thirdly, establish effective incentive mechanisms to promote supply chain coordination. To address costs arising from information sharing and collaborative innovation, consider the economic capabilities and contributions of all parties. By designing a reasonable cost-sharing mechanism, the financial burden on supply chain members can be alleviated, thereby increasing their willingness to share data resources and engage in collaborative innovation.

Current research focuses on the universal issue of the impact of data resource sharing on innovation prioritization, but different industries and different data mining approaches may have unique characteristics that affect innovation prioritization and data sharing capabilities. Meanwhile, the risk aversion of supply chain members, the long-term dynamic impact of data resource sharing, and competitor pricing decisions can be further discussed in the future.

Author Contributions: Conceptualization, J.S. and W.L.; methodology, J.S. and W.L.; software, W.L.; data curation, W.L.; writing—original draft preparation, W.L. and Y.S; validation, W.L. and Y.S.; writing—review and editing, W.L. and Y.S. All authors have read and agreed to the published version of the manuscript.

Funding: This research and the APC were funded by the National Natural Science Foundation of China (71371172), the Research and Practice on Higher Education Teaching Reform in Henan

Province (2021SJGLX016), and the Philosophy and Social Science Planning Project of Henan Province (2022BJJ066).

Data Availability Statement: Data are contained within the article.

Conflicts of Interest: The authors declare no conflicts of interest.

Appendix A

Corollary 1 Proof process

Calculating the partial derivatives of e_m^{DP*}, e_s^{DP*}, ω^{DP*}, p^{DP*}, π_m^{DP*}, π_s^{DP*} with respect to λ, μ_m, respectively.

$$
\begin{aligned}
\frac{\partial e_m^{DP*}}{\partial \lambda} &= \frac{a\mu_s(-2\beta^2\mu_m+(\lambda^2+4b\mu_m)\mu_s)}{(2\beta^2\mu_m+(\lambda^2-4b\mu_m)\mu_s)^2} > 0, &
\frac{\partial e_s^{DP*}}{\partial \lambda} &= \frac{2a\lambda\beta\mu_m\mu_s}{(2\beta^2\mu_m+(\lambda^2-4b\mu_m)\mu_s)^2} > 0, \\
\frac{\partial \omega^{DP*}}{\partial \lambda} &= \frac{2a\lambda\mu_m\mu_s^2}{(2\beta^2\mu_m+(\lambda^2-4b\mu_m)\mu_s)^2} > 0, &
\frac{\partial p^{DP*}}{\partial \lambda} &= \frac{2a\lambda\mu_m\mu_s(3b\mu_s-\beta^2)}{b(2\beta^2\mu_m+(\lambda^2-4b\mu_m)\mu_s)^2} > 0, \\
\frac{\partial \pi_m^{DP*}}{\partial \lambda} &= \frac{4a^2\lambda\mu_m\mu_s^2}{(-4\beta^2\mu_m-2(\lambda^2-4b\mu_m)\mu_s)^2} > 0, &
\frac{\partial \pi_s^{DP*}}{\partial \lambda} &= \frac{2a^2\lambda\mu_m^2\mu_s^2(\beta^2-2b\mu_s)}{(2\beta^2\mu_m+(\lambda^2-4b\mu_m)\mu_s)^3} > 0; \\
\frac{\partial e_m^{DP*}}{\partial \mu_m} &= \frac{a\lambda\mu_s(2\beta^2-4b\mu_s)}{(2\beta^2\mu_m+(\lambda^2-4b\mu_m)\mu_s)^2} < 0, &
\frac{\partial e_s^{DP*}}{\partial \mu_m} &= \frac{-a\lambda^2\beta\mu_s}{(2\beta^2\mu_m+(\lambda^2-4b\mu_m)\mu_s)^2} < 0, \\
\frac{\partial \omega^{DP*}}{\partial \mu_m} &= \frac{-a\lambda^2\mu_s^2}{(2\beta^2\mu_m+(\lambda^2-4b\mu_m)\mu_s)^2} < 0, &
\frac{\partial p^{DP*}}{\partial \mu_m} &= \frac{a\lambda^2\mu_s(\beta^2-3b\mu_s)}{b(2\beta^2\mu_m+(\lambda^2-4b\mu_m)\mu_s)^2} < 0, \\
\frac{\partial \pi_m^{DP*}}{\partial \mu_m} &= \frac{-a^2\lambda^2\mu_s^2}{2(2\beta^2\mu_m+(\lambda^2-4b\mu_m)\mu_s)^2} < 0, &
\frac{\partial \pi_s^{DP*}}{\partial \mu_m} &= \frac{a^2\lambda^2\mu_m\mu_s^2(2b\mu_s-\beta^2)}{(2\beta^2\mu_m+(\lambda^2-4b\mu_m)\mu_s)^3} < 0
\end{aligned}
$$

Corollary 2 Proof process

Calculating the partial derivatives of e_m^{SP*}, e_s^{SP*}, ω^{SP*}, p^{SP*}, π_m^{SP*}, π_s^{SP*} with respect to β, μ_s, respectively.

$$
\begin{aligned}
\frac{\partial e_m^{SP*}}{\partial \beta} &= \frac{2a\lambda\beta\mu_m\mu_s}{(\beta^2\mu_m+2(\lambda^2-2b\mu_m)\mu_s)^2} > 0, &
\frac{\partial e_s^{SP*}}{\partial \beta} &= \frac{a\mu_m(\beta^2\mu_m+2(\lambda^2-2b\mu_m)\mu_s)}{(\beta^2\mu_m+2(\lambda^2-2b\mu_m)\mu_s)^2} > 0, \\
\frac{\partial \omega^{SP*}}{\partial \beta} &= \frac{2a\beta\mu_m(2b\mu_m-\lambda^2)\mu_s}{b(\beta^2\mu_m+2(\lambda^2-2b\mu_m)\mu_s)^2} > 0, &
\frac{\partial p^{SP*}}{\partial \beta} &= \frac{2a\beta\mu_m(3b\mu_m-\lambda^2)\mu_s}{b(\beta^2\mu_m+2(\lambda^2-2b\mu_m)\mu_s)^2} > 0, \\
\frac{\partial \pi_m^{SP*}}{\partial \beta} &= \frac{2a^2\beta\mu_m^2(2b\mu_m-\lambda^2)\mu_s^2}{-(\beta^2\mu_m+2(\lambda^2-2b\mu_m)\mu_s)^3} > 0, &
\frac{\partial \pi_s^{SP*}}{\partial \beta} &= \frac{4a^2\beta\mu_m^2\mu_s}{(-2\beta^2\mu_m-4(\lambda^2-2b\mu_m)\mu_s)^2} > 0; \\
\frac{\partial e_m^{SP*}}{\partial \mu_s} &= \frac{-a\lambda\beta^2\mu_m}{(\beta^2\mu_m+2(\lambda^2-2b\mu_m)\mu_s)^2} < 0, &
\frac{\partial e_s^{SP*}}{\partial \mu_s} &= \frac{a\beta\mu_m(2\lambda^2-4b\mu_m)}{(\beta^2\mu_m+2(\lambda^2-2b\mu_m)\mu_s)^2} < 0, \\
\frac{\partial \omega^{SP*}}{\partial \mu_s} &= \frac{a\beta^2\mu_m(\lambda^2-2b\mu_m)}{b(\beta^2\mu_m+2(\lambda^2-2b\mu_m)\mu_s)^2} < 0, &
\frac{\partial p^{SP*}}{\partial \mu_s} &= \frac{a\beta^2\mu_m(\lambda^2-3b\mu_m)}{b(\beta^2\mu_m+2(\lambda^2-2b\mu_m)\mu_s)^2} < 0 \\
\frac{\partial \pi_m^{SP*}}{\partial \mu_s} &= \frac{a^2\beta^2\mu_m^2(2b\mu_m-\lambda^2)\mu_s}{(\beta^2\mu_m+2(\lambda^2-2b\mu_m)\mu_s)^3} < 0, &
\frac{\partial \pi_s^{SP*}}{\partial \mu_s} &= \frac{-a^2\beta^2\mu_m^2}{2(\beta^2\mu_m+2(\lambda^2-2b\mu_m)\mu_s)^2} < 0
\end{aligned}
$$

Corollary 3 Proof process

Calculating the partial derivatives of e_m^{DPI*}, e_s^{DPI*}, ω^{DPI*}, p^{DPI*}, π_m^{DPI*}, π_s^{DPI*} with respect to k_s, η_s, respectively.

$$
\begin{aligned}
\frac{\partial e_m^{DPI*}}{\partial k_s} &= \frac{b\lambda\mu_s}{4b\mu_m\mu_s-2\beta^2\mu_m-\lambda^2\mu_s} > 0, &
\frac{\partial e_s^{DPI*}}{\partial k_s} &= \frac{b\beta\mu_m}{4b\mu_m\mu_s-2\beta^2\mu_m-\lambda^2\mu_s} > 0, \\
\frac{\partial \omega^{DPI*}}{\partial k_s} &= \frac{b\mu_m\mu_s}{4b\mu_m\mu_s-2\beta^2\mu_m-\lambda^2\mu_s} > 0, &
\frac{\partial p^{DPI*}}{\partial k_s} &= \frac{\beta^2\mu_m+\lambda^2\mu_s-b\mu_m\mu_s}{4b\mu_m\mu_s-2\beta^2\mu_m-\lambda^2\mu_s}, \\
\frac{\partial \pi_m^{DPI*}}{\partial k_s} &= \frac{2b(a+b(k_s-\eta_s))\mu_m\mu_s}{8b\mu_m\mu_s-4\beta^2\mu_m-2\lambda^2\mu_s} > 0, &
\frac{\partial \pi_s^{DPI*}}{\partial k_s} &= \frac{b(a+b(k_s-\eta_s))\mu_m^2\mu_s(2b\mu_s-\beta^2)}{(\beta^2\mu_m+\lambda^2\mu_s-4b\mu_m\mu_s)^2} > 0; \\
\frac{\partial e_m^{DPI*}}{\partial \eta_s} &= \frac{-b\lambda\mu_s}{4b\mu_m\mu_s-2\beta^2\mu_m-\lambda^2\mu_s} < 0, &
\frac{\partial e_s^{DPI*}}{\partial \eta_s} &= \frac{-b\beta\mu_m}{4b\mu_m\mu_s-2\beta^2\mu_m-\lambda^2\mu_s} < 0,
\end{aligned}
$$

$\frac{\partial \omega^{DPI*}}{\partial \eta_s} = \frac{3b\mu_m\mu_s-2\beta^2\mu_m-\lambda^2\mu_s}{4b\mu_m\mu_s-2\beta^2\mu_m-\lambda^2\mu_s}$. It can be shown that, when $3b\mu_m\mu_s - 2\beta^2\mu_m - \lambda^2\mu_s > 0$, $\frac{\partial \omega^{DPI*}}{\partial \eta_s} > 0$, ω^{DPI*} monotonically increases with η_s. When $3b\mu_m\mu_s - 2\beta^2\mu_m - \lambda^2\mu_s < 0$, $\frac{\partial \omega^{DPI*}}{\partial \eta_s} < 0$. ω^{DPI*} monotonically decreases with η_s.

$\frac{\partial p^{DPI*}}{\partial \eta_s} = \frac{b\mu_m\mu_s - \beta^2\mu_m - \lambda^2\mu_s}{4b\mu_m\mu_s - 2\beta^2\mu_m - \lambda^2\mu_s}$. It can be shown that, when $b\mu_m\mu_s - \beta^2\mu_m - \lambda^2\mu_s > 0$, $\frac{\partial p^{DPI*}}{\partial \eta_s} > 0$. p^{DPI*} monotonically increases with η_s. When $b\mu_m\mu_s - \beta^2\mu_m - \lambda^2\mu_s < 0$, $\frac{\partial p^{DPI*}}{\partial \eta_s} < 0$. p^{DPI*} monotonically decreases with η_s.

$$\frac{\partial \pi_m^{DPI*}}{\partial \eta_s} = \frac{-2b(a + b(k_s - \eta_s))\mu_m\mu_s}{8b\mu_m\mu_s - 4\beta^2\mu_m - 2\lambda^2\mu_s} < 0, \quad \frac{\partial \pi_s^{DPI*}}{\partial \eta_s} = \frac{-b(a + b(k_s - \eta_s))\mu_m^2\mu_s(2b\mu_s - \beta^2)}{(2\beta^2\mu_m + (\lambda^2 - 4b\mu_m)\mu_s)^2} < 0$$

Corollary 4 Proof process

Calculating the partial derivatives of e_m^{SPI*}, e_s^{SPI*}, ω^{SPI*}, p^{SPI*}, π_m^{SPI*}, π_s^{SPI*} with respect to k_m, η_s, respectively.

$$\frac{\partial e_m^{SPI*}}{\partial \eta_s} = \frac{\lambda(bk_m+\beta)}{(bk_m+\beta)^2\mu_m+2(\lambda^2-2b\mu_m)\mu_s} < 0,$$
$$\frac{\partial e_s^{SPI*}}{\partial \eta_s} = \frac{2(2b\mu_m-\lambda^2)}{(bk_m+\beta)^2\mu_m+2(\lambda^2-2b\mu_m)\mu_s} < 0,$$
$$\frac{\partial \omega^{SPI*}}{\partial \eta_s} = \frac{(bk_m+\beta)(2b\mu_m-\lambda^2)}{b(bk_m+\beta)^2\mu_m+2b(\lambda^2-2b\mu_m)\mu_s} < 0,$$

$\frac{\partial p^{SPI*}}{\partial \eta_s} = \frac{-\lambda^2\beta - b^2k_m\mu_m + b(k_m\lambda^2+3\beta\mu_m)}{b(bk_m+\beta)^2\mu_m+2b(\lambda^2-2b\mu_m)\mu_s}$. Set $\delta(k_m) = -\lambda^2\beta - b^2k_m\mu_m + b(k_m\lambda^2 + 3\beta\mu_m)$ and make it equal to 0 to obtain $k_m = \frac{3b\beta\mu_m - \beta\lambda^2}{b(b\mu_m-\lambda^2)}$. Solve for the partial derivative of $\delta(k_m)$ with respect to k_m to obtain $\delta'(k_m) = b\lambda^2 - b^2\mu_m$. When $\lambda^2 < b\mu_m$, $\delta(k_m)$ is monotonically decreasing with respect to k_m. At this point, the zeros satisfy $k_m = \frac{3b\beta\mu_m - \beta\lambda^2}{b(b\mu_m-\lambda^2)} > \frac{-b\beta\mu_m + \sqrt{2}\sqrt{b^2\mu_m(2b\mu_m-\lambda^2)\mu_s}}{b^2\mu_m}$. When $0 < k_m < \frac{-b\beta\mu_m + \sqrt{2}\sqrt{b^2\mu_m(2b\mu_m-\lambda^2)\mu_s}}{b^2\mu_m}$, $\delta(k_m) > 0$, then $\frac{\partial p^{SPI*}}{\partial \eta_s} < 0$. When $\lambda^2 > b\mu_m$, $\delta(k_m)$ is monotonically increasing with respect to k_m. At this point, the zero point satisfies $k_m = \frac{3b\beta\mu_m - \beta\lambda^2}{b(b\mu_m-\lambda^2)} < 0$. When $0 < k_m < \frac{-b\beta\mu_m + \sqrt{2}\sqrt{b^2\mu_m(2b\mu_m-\lambda^2)\mu_s}}{b^2\mu_m}$, then $\frac{\partial p^{SPI*}}{\partial \eta_s} < 0$.

$$\frac{\partial \pi_m^{SPI*}}{\partial \eta_s} = \frac{(bk_m + \beta)\mu_m(2b\mu_m - \lambda^2)(b\eta_s k_m + \eta_s\beta - a\mu_s)}{((bk_m + \beta)^2\mu_m + 2(\lambda^2 - 2b\mu_m)\mu_s)^2}$$

$\frac{\partial \pi_m^{SPI*}}{\partial \eta_s^2} = \frac{(bk_m+\beta)^2\mu_m(2b\mu_m-\lambda^2)}{((bk_m+\beta)^2\mu_m+2(\lambda^2-2b\mu_m)\mu_s)^2}$, where the second-order derivative is greater than 0 and is a concave function. The first-order derivative is equal to zero to obtain $\eta_s = \frac{a\mu_s}{bk_m+\beta}$, by condition $\eta_s < \frac{a\mu_m(bk_m+\beta)}{2(2b\mu_m-\lambda^2)} < \frac{a\mu_s}{bk_m+\beta}$. When $\eta_s < \frac{a\mu_m(bk_m+\beta)}{2(2b\mu_m-\lambda^2)}$, $\frac{\partial \pi_m^{SPI*}}{\partial \eta_s} < 0$, π_m^{SPI*} monotonically decreases with η_s.

$$\frac{\partial \pi_s^{SPI*}}{\partial \eta_s} = \frac{2\eta_s(\lambda^2 - 2b\mu_m) + 2(a(bk_m + \beta)\mu_m + \eta_s(\lambda^2 - 2b\mu_m))}{2(bk_m + \beta)^2\mu_m + 4(\lambda^2 - 2b\mu_m)\mu_s},$$

$\frac{\partial \pi_m^{SPI*}}{\partial \eta_s^2} = \frac{(bk_m+\beta)^2\mu_m(2b\mu_m-\lambda^2)}{((bk_m+\beta)^2\mu_m+2(\lambda^2-2b\mu_m)\mu_s)^2}$, where the second-order derivative is greater than 0 and is a concave function. The first-order derivative is equal to zero to obtain $\eta_s = \frac{a\mu_m(bk_m+\beta)}{2(2b\mu_m-\lambda^2)}$. When $\eta_s < \frac{a\mu_m(bk_m+\beta)}{2(2b\mu_m-\lambda^2)}$, $\frac{\partial \pi_s^{SPI*}}{\partial \eta_s} < 0$, π_s^{SPI*} monotonically decreases with η_s.

$\frac{\partial e_m^{SPI*}}{\partial k_m} = \frac{b\lambda(\eta_s(bk_m+\beta)^2\mu_m + 2\eta_s(\lambda^2-2b\mu_m)\mu_s - 2(bk_m+\beta)\mu_m(b\eta_s k_m + \eta_s\beta - a\mu_s))}{((bk_m+\beta)^2\mu_m+2(\lambda^2-2b\mu_m)\mu_s)^2}$, and it is not possible to determine the positive or negative of the derivative function directly. Therefore, make it equal to 0 to obtain the only zero point, $\eta_s = \frac{2a(bk_m+\beta)\mu_m\mu_s}{(bk_m+\beta)^2\mu_m - 2(\lambda^2-2b\mu_m)\mu_s}$, and $\frac{\partial e_m^{SPI*}}{\partial k_m}$ solves the partial derivative with respect to η_s to obtain $\frac{\partial e_m^{SPI*}}{\partial k_m \partial \eta_s} = \frac{b\lambda(-(bk_m+\beta)^2\mu_m+2(\lambda^2-2b\mu_m)\mu_s)}{((bk_m+\beta)^2\mu_m+2(\lambda^2-2b\mu_m)\mu_s)^2} < 0$, indicating that $\frac{\partial e_m^{SPI*}}{\partial k_m}$ monotonically decreases with respect to η_s. As η_s needs to meet the $0 < \eta_s < \frac{a\mu_m(bk_m+\beta)}{2(2b\mu_m-\lambda^2)}$ condition, the comparison finds $\frac{2a(bk_m+\beta)\mu_m\mu_s}{(bk_m+\beta)^2\mu_m-2(\lambda^2-2b\mu_m)\mu_s} > \frac{a\mu_m(bk_m+\beta)}{2(2b\mu_m-\lambda^2)}$, then indicates that, when $0 < \eta_s < \frac{a\mu_m(bk_m+\beta)}{2(2b\mu_m-\lambda^2)}$, $\frac{\partial e_m^{SPI*}}{\partial k_m} > 0$, that is, e_m^{SPI*} monotonically increases with k_m. Likewise, this is the case for $\frac{\partial e_s^{SPI*}}{\partial k_m} > 0$, $\frac{\partial \omega^{SPI*}}{\partial k_m} > 0$, $\frac{\partial p_s^{SPI*}}{\partial k_m} > 0$, $\frac{\partial \pi_m^{SPI*}}{\partial k_m} > 0$, $\frac{\partial \pi_s^{SPI*}}{\partial k_m} > 0$.

Corollary 5 Proof process

Set $\gamma_1(\eta_s) = e_m^{DPI*} - e_m^{DP*}$, for γ_1, solving the derivative function $\gamma_1'(\eta_s)$ on η_s, can be obtained, $\gamma_1'(\eta_s) = \frac{-b\lambda\mu_s}{4b\mu_m\mu_s-2\beta^2\mu_m-\lambda^2\mu_s} < 0$, and then γ_1 on η_s is monotonically decreasing. Make $\gamma_1(\eta_s)$ equal to 0, and $\eta_s = k_s$ can be obtained. By Proposition 3, it can be seen that, when the analytical solution is meaningful, to meet the $0 < \eta_s < \frac{(-a+bk_s)\beta^2\mu_m+b\mu_s(2a\mu_m+k_s(\lambda^2-2b\mu_m))}{b\mu_m(2b\mu_s-\beta^2)} = \overline{\eta_{sd}}$ condition, set $\gamma_2 = \overline{\eta_{sd}} - k_s$, so that it is equal to zero and $k_s = \frac{a\mu_m(\beta^2-2b\mu_s)}{b(2\beta^2\mu_m+(\lambda^2-4b\mu_m)\mu_s)} > 0$ can be obtained. Solving the derivative function with respect to k_s for γ_2 yields $\gamma_2'(k_s) = \frac{2b\beta^2\mu_m+b\lambda^2\mu_s-4b^2\mu_m\mu_s}{b\mu_m(2b\mu_s-\beta^2)} < 0$. Then, γ_2 is monotonically decreasing with respect to k_s. Then, it can show that, when $0 < k_s < \frac{a\mu_m(\beta^2-2b\mu_s)}{b(2\beta^2\mu_m+(\lambda^2-4b\mu_m)\mu_s)}$, $\overline{\eta_{sd}} > k_s$. At this time, if $0 < \eta_s < k_s$, then $\gamma_1(\eta_s) > 0$, that is, $e_m^{DPI*} > e_m^{DP*}$. If $k_s < \eta_s < \frac{(-a+bk_s)\beta^2\mu_m+b\mu_s(2a\mu_m+k_s(\lambda^2-2b\mu_m))}{b\mu_m(2b\mu_s-\beta^2)}$, then $\gamma_1(\eta_s) < 0$, that is, $e_m^{DPI*} < e_m^{DP*}$. When $k_s > \frac{a\mu_m(\beta^2-2b\mu_s)}{b(2\beta^2\mu_m+(\lambda^2-4b\mu_m)\mu_s)}$, $\overline{\eta_{sd}} < k_s$. At this time, if $\eta_s < \frac{(-a+bk_s)\beta^2\mu_m+b\mu_s(2a\mu_m+k_s(\lambda^2-2b\mu_m))}{b\mu_m(2b\mu_s-\beta^2)}$, then $\gamma_1(\eta_s) > 0$, that is, $e_m^{DPI*} > e_m^{DP*}$. The same reasoning can be used to compare the size relationships of other decisions. Similarly to the above approach, (ii) is proven.

Corollary 6 Proof process

Set $\gamma_3(\eta_s) = e_m^{SPI*} - e_m^{SP*}$. For γ_3, solving the derivative function $\gamma_3'(\eta_s)$ with respect to η_s can obtain $\gamma_3'(\eta_s) = \frac{\lambda(bk_m+\beta)}{(bk_m+\beta)^2\mu_m+2(\lambda^2-2b\mu_m)\mu_s} < 0$. Then, γ_3 with respect to η_s is monotonically decreasing. Make $\gamma_3(\eta_s)$ equal to 0 to obtain its only zero point for $\eta_s = \frac{abk_m(bk_m+2\beta)\mu_m\mu_s}{(bk_m+\beta)(-\beta^2\mu_m-2\lambda^2\mu_s+4b\mu_m\mu_s)}$. From Proposition 4, it can be seen that the analytic solution has meaning when η_s needs to satisfy $0 < \eta_s < \frac{a\mu_m(bk_m+\beta)}{2(2b\mu_m-\lambda^2)} = \overline{\eta_{ss}}$. For this, there is a need to further judge the size of the zero point and the constraints. Let $\gamma_4(\eta_s) = \frac{a\mu_m(bk_m+\beta)}{2(2b\mu_m-\lambda^2)} - \frac{abk_m(bk_m+2\beta)\mu_m\mu_s}{(bk_m+\beta)(-\beta^2\mu_m-2\lambda^2\mu_s+4b\mu_m\mu_s)}$. The simplification $\gamma_4(\eta_s) = \frac{a\beta^2\mu_m((bk_m+\beta)^2\mu_m+2(\lambda^2-2b\mu_m)\mu_s)}{2(bk_m+\beta)(2b\mu_m-\lambda^2)(\beta^2\mu_m+2(\lambda^2-2b\mu_m)\mu_s)} > 0$ can be obtained, that is, denoted as $\frac{a\mu_m(bk_m+\beta)}{2(2b\mu_m-\lambda^2)} > \frac{abk_m(bk_m+2\beta)\mu_m\mu_s}{(bk_m+\beta)(-\beta^2\mu_m-2\lambda^2\mu_s+4b\mu_m\mu_s)}$. Since γ_3 with respect to η_s is monotonically decreasing, therefore, when $0 < \eta_s < \frac{abk_m(bk_m+2\beta)\mu_m\mu_s}{(bk_m+\beta)(-\beta^2\mu_m-2\lambda^2\mu_s+4b\mu_m\mu_s)}$, $e_m^{SPI*} > e_m^{SP*}$. When $\frac{abk_m(bk_m+2\beta)\mu_m\mu_s}{(bk_m+\beta)(-\beta^2\mu_m-2\lambda^2\mu_s+4b\mu_m\mu_s)} < \eta_s < \frac{a\mu_m(bk_m+\beta)}{2(2b\mu_m-\lambda^2)}$, $e_m^{SPI*} < e_m^{SP*}$. Similar to the above method, (ii), (iii), and (iv) are proved.

Corollary 7 Proof process

The quotient method compares the market demand under the SP and DP models to obtain $\chi = \frac{D^{DP*}}{D^{SP*}} = \frac{4b\mu_m\mu_s-\beta^2\mu_m-2\lambda^2\mu_s}{4b\mu_m\mu_s-2\beta^2\mu_m-\lambda^2\mu_s}$. Two covariates, $\varphi_m = \frac{\lambda^2}{\mu_m}$ and $\varphi_s = \frac{\beta^2}{\mu_s}$, are introduced, which represent the relationship between the service innovation effect and the service innovation cost and the relationship between the product innovation effect and the product innovation cost, respectively, and are called the product innovation efficiency and the service innovation efficiency. As a result, the market demand ratio can be reduced to $\chi = \frac{4b-\varphi_s-2\varphi_m}{4b-2\varphi_s-\varphi_m}$. When $\varphi_m = \varphi_s$, $\chi = 1$, the market demand is equal under the two innovation modes. When $\chi = \frac{4b-\varphi_s-2\varphi_m}{4b-2\varphi_s-\varphi_m} > 1$, $\varphi_s > \varphi_m$ can be obtained, that is, when the efficiency of service innovation is higher than the efficiency of product innovation, the market demand for the product innovation priority mode is higher than the market demand of service innovation. When $\chi = \frac{4b-\varphi_s-2\varphi_m}{4b-2\varphi_s-\varphi_m} < 1$, $\varphi_m > \varphi_s$ can be obtained, that is, when the efficiency of product innovation is higher than the efficiency of service innovation, the market demand of the service innovation priority mode is higher than the market demand of the product innovation priority. This leads to the conclusion that the party with high innovation efficiency adopts the sub-priority innovation strategy, which can lead to a larger market share for the innovation results.

Corollary 8 Proof process

Set $\gamma_5(\eta_s) = D^{SPI*} - D^{DPI*}$, and, for γ_5, solve the derivative function $\gamma_5'(\eta_s)$ with respect to η_s to obtain $\gamma_5'(\eta_s) = \frac{b^2\mu_m\mu_s}{-2\beta^2\mu_m-\lambda^2\mu_s+4b\mu_m\mu_s} + \frac{b(bk_m+\beta)\mu_m}{(bk_m+\beta)^2\mu_m+2(\lambda^2-2b\mu_m)\mu_s}$, and make it equal to 0 to obtain two zeros:

$$k_{m11} = \frac{\lambda^2\mu_s+2\mu_s(\beta^2-b(2+\beta)\mu_s)+\sqrt{4\beta^4\mu_m^4+4\beta^2\mu_m\mu_s(\lambda^2-4b\mu_m)+\mu_s^2((\lambda^2-4b\mu_m)^2-8b^2\mu_m\mu_s(\lambda^2-2b\mu_m))}}{2b^2\mu_m\mu_s}$$

$$k_{m12} = \frac{\lambda^2\mu_s+2\mu_s(\beta^2-b(2+\beta)\mu_s)-\sqrt{4\beta^4\mu_m^4+4\beta^2\mu_m\mu_s(\lambda^2-4b\mu_m)+\mu_s^2((\lambda^2-4b\mu_m)^2-8b^2\mu_m\mu_s(\lambda^2-2b\mu_m))}}{2b^2\mu_m\mu_s}$$

k_{m12} does not meet the constraints. Solve $\gamma_5'(\eta_s)$ with respect to the k_m, the first-order partial derivative, and the second-order partial derivative to obtain $\frac{\partial\gamma_5'(\eta_s)}{\partial k_m} = \frac{b^2\mu_m(-(bk_m+\beta)^2\mu_m+2(\lambda^2-2b\mu_m)\mu_s)}{((bk_m+\beta)^2\mu_m+2(\lambda^2-2b\mu_m)\mu_s)^2} > 0$, $\frac{\partial\gamma_5'(\eta_s)}{\partial k_m^2} = \frac{2b^3(bk_m+\beta)^2\mu_m^2((bk_m+\beta)^2\mu_m-6(\lambda^2-2b\mu_m)\mu_s)}{((bk_m+\beta)^2\mu_m+2(\lambda^2-2b\mu_m)\mu_s)^3} < 0$. It can be seen that, when $k_m < k_{m11}$ or $k_m > k_{m12}$, γ_5 monotonically decreases with respect to η_s. When $k_{m11} < k_m < k_{m12}$, γ_5 is monotonically increasing with respect to η_s. Make $\gamma_5(\eta_s)$ equal to 0 to obtain the unique zero point:

$$\eta_{s5} = -\frac{\mu_s(a(b^2k_m^2\mu_m + 2bk_m\beta\mu_s - \beta^2\mu_m + \lambda^2\mu_s) + bk_s((bk_m+\beta)^2\mu_m + 2(\lambda^2 - 2b\mu_m)\mu_s))}{2\beta^2(bk_m+\beta)\mu_m - (bk_m+\beta)(-\lambda^2 + b(4 + bk_m + \beta)\mu_m)\mu_s + 2b(2b\mu_m - \lambda^2)\mu_s^2}$$

Make $\eta_{s5}(k_s)$ equal to 0 to obtain $k_{s3} = \frac{a(b^2k_m^2\mu_m+2bk_m\beta\mu_m-\beta^2\mu_m+\lambda^2\mu_s)}{b(4b\mu_m\mu_s-(bk_m+\beta)^2\mu_m-2\lambda^2\mu_s)}$. Solve the derivative function $\eta_{s5}'(k_s)$ of $\eta_{s5}(k_s)$ with respect to k_s to obtain $\eta_{s5}'(k_s) = \frac{b\mu_s((bk_m+\beta)^2\mu_m+2(\lambda^2-2b\mu_m)\mu_s)}{2\beta^2(bk_m+\beta)\mu_m-\mu_s(bk_m+\beta)(-\lambda^2+b(4+bk_m+\beta)\mu_m)+2b(2b\mu_m-\lambda^2)\mu_s^2}$. Analysis shows that, when $k_m < k_{m11}$ or $k_m > k_{m12}$, $\eta_{s5}'(k_s)$ is monotonically decreasing with respect to k_m, and, when $k_{m11} < k_m < k_{m12}$, $\eta_{s5}'(k_s)$ is monotonically increasing with respect to k_m. Make $k_{s3}(k_m)$ equal to 0 to obtain $k_{m21} = \frac{-b\beta\mu_m+\sqrt{2b^2\beta^2\mu_m^2-b^2\lambda^2\mu_m\mu_s}}{b^2\mu_m}$, $k_{m22} = \frac{-b\beta\mu_m-\sqrt{2b^2\beta^2\mu_m^2-b^2\lambda^2\mu_m\mu_s}}{b^2\mu_m}$, where k_{m22} does not meet the constraints. From proposition 4, it is known that $k_m < \frac{-b\beta\mu_m+\sqrt{2}\sqrt{b^2\mu_m(2b\mu_m-\lambda^2)\mu_s}}{b^2\mu_m}$ needs to be satisfied. So, it needs to be verified that k_{m11} and k_{m21} meet the conditions, and, through the calculation, it can be known that $k_{m11}, k_{m21} < \frac{-b\beta\mu_m+\sqrt{2}\sqrt{b^2\mu_m(2b\mu_m-\lambda^2)\mu_s}}{b^2\mu_m}$ when $b^2\mu_m\mu_s - \sqrt{b^2\mu_m(2\beta^2\mu_m - \lambda^2\mu_s)} < 0$, $k_{m11} < k_{m21}$. When $b^2\mu_m\mu_s - \sqrt{b^2\mu_m(2\beta^2\mu_m - \lambda^2\mu_s)} > 0$, $k_{m11} > k_{m21}$. Therefore, when $b^2\mu_m\mu_s - \sqrt{b^2\mu_m(2\beta^2\mu_m - \lambda^2\mu_s)} > 0$, $k_{m11} < k_{m21}$. The reverse derivation is obtained as follows: When $0 < k_m < k_{m11}$, $k_{s3} < 0$ and γ_5 are monotonically increasing with respect to η_s. If $k_s > 0$ and $\eta_{s5} > 0$, when $0 < \eta_s < \eta_{s5}$, $D^{SPI*} < D^{DPI*}$. When $\eta_s > \eta_{s5}$, $D^{SPI*} > D^{DPI*}$. When $k_{m11} < k_m < k_{m21}$, $k_{s3} < 0$. When $k_{s3} > 0$, $\eta_{s5} < 0$. When γ_5 is monotonically decreasing with respect to η_s, then $D^{SPI*} < D^{DPI*}$. When $k_{m21} < k_m < \frac{-b\beta\mu_m+\sqrt{2}\sqrt{b^2\mu_m(2b\mu_m-\lambda^2)\mu_s}}{b^2\mu_m}$, $k_{s3} > 0$ and η_{s5} is monotonously decreasing with respect to k_s. At this point, if $0 < k_s < k_{s3}$, then $\eta_{s5} > 0$ and γ_5 is monotonically decreasing with respect to η_s. When $0 < \eta_s < \eta_{s5}$, $D^{SPI*} > D^{DPI*}$.

When $\eta_s > \eta_{s5}$, $D^{SPI*} < D^{DPI*}$. If $k_s > k_{s3}$, then $\eta_{s5} < 0$ and γ_5 is monotonically decreasing with respect to η_s. When $\eta_s > 0$, $D^{SPI*} < D^{DPI*}$. Similarly, the conclusion when $b^2\mu_m\mu_s - \sqrt{b^2\mu_m(2\beta^2\mu_m - \lambda^2\mu_s)} > 0$ can be proved.

References

1. Li, Y.; Luo, J. Diffusion Mechanism of Service Innovation Back-Feeding Product Innovation for Manufacturing Enterprises in the Digital Environment. *J. Syst. Manag.* **2023**, *32*, 995–1008.
2. Li, Y.; Gao, N.; Yi, Q.; Tan, R. Research on the Relationship between Innovation Network Embeddedness and Innovation Performance of High-tech Enterprise R&D Personnel. *J. Manag. Sci.* **2018**, *31*, 3–19.
3. Liu, W.; Wang, D.; Shen, X.; Yan, X.; Wei, W. The impacts of distributional and peer-induced fairness concerns on the decision-making of order allocation in logistics service supply chain. *Transp. Res. Part E Logist. Transp. Rev.* **2018**, *116*, 102–122. [CrossRef]
4. Wang, Q.; Wang, Z.; Zhao, X. Strategic orientations and mass customisation capability: The moderating effect of product life cycle. *Int. J. Prod. Res.* **2015**, *53*, 5278–5295. [CrossRef]
5. Vargo, S.L. From promise to perspective: Reconsidering value propositions from a service-dominant logic orientation. *Ind. Mark. Manag.* **2020**, *87*, 309–311. [CrossRef]
6. Ulaga, W.; Reinartz, W.J. Hybrid offerings: How manufacturing firms combine goods and services successfully. *J. Mark.* **2011**, *75*, 5–23. [CrossRef]
7. Luo, J.; Jiang, Q. Priority evolution of product and service innovation under digital transformation—Based on the case of Haier smart home. *Stud. Sci. Sci.* **2022**, *40*, 1710–1720.
8. Zhou, X.; Ye, W.; Li, X. Digit-intellectualized Knowledge Orchestration and the Evolution of Organizational Dynamic Capabilities: A Case Study of Xiaomi Technology. *J. Manag. World* **2023**, *39*, 138–157.
9. Treleaven, P.C. Control-driven, data-driven and demand-driven computer architecture. *Parallel Comput.* **1985**, *2*, 287–288. [CrossRef]
10. Lapalme, E.; Lina, J.-M.; Mattout, J. Data-driven parceling and entropic inference in MEG. *NeuroImage* **2006**, *30*, 160–171. [CrossRef]
11. Wang, D.; Liu, W.; Liang, Y.; Wei, S. Decision optimization in service supply chain: The impact of demand and supply-driven data value and altruistic behavior. *Ann. Oper. Res.* **2023**, *324*, 971–992. [CrossRef]
12. Grover, V.; Chiang, R.H.; Liang, T.-P.; Zhang, D. Creating strategic business value from big data analytics: A research framework. *J. Manag. Inf. Syst.* **2018**, *35*, 388–423. [CrossRef]
13. Benlian, A.; Kettinger, W.J.; Sunyaev, A.; Winkler, T.J.; Editors, G. The transformative value of cloud computing: A decoupling, platformization, and recombination theoretical framework. *J. Manag. Inf. Syst.* **2018**, *35*, 719–739. [CrossRef]
14. Xie, X.; Wang, H.; García, J.S. How does customer involvement in service innovation motivate service innovation performance? The roles of relationship learning and knowledge absorptive capacity. *J. Bus. Res.* **2021**, *136*, 630–643. [CrossRef]
15. Chen, W.; Wang, J. Platform-dependent Upgrade: Digital Transformation Strategy of Complementors in Platform-based Ecosystem. *J. Manag. World* **2021**, *37*, 195–214.
16. Aksoy, L.; Alkire, L.; Choi, S.; Kim, P.B.; Zhang, L. Social innovation in service: A conceptual framework and research agenda. *J. Serv. Manag.* **2019**, *30*, 429–448. [CrossRef]
17. Gunasekaran, A.; Papadopoulos, T.; Dubey, R.; Wamba, S.F.; Childe, S.J.; Hazen, B.; Akter, S. Big data and predictive analytics for supply chain and organizational performance. *J. Bus. Res.* **2017**, *70*, 308–317. [CrossRef]
18. Ugray, Z.; Paper, D.; Johnson, J. How Business Value Is Extracted from Operational Data: A Case Study. In *Digital Business Models: Driving Transformation Innovation*; Springer: Cham, Switzerland, 2019; pp. 117–145.
19. Du, X.; Li, L.; Sang, Z.; Zhan, H. Supply Chain Upstream Innovation Decisions under Asymmetric Effectiveness with Overconfident Retailer. *Ind. Eng. Manag.* **2023**, *28*, 67–79.
20. Liu, K.; Li, Q.; Zhang, H. Analysis of the impact of remanufacturing process innovation on closed-loop supply chain from the perspective of government subsidy. *Sustainability* **2022**, *14*, 11333. [CrossRef]
21. Qu, J.; Hu, B.; Meng, C. Joint Innovation Investment and Pricing Decisions in Retail Supply Chains with Customer Value. *Sustainability* **2021**, *13*, 1309. [CrossRef]
22. Xing, P.; Zhou, C.; Li, C. Innovation effort strategy of green supply chain considering government subsidies. *Comput. Integr. Manuf. Syst.* **2024**, *30*, 1889–1907.
23. Yi, H.; Tang, L. Differential strategies for collaborative product innovation in supply chain under decision makers' disappointed circumvention. *Control. Decis.* **2024**, *39*, 271–280.
24. Liu, Z.; Guo, Q.; Nie, J. Retailer's Optimal Extended Warranty Strategy? Collaboration with Manufacturer or Not. *Chin. J. Manag. Sci.* **2024**, *32*, 268–275.
25. Yu, H.; Li, Y. Supplier's product innovation strategy choice from the perspective of supply chain: Financing VS not financing. *J. Ind. Eng. Eng. Manag.* **2021**, *35*, 172–180.
26. Li, M.; Chen, L.; Guo, G. Supply chain decision considering lead user innovation. *Comput. Integr. Manuf. Syst.* **2024**, 1–25.
27. Lin, Z.; Cai, W. Green technological innovation decision-making of risk averse manufacturer with uncertain technical efficiency. *Comput. Integr. Manuf. Syst.* **2024**, 1–20.

28. Jiang, Z.; Hou, X.; Li, K. Process innovation or mode innovation? Innovation strategies of new energy vehicle (NEV) companies. *Syst. Eng. Theory Pract.* **2024**, *44*, 947–974.
29. Brinch, M. Understanding the value of big data in supply chain management and its business processes: Towards a conceptual framework. *Int. J. Oper. Prod. Manag.* **2018**, *38*, 1589–1614. [CrossRef]
30. Zhang, L.; Zhang, P. Research on Selection of Process and Product Innovation Modes in Supply Chains under Asymmetric Demand Information. *Oper. Res. Manag. Sci.* **2022**, *31*, 128–134.
31. Shi, C.; Nie, J. Sharing Information in a Dual-channel Supply Chain under Network Externality. *Chin. J. Manag. Sci.* **2019**, *27*, 142–150.
32. Wang, F.; Li, H.; Cao, Y.; Zhang, C.; Ran, Y. Knowledge sharing strategy and emission reduction benefits of low carbon technology collaborative innovation in the green supply chain. *Front. Environ. Sci.* **2022**, *9*, 783835. [CrossRef]
33. Tang, C.; Gao, P.; Xue, J. Green supply chain information sharing strategy of manufacturing enterprises under different product innovation modes. *J. Cent. South Univ. (Soc. Sci.)* **2023**, *29*, 96–108.
34. Nie, J.; Wang, Q.; Li, G.; Liu, D. To share or not to share? When information sharing meets remanufacturing. *Ann. Oper. Res.* **2023**, *329*, 815–846. [CrossRef]
35. Wei, J.; Wang, Y.; Lu, J. Information sharing and sales patterns choice in a supply chain with product's greening improvement. *J. Clean. Prod.* **2021**, *278*, 123704. [CrossRef]
36. Hwang, B.-N.; Hsu, M.-Y. The impact of technological innovation upon servitization: Evidence from Taiwan community innovation survey. *J. Manuf. Technol. Manag.* **2019**, *30*, 1097–1114. [CrossRef]
37. Liu, C.; Huang, W.; Zheng, B.; Yang, C. The decision-making of manufacturer stimulating supplier innovation considering marketing effort and innovation capability. *Syst. Eng. Theory Pract.* **2017**, *37*, 3040–3051.
38. Song, J.; Li, F.; Wu, D.D.; Liang, L.; Dolgui, A. Supply chain coordination through integration of innovation effort and advertising support. *Appl. Math. Model.* **2017**, *49*, 108–123. [CrossRef]
39. Lu, X.; Li, Y.; Wang, J.; Yu, S. Differential Game Analysis of Technological Innovation and Dynamic Pricing in the Cloud Service Supply Chain. *Oper. Res. Manag. Sci.* **2020**, *29*, 49–57.
40. Zhu, L.; Sun, S. Manufacturer's Innovation Invest Decision Influence Analysis of Three Manufacturer's Innovation Invest Decision Influence Analysis of Three. *Syst. Eng.* **2018**, *36*, 136–140.
41. Chen, H.; Lu, Q.; Huan, X. Research on Coordination Strategies of Cruise Industry Supply Chain Considering Sales Effort and Innovation Effort. *Syst. Eng.* **2024**, 1–12.
42. Ma, P.; Shang, J.; Wang, H. Enhancing corporate social responsibility: Contract design under information asymmetry. *Omega* **2017**, *67*, 19–30. [CrossRef]
43. Correani, A.; De Massis, A.; Frattini, F.; Petruzzelli, A.M.; Natalicchio, A. Implementing a digital strategy: Learning from the experience of three digital transformation projects. *Calif. Manag. Rev.* **2020**, *62*, 37–56. [CrossRef]
44. Wang, G.; Gunasekaran, A.; Ngai, E.W.; Papadopoulos, T. Big data analytics in logistics and supply chain management: Certain investigations for research and applications. *Int. J. Prod. Econ.* **2016**, *176*, 98–110. [CrossRef]
45. Xie, K.; Xia, Z.; Xiao, J. The Enterprise Realization Mechanism of Big Data Becoming a Real Production Factor: From the Product Innovation Perspective. *China Ind. Econ.* **2020**, *5*, 42–60.
46. Yang, X.; Li, Y. Study on the Willingness of Sharing Multi-agent Data of Supply Chain from the Perspective of Blockchain Technology. *Sci. Technol. Manag. Res.* **2021**, *41*, 181–192.
47. Ma, R.; Jiang, L.; Wang, T.; Wang, X.; Ruan, J. How do manufacturing companies and service providers share knowledge in the context of servitization? An evolutionary-game model of complex networks. *Int. J. Prod. Res.* **2023**, *61*, 4279–4301. [CrossRef]
48. Liu, D.; Chen, H. Pricing Decision of Product Service Supply Chain: Impact of Data Resource Mining and Sharing Strategies. *Chin. J. Manag. Sci.* **2024**, *32*, 129–140.
49. Liu, W.; Long, S.; Liang, Y.; Wang, J.; Wei, S. The influence of leadership and smart level on the strategy choice of the smart logistics platform: A perspective of collaborative innovation participation. *Ann. Oper. Res.* **2023**, *324*, 893–935. [CrossRef]

mathematics

MDPI

Article

Insider Trading with Semi-Informed Traders and Information Sharing: The Stackelberg Game

Wassim Daher [1,2,*,†], Fida Karam [3,†] and Naveed Ahmed [1,2,†]

1 Department of Mathematics and Natural Sciences, Gulf University for Science and Technology, P.O. Box 7207, Hawally 32093, Kuwait; ahmed.n@gust.edu.kw
2 Center of Applied Mathematics and Bioinformatics (CAMB), Gulf University for Science and Technology, P.O. Box 7207, Hawally 32093, Kuwait
3 Department of Economics and Finance, Gulf University for Science and Technology, P.O. Box 7207, Hawally 32093, Kuwait; karam.f@gust.edu.kw
* Correspondence: daher.w@gust.edu.kw
† These authors contributed equally to this work.

Abstract: This paper presents a financial Stackelberg game model with two partially informed risk neutral insiders. Each insider receives a private signal about the stock value and competes with the other insider under a Stackelberg setting. Linear strategies for the game's participants are considered and normal distributions for the fundamentals are assumed. Based on the Stackelberg game and the Backward Induction theory, the unique linear equilibrium is characterized. The findings reveal that the level of partial information might increase/decrease the insiders' profits as well as the market parameter in the Stackelberg setting relative to the Cournot setting. Additionally, this paper considers the information sharing scenario between the two insiders competing in this Stackelberg game. The results show that multiple equilibria exist in contrast to the information sharing scenario in the Cournot game where the Nash equilibrium is unique.

Keywords: insider trading; risk neutrality; partial information; Stackelberg structure; Kyle model

MSC: 91A10; 91A27; 91A80; 91B44

JEL Classification: G14; D82

1. Introduction

Asymmetric information refers to one party having more information than other market participants, leading to an imbalance in the market. In the case of insider trading, where insiders have access to material non-public information, asymmetric information can distort the efficient allocation of resources, where those may be allocated towards investments that are not necessarily the most productive or deserving, as prices may not accurately reflect the true value of the securities. In addition, investors lose confidence in the fairness and integrity of the market and are discouraged to trade, leading to reduced liquidity and potentially hindering market efficiency.

Owing to their position within the firm, corporate insiders, such as corporate executives or board members, are more informed than outsiders about any event involving the firm they work for and whose stock is publicly traded. They can use this information to make profitable trades at the expense of other market participants who do not have access to the same information. Typically, this private, price-sensitive information is expected to alter the behavior of the company's stock and reduce market efficiency.

With the recurrence of insider trading scandals, academic research has consistently renewed its interest in this topic. The innovative work of [1] on insider trading is very famous in explaining how markets can incorporate private information. Ref. [1] studies the single auction model where three kinds of traders exchange one risky asset for a riskless asset. The traders are a fully informed risk neutral insider who has access to private observation of the ex-post liquidation value of the risky asset of the firm, uninformed noise traders, and market makers who set prices conditional on the information they have about the quantities traded by others. Then, many other scholars extended the Kyle one-shot game model in several directions. One direction investigated the effect of risk attitudes of fully informed insiders on the equilibrium outcomes. For example, [2] studies the case of risk averse insiders. Ref. [3] consider the case of one risk averse and one risk neutral insiders. Ref. [4] studies the case of two insiders in which the first insider is risk-neutral while the second insider is overconfident.

A second direction extended the static Kyle model to the case in which signals are correlated. For example, in [5]'s model, the market maker observed correlated signals about the stock value. Ref. [6] studied the Kyle model in which the noise traders are able to correlate their trade with the true price.

A third extension of the Kyle model, studied the case of partially informed insiders (see [7,8]). Moreover [9] extended the Kyle model to incorporate a finite number of insiders, all knowing perfectly the value of the risky asset. Recently, ref. [10] investigated the case of two partially risk neutral informed traders.

A fourth extension of the Kyle model investigated the existence and/or uniqueness of equilibria in the Kyle-type model. For instance, ref. [11] provided conditions under which normality is necessary for the existence of a linear equilibrium (Ref. [12] characterized the distributions that imply existence of linear equilibria in the Kyle-model. Ref. [13] discussed the existence of equilibrium when the market maker is risk averse and does not set the price according to semi-strong form efficiency. Ref. [14] analyzed the closed form solution of the Kyle model. Ref. [15] allowed for a more complex environment of the information structure of the model).

A very interesting extension of the static Kyle model resides in the investigation of the effect of different market structures on the equilibrium outcomes. Indeed, there are different types of insiders in the firm, some without any managerial responsibilities (the president and the members of the board of directors, for example), with the objective of maximizing their profits from trading the stock of the firm whose inside information they possess. Therefore, competition among fully informed insiders is another form of competition that influences the amount of information disseminated in the stock price. Fully informed insiders could compete in a Cournot duopoly model or in a Stackelberg model in the financial market (see [16]).

This paper adds to the literature on insider trading in the spirit of [1] by modeling Stackelberg competition among two insiders who are partially informed. The owner acts as a Stackelberg leader in the financial market, and the manager acts as a Stackelberg-follower in the financial sector to the owner who knows his reaction function before making any decision. The Stackelberg structure in the financial market with the owner serving as the Stackelberg leader seems very realistic. Indeed, the owner designs the manager's compensation mechanism and chooses a capable manager to serve his purpose; therefore, he should have information on the manager's reaction. In addition, feedback from staff of various levels can tell the owner about any improper behavior of the manager, so the owner can frequently adjust his perception. The results show that in the presence of partial information, the follower makes more profits than the leader. Moreover, when the insiders' signals have the same precision, the insiders' profits are higher under Stackelberg setting than in Cournot setting.

Although insiders tend to compete in the financial markets, they also tend to share trading information. For instance, evidence suggests that mutual fund managers trade based on local word-of-mouth communication in the asset-management community. Sharing information reduces the informational advantage of the insider and therefore it is important to understand what benefit investors obtain from sharing information. Ref. [17] study information sharing between strategic investors who are informed about asset fundamentals. They show that a coarsely informed investor optimally chooses to share information if his counterpart investor is well informed. By doing so, the coarsely informed investor invites the other investor to trade against his information, thereby reducing his price impact. Paradoxically, the well-informed investor loses from receiving information because of the resulting worsened market liquidity and the more aggressive trading by the coarsely informed investor. Another contribution of the current paper is to investigate the effect of information sharing among partially informed insiders competing in a Stackelberg game on the dissemination of information. The results show that the follower has no incentive to share information, while the leader is indifferent between sharing and not sharing information, leading to multiple equilibria.

The paper is organized as follows: Section 2 contains to the literature review. Section 3 presents the model and methodology. Section 4 displays the results and analysis. Finally, Section 5 presents the discussion and conclusion.

2. Literature Review

The pioneering paper of [1] on insider trading is considered as the baseline model that explains how an insider uses his private information to take advantage of noise traders in a security market. A large body of the literature has since then been devoted to extend the Kyle one-shot game model. Ref. [18] extend Kyle's model to the case where the insider may be subject to an additional trading penalty, increasing in the size of their trades, and characterize the set of insider trading penalties which are efficient from the point of view of a regulator who cares both about market liquidity and price informativeness.

Some papers extended the Kyle model to include different risk attitude traders. Ref. [2] shows that competition among multiple risk-averse traders using a dynamic setting, increases the information rapidly. Ref. [19] studies a continuous time version of [2] and shows that the market depth decreases with time. Ref. [20] shows that the information is revealed slowly under disclosure requirements. Recently, ref. [3] extend the Kyle model to the case of two insiders, one risk-neutral and one risk-averse. They find that the market depth depends crucially on the degree of risk aversion, and show that regardless of the degree of risk aversion, the stock price reveals more information than the stock price in [2]. Ref. [4] study the case of two insiders in which the first insider is risk-neutral while the second insider is overconfident. They show that the market depth parameter and price information revelation vary with the degree of overconfidence.

Another direction of the extension of [1] is in [5] who allow for the market maker to observe a second signal that is correlated with the order flow. They find that the stock price is more informative and that the insider's profits are lower than in [1], when the market maker only observes the order flow.

Subsequent works have also studied the relationship between the financial decisions and real decisions of the firm in a static Kyle model. In [21]'s model, the insider is considered the manager of the firm that profits from a monopoly position on the real market. Therefore, the insider chooses both the amount of the real output to be produced and the amount of the stock of the firm to trade. In the model, the market maker observes two correlated signals: the total order flow and the market price of the real good. They show that the real output chosen by the insider/manager is greater than it would be in the absence of insider trading, and that the stock price reveals more information than in [1]. In an extension of this model, ref. [22] assume that the publicly-owned firm competes in the real market, like Cournot, with another privately-owned firm. Their findings highlight the effect Cournot duopoly in the real market on the stock price coefficients relative to the monopoly case, as well as the negative effect on the insider's profits and his compensation scheme relative to monopoly, but do not show any effect on the amount of information incorporated in the stock price. Ref. [23] suggest a two-tier model where the production of the real good by the publicly-owned firm involves constant labor costs, in addition to the intermediary products produced by another privately-owned firm. Their findings highlight the influence of the insider on the market of the final good and the stock price coefficients. However, this two-tier market structure does not change the amount of information disseminated in the stock price relatively to [21,22]. Ref. [24] study the impact of product differentiation on the real and financial decisions of a publicly-owned firm, competing like Cournot with another privately-owned firm. They show that the degree of product differentiation alters the output of the publicly-owned firm and the stock price coefficients, in addition to having a detrimental effect on the insider/manager's profits and compensation scheme.

As different types of insiders with different managerial responsibilities exist in the firms, another strand of the literature extended the Kyle model to investigate the effect of competition among two insiders (the owner and the manager) on the amount of information disseminated in the stock price. Refs. [25,26] extend, respectively, refs. [21,22] and model Cournot competition between insiders in the financial market. Their findings show that competition between insiders increases the amount of information disseminated in the stock price. Ref. [27]'s model incorporates Cournot competition in the real market, with Stackelberg competition in the financial market where the owner is considered as the Stackelberg leader. They find that the owner's profits as well as the amount of information revealed in the stock price increase in comparison with [25]. Ref. [16] extend the Kyle-type model of [5] with two signals to include Cournot duopoly among insiders in the financial market and find that each insider loses market power and partially controls the stock price. Hence, the stock price reveals more information with respect to [5]. The unconditional profits of each insider also decrease. Ref. [16] also extend the model of [5] with two signals to investigate the effect Stackelberg competition in the financial market on information revelation. They assume that one of the insiders, the owner, is high in the organizational hierarchy and chooses the second insider, the manager, to serve his purpose. In other words, the owner is the leader and knows the reaction function of the manager. They show that with Stackelberg competition in the financial market, the manager trades less and hence earns less than in the Cournot case. However, the owner, due to their role as a leader, trades more than in the Cournot and earns more profits. In addition, the price reveals more information in the Stackelberg than in the Cournot structure. Ref. [28] develops a sequential fair Stackelberg auction model in the spirit of [1] in which each of the two risk-seeking insiders has an equal chance to be a leader or a follower at each auction stage. The authors establish the existence and uniqueness of sequential fair Stackelberg equilibria when both insiders adopt linear strategies, and find that, at the sequential equilibria, these two insiders compete aggressively and cause the liquidity of the market to drop, the information to be revealed, and the profit to go down very rapidly while the trading intensity becomes substantially high.

While the above-mentioned works study the behavior of fully informed insiders, the originality of the current paper stems from adding to the literature on the Kyle model by modeling Stackelberg competition among two partially informed insiders in the financial market. It also adds to the literature on information sharing [17] by investigating the effect of information sharing among partially informed insiders on the dissemination of information.

3. Model and Methodology

Let $(\Omega, \mathcal{F}, \mathbb{P})$ be a probability space, all the random variables are defined with respect to this probability space. This model is a modified version of [1] with two insiders, each of them being partially informed about the underlying value of the stock. Consider an economy with one financial asset, the stock of the firm. There are three types of agents trading in the financial market. First, there are two risk-neutral rational traders (the insiders), each of them is partially informed about the the realization z of $\tilde{z}$, the value of the stock. Specifically, it is assumed that the partially informed trader i, $(i = 1, 2)$ observes the signal $\tilde{s}_i = \tilde{z} + \tilde{\varepsilon}_i$ correlated with $\tilde{z}$ where $\tilde{\varepsilon}_i$ is normally distributed with zero mean and variance σ_i^2. Second, there are (non-rational) noise traders, representing small investors with no information on z. The aggregate noise trade is assumed to be a random variable $\tilde{u}$, which is normally distributed with mean zero and variance σ_u^2. Finally, there are $K (K \geq 2)$

risk-neutral market makers who act like Bertrand competitors. Each of the variables $\tilde{z}$ and $\tilde{\varepsilon}_i(i = 1,2)$, is independent from $\tilde{u}$. (Random variables are denoted with a tilde. Realized values lack the tilde. The mean of the random variable is denoted with bar).

Following [1], the trading mechanism is organized in two steps. In step one, a linear pricing rule and optimal order rule are determined by the market makers and the insiders, respectively, as a Bayesian–Nash equilibrium. The market makers determine a (linear) pricing rule p, based on their a priori beliefs, where p is a measurable function $p : \mathbb{R} \longrightarrow \mathbb{R}$. Each insider $i = 1,2$ chooses a stock trade function $\tilde{x}_i$, where $x_i : \mathbb{R} \longrightarrow \mathbb{R}$ is a measurable function. In the second step, the insiders observe the realization of the signals, and submit their stock order to the market makers based on the equilibrium stock trade functions. The market makers also receive orders from the noise traders, all these orders arrive as a total order flow signal $\tilde{r} = \tilde{x}_1 + \tilde{x}_2 + \tilde{u}$. The order flow signal is used by the market makers to set the price $\tilde{p} = p(\tilde{r})$, based on the equilibrium price function, to clear the market. Each partially informed insider $i, i = 1,2$ knows only the realization s_i of $\tilde{s}_i$ and does not know the values of $\tilde{u},, \tilde{r}$, before the order flow decisions is made. Moreover, each market maker does not know neither the realization z of $\tilde{z}$ nor the realization s_i of $\tilde{s}_i$ but only knows their distribution. Finally, the market makers cannot observe either x_i or u.

In this model, Stackelberg competition is introduced among the two insiders in the financial market. Specifically, we assume that one of the insiders, the owner (insider 1), is high on the organizational hierarchy and acts as a Stackelberg leader in the financial market. The second insider (insider 2), the manager, is in the lower ladder of the organizational hierarchy and acts as a Stackelberg follower in the financial sector to the owner who knows his reaction function before making any decision (e.g., see [29]). The Stackelberg structure in the financial market with the owner serving as the Stackelberg leader seems very realistic. Indeed, the owner designs the manager's compensation mechanism and chooses a capable manager to serve their purpose; therefore, they should have information on the manager's reaction.

The theoretical methods of economic theory are used to characterize and analyze the main results of the model. Specifically, this is a game of incomplete information and we seek for a Bayesian–Nash equilibrium. A Bayesian–Nash equilibrium is a vector of three functions $[x_1(.), x_2(.), p(.)]$ such that:

(a) Profit maximization of insider 1,

$$E[\tilde{z} - p(\tilde{x}_1 + \tilde{x}_2 + \tilde{u}))\tilde{x}_1|\tilde{s}_1] \geq E[\tilde{z} - p\tilde{x}'_1 + \tilde{x}_2 + \tilde{u}))\tilde{x}'_1|\tilde{s}_1] \tag{1}$$

for any level of trading order $\tilde{x}'_1$ decided by the insider;

(b) Profit maximization of insider 2,

$$E[\tilde{z} - p(\tilde{x}_1 + \tilde{x}_2 + \tilde{u}))\tilde{x}_1|\tilde{x}_1, \tilde{s}_2] \geq E[\tilde{z} - p(\tilde{x}_1 + \tilde{x}'_2 + \tilde{u}))\tilde{x}'_2|\tilde{x}_1, \tilde{s}_2] \tag{2}$$

for any level of trading order $\tilde{x}'_2$ decided by the insider;

(c) Semi-strong market efficiency: The pricing rule $p(.)$ satisfies,

$$p(\tilde{r}) = E[\tilde{z}|\tilde{r}]. \tag{3}$$

4. Results and Analysis

Before characterizing the equilibrium of the model, it should be pointed out to the reader that a linear linear Bayesian equilibrium is being sought. Formally, an equilibrium is linear if the insiders' strategies are linear with respect to their observed signals and the pricing rule is linear with respect to the order flow signal. In other words, there are constants $a_0, a_1, b_0, b_1, b_2, \mu, \lambda$ such that,

$$x_1(s_1) = a_0 + a_1 s_1 \quad x_2(s_2, x_1) = b_0 + b_1 x_1 + b_2 s_2 \tag{4}$$

and

$$\forall r, \quad p(r) = \mu + \lambda r. \tag{5}$$

Note that conditions (1) and (2) define the optimal strategies of the insiders while condition (3) guarantees the zero expected profits for the market makers. The stock price, set by the market makers, is equal to the conditional expectation of the asset given the available information. The study is restricted to linear equilibrium. The normal distributions of the exogenous random variables, together with the particular expressions of the insiders' strategies and the market price, enable us to derive and to prove the existence of a unique linear equilibrium. In Proposition 1, the unique linear equilibrium outcomes of the model are characterized.

Proposition 1. *In the Stackelberg model with two partially informed traders, a linear equilibrium exists and it is unique. It is characterized by,*

$$\tilde{x}_1(\tilde{s}_1) = a_1(\tilde{s}_1 - \bar{s}_1) \quad \text{and} \quad \tilde{x}_2(\tilde{x}_1, \tilde{s}_2) = b_1\tilde{x}_1 + b_2(\tilde{s}_2 - \bar{s}_2) \tag{6}$$

$$a_1 = \frac{h_1 h_2}{2\lambda(1+h_1)(1+h_1+h_2)}, \; b_1 = -\frac{1}{2} + \frac{1+h_1}{h_2} \text{ and } b_2 = \frac{h_2}{2\lambda(1+h_1+h_2)} \tag{7}$$

$$\lambda = \frac{\sqrt{(1+h_1)[4h_1(1+h_1)^2 + 4(1+h_1)(1+2h_1)h_2 + (4+3h_1)h_2^2)\sigma_z^2]\sigma_u^2}}{4(1+h_1)(1+h_1+h_2)\sigma_u^2} \tag{8}$$

$$\mu = \bar{z}, \; E[\pi_1] = a_1[\gamma(1-\lambda b_2) - \lambda a_1(1+b_1)](\sigma_z^2 + \sigma_1^2) \text{ and } E[\pi_2] = \lambda E[\tilde{x}_2^2] \tag{9}$$

where $h_1 = \frac{\sigma_z^2}{\sigma_1^2}$, $h_2 = \frac{\sigma_z^2}{\sigma_2^2}$ *and* $\gamma = \frac{\sigma_z^2}{\sigma_z^2+\sigma_1^2}$,

Proof. See Appendix A. □

Proposition 1 highlights the impact of market structure (the Stackelberg setting) and information asymmetry on the equilibrium outcomes. Hence, the relationship between this model and the existing literature should be clarified. First, this paper studies the impact of partial information on the equilibrium outcomes in alignment with [16,28] which both consider a Stackelberg structure of [1] with fully informed traders. Moreover, this model differs from [8,17] which study the case of a Cournot setting with partially informed traders, by studying the impact of the Stackelberg structure in the financial sector. Consequently, this paper will be able to show the effect of market structure and information asymmetry on the equilibrium outcomes.

The sequential structure of the players' information and moves in the Stackelberg game has a direct effect on the equilibrium outcomes. Indeed, when the two insiders are fully informed, this paper reduces to the case studied by [16]. To achieve such result, a two-step procedure is implemented: in the first step, the equilibrium outcomes are characterized when the variance of the follower's signal noise is replaced by zero (equivalently, the limit when h_2 goes to infinity is considered). In the second step, the variance of the leader' signal noise is replaced by zero (equivalently, the limit when h_1 goes to infinity is considered) of the resulting outcomes. Specifically, the following result is as follows.

Lemma 1. *When the two insiders are fully informed, this paper reduces to the case studied in* [16]. *(The superscript "D" refers to* [16].*) Indeed,*

$$\tilde{x}_1^D(\tilde{z}) = \lim_{h_1\to\infty}\left(\lim_{h_2\to\infty}(a_1(\tilde{s}_1 - \bar{s}_1))\right) = \frac{1}{2\lambda^D}(\tilde{z} - \bar{z}) \tag{10}$$

$$\tilde{x}_2^D(\tilde{z}) = \lim_{h_1\to\infty}\left(\lim_{h_2\to\infty}(b_1\tilde{x}_1 + b_2(\tilde{s}_2 - \bar{s}_2))\right) = \frac{1}{2\lambda^D}(\tilde{z} - \bar{z}) \tag{11}$$

$$\mu = \bar{z} \text{ and } \lambda^D = \lim_{h_1\to\infty}\left(\lim_{h_2\to\infty}\lambda\right) = \frac{\sigma_z\sqrt{3}}{4\sigma_u} \tag{12}$$

$$E[\pi_1]^D = \lim_{h_1\to\infty}\left(\lim_{h_2\to\infty}(E[\pi_1])\right) = \frac{\sigma_z\sigma_u}{2\sqrt{3}} \tag{13}$$

$$E[\pi_2]^D = \lim_{h_1\to\infty}\left(\lim_{h_2\to\infty}(E[\pi_2])\right) = \frac{\sigma_z\sigma_u}{4\sqrt{3}} \tag{14}$$

Note that this procedure is restrictive to the Stackelberg structure. Indeed. in the Cournot model with partially informed traders studied in [8,17], the equilibrium outcomes converge to the full information case when both h_1 and h_2 go simultaneously to infinity. Moreover, it should be noted that the information structure in the Stackelberg game imposes the order of convergence. In other words, the full information case is not obtained if first h_1 goes to infinity and then h_2 goes to infinity.

5. Discussion and Conclusions

This section discusses the impact of market structure on equilibrium outcomes. On the one hand, a comparative statics analysis of this model's results is carried out with respect to the results of [8,17], and on the other hand, the impact of information sharing on the insiders' decisions is studied.

The equilibrium outcomes of this model are first compared to the ones in [8,17]. Given the complexity of the expressions of the equilibrium outcomes, the analysis is restricted to the case when the two insiders observe the same signal. In this case, Proposition 1 becomes

Proposition 2. *When the precision of the two signals are the same ($\sigma_1^2 = \sigma_2^2$), the equilibrium is characterized by,*

$$\tilde{x}_1(\tilde{s}_1) = a_1(\tilde{s}_1 - \bar{s}_1) \quad \text{and} \quad \tilde{x}_2(\tilde{x}_1, \tilde{s}_2) = b_1\tilde{x}_1 + b_2(\tilde{s}_2 - \bar{s}_2) \tag{15}$$

$$a_1 = \frac{h^2}{2\lambda(1+h)(1+2h)}, \; b_1 = \frac{1}{2} + \frac{1}{h} \text{ and } b_2 = \frac{h}{2\lambda(1+2h)} \tag{16}$$

$$\mu = \bar{z} \text{ and } \lambda = \frac{\sigma_z\sqrt{h(1+h)(8+3h(8+5h))}}{4\sigma_u(1+h)(1+2h)} \tag{17}$$

$$E[\pi_1] = \frac{h^2(2+3h)\sigma_z\sigma_u}{2(1+2h)\sqrt{h(1+h)(8+3h(8+5h))}} \tag{18}$$

$$E[\pi_2] = \frac{h(4+3h)^2\sigma_z\sigma_u}{4(1+2h)\sqrt{h(1+h)(8+3h(8+5h))}} \tag{19}$$

where $h = h_1 = h_2$.

Lemma 2 shows graphically (see Figure 1) the impact of the Stackelberg structure when compared to the Cournot setting studied in [8,17].

Lemma 2. *The relation between the Stackelberg and Cournot models when the signals' errors have the same precision is given by*

$$\lambda^S > \lambda^C, \; \pi_1^S > \pi_1^C \text{ and } \pi_2^S > \pi_2^C \tag{20}$$

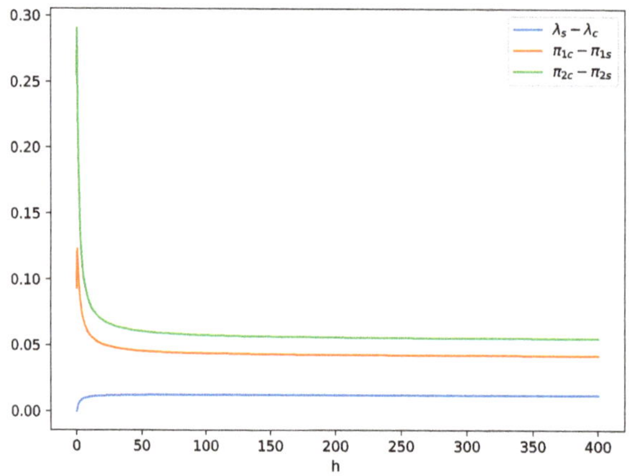

Figure 1. Market depth parameter λ and the profits of the two insiders in the Stackelberg and Cournot settings, when they observe the same signal.

First note that Stackelberg competition between the two insiders increases their unconditional profits when compared to the Cournot competition between the insiders. Second, the market depth parameter λ is higher in this model than in the Cournot case. It should be pointed out that these results do not hold in the case when both insiders are fully informed about the stock value (see [16]). Consequently, in the presence of incomplete information, the insiders benefit more in the Stackelberg setting than in the Cournot setting.

To better understand the impact the information asymmetry between the insiders on the market structure, this paper considers the case when insider 2 is fully informed about the stock value ($\tilde{s}_2 = \tilde{z}$) while insider 1 is partially informed. In this case, Proposition 1 becomes

Proposition 3. *When insider 2 is fully informed about the stock value ($\tilde{s}_2 = \tilde{z}$) and insider 1 is partially informed ($\tilde{s}_1 = \tilde{s}$ and $h_1 = h$), the equilibrium is characterized by,*

$$\tilde{x}_1(\tilde{s}) = a_1(\tilde{s}_1 - \bar{s}) \quad \text{and} \quad \tilde{x}_2(\tilde{x}_1, \tilde{z}) = b_1\tilde{x}_1 + b_2(\tilde{z} - \bar{z}) \tag{21}$$

$$a_1 = \frac{h}{2\lambda(1+h)}, \; b_1 = \frac{-1}{2} \text{ and } b_2 = \frac{1}{2\lambda} \tag{22}$$

$$\mu = \bar{z} \text{ and } \lambda = \frac{\sigma_z\sqrt{(1+h)(3h+4)}}{4\sigma_u(1+h)} \tag{23}$$

$$E[\pi_1] = \frac{h\sigma_z\sigma_u}{2\sqrt{(1+h)(3h+4)}} \tag{24}$$

$$E[\pi_2] = \frac{(4+h)\sigma_z\sigma_u}{4\sqrt{(1+h)(3h+4)}} \tag{25}$$

Lemma 3 graphically shows (See Figure 2) the impact of the Stackelberg structure when compared to the Cournot setting studied in [8,17] when insider 2 is fully informed while insider 1 is partially informed.

Lemma 3. *The relations between the Stackelberg and Cournot models when insider 2 is fully informed and insider 1 is partially informed, is given by*

$$\lambda^S < \lambda^C, \quad \pi_1^S < \pi_1^C \text{ and } \pi_2^S < \pi_2^C \tag{26}$$

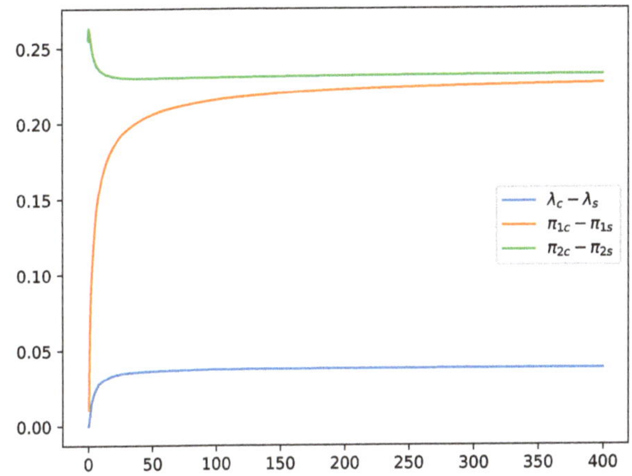

Figure 2. The market depth parameter λ and the profits of the two insiders in the Stackelberg and Cournot settings when insider 2 is fully informed and insider 1 is partially informed.

Lemma 3 reveals that Cournot structure is more beneficial for both insiders when one of the two insiders in fully informed. Moreover, a quick comparison between insiders' profits (see Equations (24) and (25)) shows that the unconditional profits of the leader (insider 1) are greater than the profits of the follower (insider 2) when the signal of insider 1 is less noisy.

Finally, it is worth noticing that Proposition 1 sheds light on the impact of partial information on the equilibrium outcomes in a Stackelberg game structure. Indeed, Figure 3 shows that the profits of the follower (insider 2) are greater than the profits of the leader (insider 1).

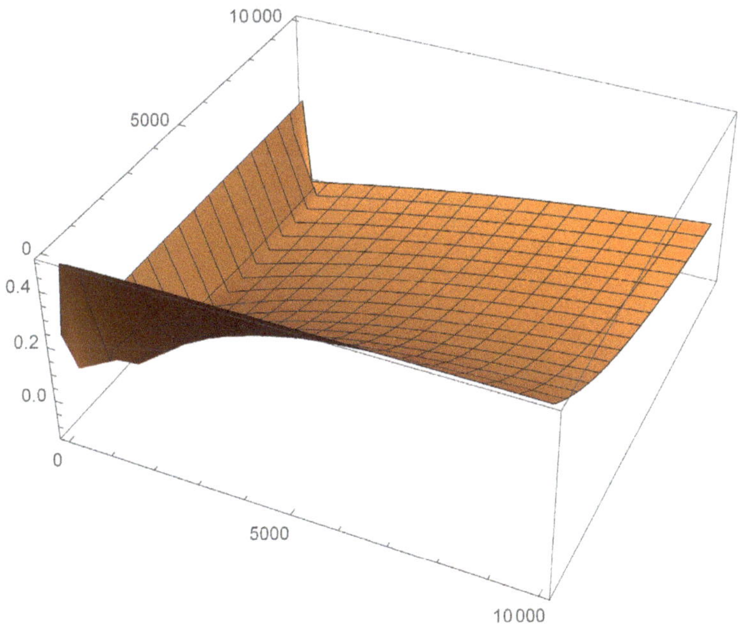

Figure 3. $\delta\Pi = \Pi_2 - \Pi_1$: the difference between the two insiders' profits in the Stackelberg setting when they observe the same signal.

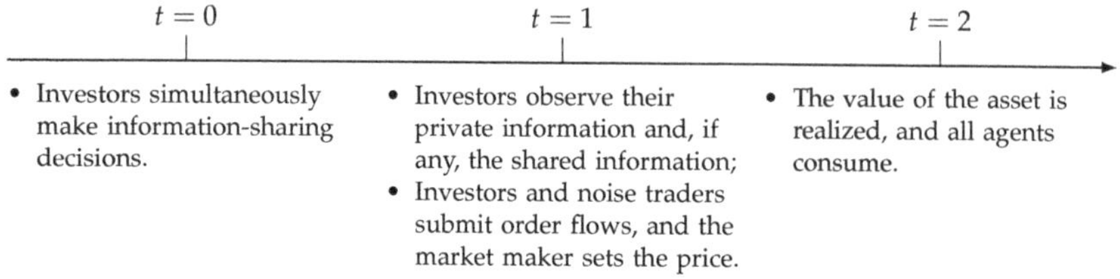

The second part of this discussion studies the impact of information sharing on the insiders' decisions. For comparison purposes, the same structure of [17] who study the impact of the information in the Cournot case is adopted. Specifically, it is assumed that insider 2 is fully informed about the asset value, i.e., Proposition 3 holds. Moreover, it is assumed as in [17], that $\bar{z} = 0$ and $\sigma_z^2 = 1$. Following [17], the fully informed insider (insider 2 in this model) is denoted by H while the semi-informed insider (insider 1 in this model) by L. At $t = 0$, H and L simultaneously decide whether to share their private information to maximize their respective expected trading profits. For investor $i \in H, L$, $A_i \in \{S, \emptyset\}$ is used to denote the information-sharing decisions, where $A_i = S$ means that the investor fully shares information with the other investor, whereas $A_i = \emptyset$ means that the investor keeps it secret. It is assumed that the date-0 information sharing decisions become common knowledge at the beginning of date 1, so that backward induction can be applied to compute the equilibrium of this economy.

Trading occurs on $t = 1$. Conditional on the endowed private information, as well as the shared information (if any), investor $i \in H, L$ places market order to maximize the expected trading profit. Let $\mathbb{F}_i$ indicates investor i's information set. For instance, if L shares information with H but H does not share information with L (i.e., $A_L = S$ and $A_H = \emptyset$), then the two investors' information sets are, respectively, $\mathbb{F}_L = \{\tilde{s}\}$ and $\mathbb{F}_H = \{\tilde{z}, \tilde{s}\}$ (see Figure 1 which describes the timeline of the economy).

Given the information-sharing decisions of the investors A_L and A_H on date 0, different trading subgames follow on date 1, depending on whether the investors share their respective private information or not. The investors' expected profits evaluated on the date-1 subgame equilibrium will serve as their payoffs of the date-0 information-sharing game. We next discuss each subgame separately.

Subgame 1: Neither Investor Shares Information ($A_L = A_H = \emptyset$).

When neither investor shares information, we obtain the expression of the expected profits stated in Proposition 3.

Subgame 2: L Shares Information but H Does Not ($A_L = \{\tilde{s}\}$ and $A_H = \emptyset$).

Since H is the follower and plays at the second stage of the game, observing the signal of L, which is linearly dependent to his strategy, will not change the profits maximization problem of each of the two players. Hence, this case will coincide with subgame 1 and the expression of the expected profits is stated in Proposition 3.

Subgame 3: H Shares Information ($A_H = S$).

There are two subgames when H shares information, depending on whether L shares his information. Nonetheless, it turns out that the equilibrium in these two subgames is the same because H owns perfect information about the asset fundamental, and once he shares it with L, L no longer uses his own noisy signal $\tilde{s}$ in predicting the asset fundamental. Thus, regardless of L's information-sharing decision, once H shares information, the trading game degenerates to the [16] with two perfectly-informed traders. Consequently, the expression of the expected profits stated in Lemma 1 is obtained.

Now, the insiders' payoffs of the date-0 information sharing game is summarized in Table 1. It can be easily noticed that this game has 2 equilibria with the same players' payoffs. Indeed, the strategy of player H, *"Not share"* is a dominant strategy. (Consider the function $f(h) = \frac{(4+h)\sigma_u}{4\sqrt{(1+h)(3h+4)}}$. It is easy to check that this function is decreasing on $(0, \infty)$ and having a horizontal asymptote with equation $y = \frac{\sigma_u}{4\sqrt{3}}$.) Taking this information into account, player L is indifferent between *"Share"* or *"Not share"* since both strategies lead to the same payoff.

Table 1. Insiders' payoffs of the date-0 information sharing game.

		H	
		Not Share	*Share*
L	*Not share*	$\frac{h\sigma_u}{2\sqrt{(1+h)(3h+4)}}, \frac{(4+h)\sigma_u}{4\sqrt{(1+h)(3h+4)}}$	$\frac{\sigma_u}{2\sqrt{3}}, \frac{\sigma_u}{4\sqrt{3}}$
	Share	$\frac{h\sigma_u}{2\sqrt{(1+h)(3h+4)}}, \frac{(4+h)\sigma_u}{4\sqrt{(1+h)(3h+4)}}$	$\frac{\sigma_u}{2\sqrt{3}}, \frac{\sigma_u}{4\sqrt{3}}$

Finally, it should be noted that the result is quite different than the result found in [17] in which the equilibrium consists of player L sharing his information whereas player H does not. This difference is due to the fact that in the Stackelberg game, on the one hand, the fully informed player is the follower and on the other hand, it is also due to the sequential structure of the game timing.

In this paper, the authors generalize the Kyle model with two risk neutral insiders to the case where each insider is partially informed about the value of the stock and compete under a Stackelberg game setting in the financial market. Based on the backward induction theory, the linear Bayesian equilibrium is first characterized. Given the information structure of the signal of the two insiders which is realistic and well used in the field of market microstructure, this paper can provide important reference for market analysts and practitioners in the field of insider trading.

However, there are still some limitations in the paper. First, the current analysis illustrates the impact and the important role that Stackelberg structure has on the insiders' decisions and the equilibrium variables. In particular, it is shown that the relation between the market depth parameters under Cournot and Stackelberg settings is quite ambiguous and depends crucially on the precision of the insiders' signals. Therefore, understanding the role of partial information and its implications to the regulatory aspects of the world of insider trading should be examined more closely. Secondly, in the case of general correlated signals (see [10]), a linear equilibrium might not be necessarily unique. This is an open question for future research.

Author Contributions: Conceptualization, W.D., F.K. and N.A.; Methodology, W.D., F.K. and N.A.; Formal analysis, W.D., F.K. and N.A.; Writing—review & editing, W.D., F.K. and N.A. All authors have read and agreed to the published version of the manuscript.

Funding: This research received no external funding.

Data Availability Statement: Not applicable.

Acknowledgments: We would like to express our sincere gratitude to the editor and the anonymous referees for their invaluable contributions to this research article. Their expertise, careful assessment, and constructive feedback have greatly enhanced the quality and rigor of our work. The authors also thank Gunar Matthies for the useful comments and discussions. The authors gratefully acknowledge technical and financial support provided by Gulf University of Science and Technology (GUST), Kuwait.

Conflicts of Interest: The authors declare no conflict of interest.

Appendix A

Proof of Proposition 1. Applying the backward induction approach to solve for the equilibrium, insider 2 solves,

$$\begin{aligned}\underset{\tilde{x}_2}{\text{Max}}\, E[(\tilde{z}-p(\tilde{r}))\tilde{x}_2|\tilde{s}_2,\tilde{x}_1] &= \underset{\tilde{x}_2}{\text{Max}}\, E[(\tilde{z}-\mu-\lambda(\tilde{x}_1+\tilde{x}_2+\tilde{u}))\tilde{x}_2|\tilde{s}_2,\tilde{x}_1] \\ &= \underset{\tilde{x}_2}{\text{Max}}(E[\tilde{z}|\tilde{s}_2,\tilde{x}_1]-\mu-\lambda\tilde{x}_1-\lambda\tilde{x}_2)\tilde{x}_2\end{aligned}$$

The F.O.C implies that

$$\tilde{x}_2 = \frac{E[\tilde{z}|\tilde{s}_2,\tilde{x}_1]-\mu-\lambda\tilde{x}_1}{2\lambda} \tag{A1}$$

and the S.O.C. requires $\lambda > 0$. Since all the variables $\tilde{z}, \tilde{s}_2$ and $\tilde{x}_1$ are all normally distributed, then by the projection Theorem, we have $E[\tilde{z}|\tilde{s}_2,\tilde{x}_1] = \alpha_0+\alpha_1\tilde{x}_1+\alpha_2\tilde{s}_2$ where α_0, α_1 and α_2 will be expressed later in the paper. Hence, Equation (A1) becomes,

$$\tilde{x}_2(\tilde{s}_2,\tilde{x}_1) = \frac{\alpha_0+\alpha_1\tilde{x}_1+\alpha_2\tilde{s}_2-\mu-\lambda\tilde{x}_1}{2\lambda} = \frac{\alpha_0-\mu}{2\lambda}+\frac{(\alpha_1-\lambda)}{2\lambda}\tilde{x}_1+\frac{\alpha_2}{2\lambda}\tilde{s}_2 \tag{A2}$$

Thus we obtain,

$$b_0 = \frac{\alpha_0-\mu}{2\lambda}, \quad b_1 = \frac{(\alpha_1-\lambda)}{2\lambda}, \quad b_2 = \frac{\alpha_2}{2\lambda} \tag{A3}$$

Now insider 1 solves

$$\underset{\tilde{x}_1}{\text{Max}}\, E[(\tilde{z}-p(\tilde{r}))\tilde{x}_1|\tilde{s}_1] = \underset{\tilde{x}_1}{\text{Max}}\, E[(\tilde{z}-\mu-\lambda(\tilde{x}_1+\tilde{x}_2+\tilde{u}))\tilde{x}_1|\tilde{s}_1]$$

Plugging Equation (A2) into the above equation, insider 1 solves,

$$\begin{aligned}&\underset{\tilde{x}_1}{\text{Max}}\, E[(\tilde{z}-\mu-\lambda\tilde{x}_1-\lambda[\frac{\alpha_0-\mu}{2\lambda}+\frac{(\alpha_1-\lambda)}{2\lambda}\tilde{x}_1+\frac{\alpha_2}{2\lambda}\tilde{s}_2])\tilde{x}_1|\tilde{s}_1] \\ &= \underset{\tilde{x}_1}{\text{Max}}\left(E[\tilde{z}|\tilde{s}_1]+\frac{\mu-\alpha_0}{2}-\frac{(\alpha_1+\lambda)}{2}\tilde{x}_1-\frac{\alpha_2}{2}E[\tilde{s}_2|\tilde{s}_1]\right)\tilde{x}_1\end{aligned}$$

The F.O.C implies that

$$\tilde{x}_1(\tilde{s}_1) = \frac{E[\tilde{z}|\tilde{s}_1]+\frac{\mu-\alpha_0}{2}-\frac{\alpha_2}{2}E[\tilde{s}_2|\tilde{s}_1]}{\alpha_1+\lambda} \tag{A4}$$

Recall that $E[\tilde{s}_2|\tilde{s}_1] = E[\tilde{z}+\tilde{\varepsilon}_2|\tilde{s}_1] = E[\tilde{z}|\tilde{s}_1]$ since $\tilde{\varepsilon}_2$ is independent of $\tilde{s}_1$. Hence, Equation (3) becomes

$$\tilde{x}_1(\tilde{s}_1) = \frac{(\mu-\alpha_0)+(2-\alpha_2)E[\tilde{z}|\tilde{s}_1]}{2(\alpha_1+\lambda)} \tag{A5}$$

Since $\tilde{z}$ and $\tilde{s}_1$ are jointly distributed,

$$E[\tilde{z}|\tilde{s}_1] = \gamma\tilde{s}_1+\eta \text{ where } \gamma = \frac{\sigma_z^2}{\sigma_z^2+\sigma_1^2} \text{ and } \eta = (1-\gamma)\bar{z} \tag{A6}$$

Thus, Equation (A5) becomes

$$\tilde{x}_1(\tilde{s}_1) = \frac{(\mu-\alpha_0)+(2-\alpha_2)(\gamma\tilde{s}_1+\eta)}{2(\alpha_1+\lambda)} \tag{A7}$$

Consequently, it is obtained that,

$$a_0 = \frac{(\mu - \alpha_0) + (2 - \alpha_2)\eta}{2(\alpha_1 + \lambda)}, \quad a_1 = \frac{(2 - \alpha_2)\gamma}{2(\alpha_1 + \lambda)} \tag{A8}$$

In order to express the coefficients of the insiders, the values of α_0, α_1 and α_2 have to be found. Recall the Theorem to find these values.

Theorem A1. *If the $p \times 1$ vector Y is normally distributed with mean U and covariance V and if the vector Y is partitioned into two subvectors such that $Y = \begin{pmatrix} Y_1 \\ Y_2 \end{pmatrix}$ and if $Y^* = \begin{pmatrix} Y_1 \\ Y_2^* \end{pmatrix}$*

$$U = \begin{pmatrix} U_1 \\ U_2 \end{pmatrix} \text{ and } V = \begin{pmatrix} V_{11} & V_{12} \\ V_{21} & V_{22} \end{pmatrix}$$

are the corresponding partitions of Y^*, U and V, then the conditional distribution of the $m \times 1$ ($m < p$) vector Y_1 given the vector $Y_2 = Y_2^*$ is the multivariate normal distribution with mean $U_1 + V_{12}V_{22}^{-1}(Y_2^*$ - $U_2)$ and covariance matrix $V_{11} - V_{12}V_{22}^{-1}V_{21}$.

Proof. See [30], Theorem 3.10, p. 63. □

Now, applying Theorem A1 to the normal random vector $B = (\tilde{z}, \tilde{x}_1, \tilde{s}_2)$ with $p = 3$ and $m = 1$. By identification, $Y_1 = \tilde{z}$ and $Y_2 = \begin{pmatrix} \tilde{x}_1 \\ \tilde{s}_2 \end{pmatrix}$. $U_1 = \bar{z}$, $U_2 = \begin{pmatrix} \bar{x}_1 \\ \bar{s}_2 \end{pmatrix}$ and

$$V = \begin{pmatrix} \sigma_z^2 & \sigma_{zx_1} & \sigma_{zs_2} \\ \sigma_{zx_1} & \sigma_{x_1}^2 & \sigma_{x_1s_2} \\ \sigma_{zs_2} & \sigma_{x_1s_2} & \sigma_{s_2}^2 \end{pmatrix} = \begin{pmatrix} V_{11} & V_{12} \\ V_{21} & V_{22} \end{pmatrix}$$

where $V_{11} = \sigma_z^2$, $V_{12} = (\sigma_{zx_1}, \sigma_{zs_2})$, $V_{21} = \begin{pmatrix} \sigma_{zx_1} \\ \sigma_{zs_2} \end{pmatrix}$ and $V_{22} = \begin{pmatrix} \sigma_{x_1}^2 & \sigma_{x_1s_2} \\ \sigma_{x_1s_2} & \sigma_{s_2}^2 \end{pmatrix}$.

Note that

$$V_{22}^{-1} = \frac{1}{D}\begin{pmatrix} \sigma_{s_1}^2 & -\sigma_{x_1s_2} \\ -\sigma_{x_1s_2} & \sigma_{s_2}^2 \end{pmatrix},$$

where D is the determinant of V_{22}, that is $D = \sigma_{x_1}^2\sigma_{s_2}^2 - \sigma_{x_1s_2}^2$.

Since it is assumed that $x_1(\tilde{s}_1) = a_0 + a_1\tilde{s}_1$, then

$$\sigma_{zs_2} = \sigma_z^2,\ \sigma_{zx_1} = a_1\sigma_z^2,\ \sigma_{x_1s_2} = a_1\sigma_z^2,\ \sigma_{s_2}^2 = \sigma_z^2 + \sigma_2^2,\ \sigma_{x_1}^2 = a_1^2(\sigma_z^2 + \sigma_1^2).$$

Since $E[\tilde{z}|\tilde{s}_2, \tilde{x}_1] = \alpha_0 + \alpha_1\tilde{x}_1 + \alpha_2\tilde{s}_2$, the values of α_0, α_1 and α_2 can be expressed as:

$$\alpha_0 = \bar{z} - \alpha_1\bar{x}_1 - \alpha_2\bar{s}_2 \tag{A9}$$

$$\alpha_1 = \frac{\sigma_{zx_1}\sigma_{s_2}^2 - \sigma_{zs_2}\sigma_{x_1s_2}}{D} \tag{A10}$$

$$\alpha_2 = \frac{\sigma_{zx_1}\sigma_{s_2}^2 - \sigma_{zs_2}\sigma_{x_1s_2}}{D} \tag{A11}$$

First note that $D = \sigma_{x_1}^2\sigma_{s_2}^2 - \sigma_{x_1s_2}^2 = a_1^2[\sigma_1^2(\sigma_z^2 + \sigma_2^2) + \sigma_z^2\sigma_2^2]$. Second, Equation (A10) becomes

$$\alpha_1 = \frac{a_1\sigma_z^2\sigma_2^2}{D} \tag{A12}$$

Similarly, Equation (A11) becomes after simplification

$$\alpha_2 = \frac{a_1^2\sigma_z^2\sigma_1^2}{D} \tag{A13}$$

Combining Equations (2), (A6), (A8), (A12), and (A13), gives

$$b_1 = \frac{2\sigma_z^2\sigma_2^2 + 2\sigma_1^2\sigma_2^2 - \sigma_z^2\sigma_1^2}{2\sigma_z^2\sigma_1^2} \text{ and } b_2 = \frac{\sigma_z^2\sigma_1^2}{2\lambda(\sigma_z^2\sigma_1^2 + \sigma_z^2\sigma_2^2 + \sigma_1^2\sigma_2^2)} \tag{A14}$$

$$b_0 = -\mu b_2 \text{ and } a_1 = \frac{\sigma_z^4\sigma_1^2}{2\lambda(\sigma_z^2 + \sigma_1^2)(\sigma_z^2\sigma_1^2 + \sigma_z^2\sigma_2^2 + \sigma_1^2\sigma_2^2)}, \tag{A15}$$

$$a_0 = \frac{-\sigma_z^2\sigma_1^2(\mu\sigma_1^2 - \bar{z}\sigma_1^2 + \mu\sigma_z^2)}{2\lambda(\sigma_z^2 + \sigma_1^2)(\sigma_z^2\sigma_1^2 + \sigma_z^2\sigma_2^2 + \sigma_1^2\sigma_2^2)}, \alpha_0 = \frac{\mu\sigma_2^2(\sigma_z^2 + \sigma_1^2)}{\sigma_z^2\sigma_1^2 + \sigma_z^2\sigma_2^2 + \sigma_1^2\sigma_2^2} \tag{A16}$$

$$\alpha_1 = \frac{2\lambda(\sigma_z^2 + \sigma_1^2)\sigma_2^2}{\sigma_z^2\sigma_1^2} \text{ and } \alpha_2 = \frac{\sigma_z^2\sigma_1^2}{\sigma_z^2\sigma_1^2 + \sigma_z^2\sigma_2^2 + \sigma_1^2\sigma_2^2} \tag{A17}$$

To find the expressions of μ and λ, first recall that the market efficiency condition together with the price linearity imply that

$$p(\tilde{r}) = E[\tilde{z}|\tilde{r}] = \mu + \lambda\tilde{r}$$

taking the expectation on both sides of the above equation,

$$\begin{aligned}\bar{z} &= \mu + \lambda\bar{r} = \mu + \lambda[\bar{x}_1 + \bar{x}_2] \\ &= \mu + \lambda[a_0 + b_0 + b_1a_0 + a_1(1 + b_1)\bar{s}_1 + b_2\bar{s}_2]\end{aligned}$$

Using the expressions of $a_0, b_0, a_1, b_1, b_2, \alpha_0, \alpha_1$, and α_2 written above, it is found that $\mu = \bar{z}$. Since the orders of the insiders and the liquidity traders are normally distributed, applying the projection theorem for normal random variables,

$$\lambda = \frac{Cov(\tilde{z}, \tilde{r})}{Var(\tilde{r})}$$

where $\tilde{r} = \tilde{x}_1 + \tilde{x}_2 + \tilde{u} = a_0 + b_0 + b_1a_0 + a_1(1 + b_1)\tilde{s}_1 + b_2\tilde{s}_2 + \tilde{u}$. Hence,

$$\lambda = \frac{Cov(\tilde{z}, \tilde{r})}{Var(\tilde{r})} = \frac{(a_1(1 + b_1) + b_2)\sigma_z^2}{a_1^2(1 + b_1)^2(\sigma_z^2 + \sigma_1^2) + b_2^2(\sigma_z^2 + \sigma_2^2) + 2a_1(1 + b_1)b_2\sigma_z^2 + \sigma_u^2} \tag{A18}$$

Substituting Equations (A14)–(A17) into (A18) and using the following change of variable $h_1 = \frac{\sigma_z^2}{\sigma_1^2}$, $h_2 = \frac{\sigma_z^2}{\sigma_2^2}$ and $\gamma = \frac{\sigma_z^2}{\sigma_z^2 + \sigma_1^2}$ to solve for the positive root λ, expressions (7) and (8) of the equilibrium outcomes as stated in Proposition 1 are found. It remains to find the expressions of the unconditional profits. Indeed,

$$\begin{aligned}E[(\tilde{z} - p(\tilde{r}))\tilde{x}_1|\tilde{s}_1] &= E[(\tilde{z} - \mu - \lambda(\tilde{x}_1 + \tilde{x}_2 + \tilde{u}))\tilde{x}_1|\tilde{s}_1] \\ &= (E[(\tilde{z}|\tilde{s}_1] - \mu - \lambda\tilde{x}_1 - \lambda E[\tilde{x}_2|\tilde{s}_1])\tilde{x}_1 \\ &= (E[(\tilde{z}|\tilde{s}_1] - \bar{z} - \lambda a_1(\tilde{s}_1 - \bar{s}_1) - \lambda E[b_1a_1(\tilde{s}_1 - \bar{s}_1) + b_2(\tilde{s}_2 - \bar{s}_2)|\tilde{s}_1])(a_1(\tilde{s}_1 - \bar{s}_1)) \\ &= (\bar{z} + \gamma) - \bar{z} - \lambda a_1(\tilde{s}_1 - \bar{s}_1) - \lambda b_1a_1(\tilde{s}_1 - \bar{s}_1) - \lambda b_2\gamma(\tilde{s}_1 - \bar{s}_1))(a_1(\tilde{s}_1 - \bar{s}_1)) \\ &= a_1(\tilde{s}_1 - \bar{s}_1)^2[\gamma(1 - \lambda b_2) - \lambda a_1(1 + b_1)]\end{aligned}$$

Taking the expectation on both sides of the above equation, expression in (9) is obtained.The conditional profits of insider 2 are now computed:

$$
\begin{aligned}
E[(\tilde{z} - p(\tilde{r}))\tilde{x}_2|\tilde{x}_1, \tilde{s}_2] = E[(\tilde{z} - \mu - \lambda(\tilde{x}_1 + \tilde{x}_2 + \tilde{u}))\tilde{x}_2|\tilde{x}_1, \tilde{s}_2] \\
= E[(\tilde{z} - \mu - \lambda\tilde{x}_1 - \lambda\tilde{x}_2)\tilde{x}_2|\tilde{x}_1, \tilde{s}_2] = (E[\tilde{z}|\tilde{x}_1, \tilde{s}_2] - \mu - \lambda\tilde{x}_1 - \lambda\tilde{x}_2)\tilde{x}_2 \\
= (\alpha_0 + \alpha_1\tilde{x}_1 + \alpha_2\tilde{s}_2 - \mu - \lambda\tilde{x}_1 - \lambda\tilde{x}_2)\tilde{x}_2 \\
= \lambda\tilde{x}_2^2
\end{aligned}
$$

where the last equality is based on Equations (A3) and (A15). □

References

1. Kyle, A. Continuous Auctions and Insider Trading. *Econometrica* **1985**, *53*, 1315–1335. [CrossRef]
2. Holden, C.; Subrahmanyam, A. Risk-Aversion, Imperfect Competition and Long-Lived Information. *Econ. Lett.* **1994**, *44*, 181–190. [CrossRef]
3. Daher, W.; Aydilek, A.; Saleeby, E.G. Insider Trading With Different Risk Attitudes. *J. Econ.* **2020**, *131*, 123–147. [CrossRef]
4. Jiang, Y.; Liu, H. Insider Trading, Overconfidence, and Private Information Flow. *N. Am. J. Econ. Financ.* **2022**, *60*, 101664. [CrossRef]
5. Jain, N.; Mirman, L.J. Insider Trading with Correlated Signals. *Econ. Lett.* **1999**, *65*, 105–113. [CrossRef]
6. Aase, K.K.; Bjuland, T.; Øksendal, B. Partially Informed Noise Traders. *Math. Financ. Econ.* **2012**, *6*, 93–104. [CrossRef]
7. Holden, C.; Subrahmanyam, A. Long-Lived Private Information and Imperfect Competition. *J. Financ.* **1992**, *47*, 247–270.
8. Tighe, C.; Michener, R. The Political Economy of Insider Trading Laws. *Am. Econ. Rev.* **1994**, *84*, 164–168.
9. Foster, F.D.; Viswanathan, S. Strategic Trading with Asymmetrically Informed Traders and Long-Lived Information. *J. Financ. Quant. Anal.* **1994**, *29*, 499–518. [CrossRef]
10. Daher, W.; Saleeby, E.G. Existence of Linear Equilibria in The Kyle Model with Partial Correlation and Two Risk Neutral Traders. *Heliyon* **2023**, *9*, e17574. [CrossRef]
11. Nöldeke, G.; Tröger, T. Existence of Linear Equilibria in The Kyle Model with Multiple Informed Traders. *Econ. Lett.* **2001**, *72*, 159–164. [CrossRef]
12. Nöldeke, G.; Tröger, T. A Characterization of the Distributions That Imply Existence of Linear Equilibria in the Kyle Model. *Ann. Financ.* **2006**, *2*, 73–85. [CrossRef]
13. Vitale, P. *Risk-Averse Trader with Insider Information*; Discussion Paper 9504; Department of Applied Economics, University of Cambridge: Cambridge, UK, 1995.
14. McLennan, A.; Monteiro, P.K.; Tourky, R. On Uniqueness of Equilibrium in the Kyle Model. *Math. Financ. Econ.* **2017**, *11*, 161–172. [CrossRef]
15. Lambert, N.S.; Ostrovsky, M.; Panov, M. Strategic Trading in Informationally Complex Environments. *Econometrica* **2018**, *86*, 1119–1157. [CrossRef]
16. Daher, W.; Karam, F.; Mirman, L.J. Insider Trading with Different Market Structures. *Int. Rev. Econ. Financ.* **2012**, *24*, 143–154. [CrossRef]
17. Goldstein, I.; Xiong, Y.; Yang, L. Information Sharing in Financial Markets. Working Paper. 2023 . Available online: http://dx.doi.org/10.2139/ssrn.3632315 (accessed on 1 June 2023).
18. Carré, S.; Collin-Dufresne, P.; Gabriel, F. Insider Trading with Penalties. *J. Econ. Theory* **2022**, *203*, 105461. [CrossRef]
19. Baruch, S. Insider Trading and Risk Aversion. *J. Financ. Mark.* **2002**, *5*, 451–564. [CrossRef]
20. Zhang, W.D. Risk Aversion, Public Disclosure and Long-Lived Information. *Econ. Lett.* **2004**, *85*, 327–334. [CrossRef]
21. Jain, N.; Mirman, L.J. Real and Financial Effects of Insider Trading with Correlated Signals. *Econ. Theory* **2000**, *16*, 333–353. [CrossRef]
22. Jain, N.; Mirman, L.J. Effects of Insider Trading Under Different Market Structures Signals. *Q. Rev. Econ. Financ.* **2002**, *42*, 19–39. [CrossRef]
23. Karam, F.; Daher, W. Insider Trading in a Two-Tier Real Market Structure Model. *Q. Rev. Econ. Financ.* **2013**, *53*, 44–52. [CrossRef]
24. Daher, W.; Aydilek, A.; Karam, F.; Aydilek, A. Insider Trading with Product Differentiation. *J. Econ.* **2014**, *11*, 173–201. [CrossRef]
25. Daher, W.; Mirman, L.J. Cournot Duopoly and Insider Trading with Two Insiders. *Q. Rev. Econ. Financ.* **2006**, *46*, 530–551. [CrossRef]
26. Daher, W.; Mirman, L.J. Market Structure and Insider Trading. *Int. Rev. Econ. Financ.* **2007**, *16*, 306–331. [CrossRef]
27. Wang, L.F.S.; Wang, Y.-C.; Ren, S. Stackelberg Financial-Leader in Insider Trading Model. *Int. Rev. Econ. Financ.* **2009**, *18*, 123–131. [CrossRef]
28. Gong, F.; Yonghui, Z. Sequential Fair Stackelberg Equilibria of Linear Strategies in Risk-Seeking Insider Trading. *J. Syst. Sci. Complex.* **2018**, *31*, 1302–1328. [CrossRef]

29. Nakamura, T. One-leader and multiple-follower Stackelberg games with private information. *Econ. Lett.* **2015**, *127*, 27–30. [CrossRef]
30. Graybill, F. *An Introduction to Linear Statistical Models: Volume I*; McGraw Hill: New York, NY, USA, 1961.

mathematics

MDPI

Article

Dynamic Game Analysis on Cooperative Advertising Strategy in a Manufacturer-Led Supply Chain with Risk Aversion

Jia Liu [1,2] and Cuixia Li [1,*]

[1] College of Economics and Management, Northeast Agricultural University, Harbin 150030, China
[2] School of Management, Heilongjiang University of Science and Technology, Harbin 150020, China
* Correspondence: licuixia.883@163.com

Abstract: This paper considers a dynamic Stackelberg game model for a manufacturer-led supply chain with risk aversion. Cooperative advertising strategy is applied to the marketing decisions of supply chain participants. Based on Stackelberg game and system dynamic theory, the game and complex dynamical behaviors are studied through the use of several methods, such as the stability region of the system, bifurcation diagram, attractor diagram, and the largest Lyapunov exponent diagram. The expected utilities of participants are given and compared by numerical simulation. The results illustrate that a series of variations in adjustment speed of advertising expenditure, participation rate of local advertising expenditure by manufacturer, risk tolerance levels, and the effect coefficient of advertising expenditure may cause a loss of stability to the system and evolve into chaos. Meanwhile, the Nash equilibrium point and the expected utility of the manufacturer and retailer will change greatly. The parameter control method is further applied to control the chaos phenomenon of the system effectively. By means of analyzing the impact of relevant factors on the game model, the manufacturer and retailer can make optimal strategy decisions in the supply chain competition. The findings of this study mainly include the following three aspects. Firstly, for market stability and maximizing revenue, the manufacturer adjusts the participation rate appropriately, avoiding too high or too low values. Secondly, the manufacturer will try to reduce their own risk tolerance level for the economic revenue, and the retailer appropriately adjust the risk tolerance level to adapt to their own development according to their own enterprise strategy. Finally, both the manufacturer and retailer reduce their own effect coefficients of advertising expenditure. Meanwhile, they will attempt to increase their opponent's effect coefficient to gain the most revenue. The research results of this study can provide important reference for the advertising expenditure decision and revenue maximization of participants in the context of risk aversion.

Keywords: manufacturer-led supply chain; cooperative advertising strategy; risk aversion; dynamic game; chaos

MSC: 91-10

1. Introduction

Berger [1] first discussed cooperative advertising in a manufacturer–retailer channel, proposing a financing of advertising expenditures through a discount on the wholesale price by the retailer. Cooperative advertising is an interactive scheme in a supply chain system in which the manufacturer pays a part of the advertising expenditure incurred by the retailer in local advertisement. Fulop [2] considered that cooperative advertising was an important aspect of promotional activities by many manufacturers. Manufacturers can enhance brand image and boost sales through a cooperative advertising strategy. As distinguished by Bergen and John [3], advertising could be divided into national advertising and local advertising. For retailers, local advertising investment can attract potential customers to buy products. The goal of participating in the local advertising expenditure by the manufacturer is to influence the retailer to invest more, which can generate more sales [4,5].

The concept of risk management was first put forward by Nobel Prize winner Harry Markowitz in the 1950s [6] and later became a field of enterprise management. Early research on risk management mainly focused on the field of financial investment. Markowitz [7] was the first to introduce the statistical concepts of expectation and variance into the portfolio problem and used the variance of investment revenue as a measure of risk. Later, on this basis, many scholars put forward other financial risk measurement tools and extended them to inventory control, supply chain management, and other fields [8–11]. Previous literature on the supply chain considered that enterprises were risk-neutral. Early studies on supply chain models were usually based on the assumption of risk neutrality of decision makers, and the decision-making goals were generally to maximize expected profits or minimize expected costs. However, subsequent studies by experimental economist Marshall Fisher [12] showed that most decision makers were not risk-neutral. Anupindi and Bassok [13] pointed out that relaxation of the risk-neutral hypothesis would enable a better understanding of the behavior of supply chain decision makers. With the market uncertainty and intense competition, the ability of enterprises to take risks was seen as progressively affecting its overall revenue. Some enterprises preferred to sacrifice some revenue to avoid risk.

Researchers studied the effect of risk attitude on the decision-making behavior of the supply chain recently. In the field of enterprise management, most decision makers tended to be risk-averse [14–16]. Therefore, most scholars researched the decision-making issues of supply chains that consisted of one or more risk-averse participants. They mainly researched the effect of risk attitude on the production decisions, price decisions, and corporation strategy of supply chain participants. In the context of supply chain management, both the utility function approach and the mean-risk model had been widely applied to risk-averse decision-making problems [17]. Eeckhoudt and Gollier [8] explored the effect of risk attitude on decision making when they decided inventory levels without knowing market demand. Xiao and Yang [18] proposed the price service competition model of two supply chains. The study found that the higher the risk sensitivity of the retailer, the lower the optimal solution of service level and retail price. Zhang and Wang [19] mainly studied the coordination problem on dual channels of retailers in the case of short life cycles. With the increase of market demand uncertainty, enterprises must consider the risk tolerance of enterprises while pursuing the maximum profit. Therefore, the attitudes of enterprises towards risk affected the final profit of enterprises.

Beijing Automotive Industry Corporation (BAIC), which makes new energy vehicles, and vehicle retailers of Sale Service Sparepart and Survey (4S) constitute a typical manufacturer–retailer supply chain. Due to the impact of COVID-19, BAIC was faced with environmental risks, which posed a great risk to the stability of the entire supply chain. BAIC formulated rapid response strategies to improve the level of risk aversion and the risk management system for minimizing the impact of the sudden epidemic on the enterprises. By providing customers with a series of service-type advertising, including a sufficient supply of spare parts and timely tracking system, the 4S automobile retailers made customers trust the brand and expanded sales. In order to celebrate the cumulative sales volume of 100,000, BAIC cooperated with 15 retailers in Beijing, and then launched a special campaign for 10,000 people to buy cars as a reward to consumers. BAIC completed a reasonable share of the local advertising expenditure by retailers. The overall benefits of the supply chain were optimized.

Complexity theory was widely used to study the stability of various systems, such as mechanics, meteorology, economics, and supply chain systems. Based on bounded rational expectations, complexity theory discussed the characteristics of the system complexity produced by enterprise decision making, and analyzed the effect of dynamic adjustment on the supply chain. Ma and Wang [20] considered two non-cooperative game models using the bounded rational expectations theory. They found that the closed-loop supply chain system would be more likely to fall into chaos as the retailer's position in the game increased. In addition, the decision-making stage of the decision-making body in the supply chain was a dynamic process. Based on this fact, many scholars introduced the nonlinear dynamics theory into the related problems of supply chain management and carried out a lot of research [21–23]. Sun and Ma [24] combined this with a pricing model of risk aversion in the supply chain and carried out the study of dynamic complexity by discussing the game process in the environment of incomplete information. Nonlinear dynamics was widely used as an effective tool theory and made a great impact in many research fields.

Combing the existing literature, this paper applies both risk aversion and cooperative advertising theory to the supply chain game model. Researchers consider a cooperative advertising supply chain as being composed of a manufacturer and a retailer which are both risk-averse participants. The research emphasis of this paper is the impact of the supply chain relevant factors, containing the adjustment speed of advertising expenditure, participation rate of local advertising expenditure by manufacturer, risk tolerance levels, and advertising expenditure effect coefficient. Based on the Stackelberg game model and system dynamic theory, this paper researches the effects of advertising investment and risk aversion on the stability of the supply chain and revenue of participants. There are three primary innovations in our paper. Firstly, researchers build a dynamical Stackelberg game model of cooperative advertising based on the supply chain with risk aversion. Secondly, effects of relevant factors such as adjustment speeds, risk tolerance levels, participation rate of manufacturer local advertising expenditure, and advertising expenditure effect coefficient are mainly analyzed on the stability and complex dynamic behavior of the supply chain. Thirdly, expected utilities of the participants are given and compared by the numerical simulation method that can provide guidance for the manufacturer and retailer to make optimal decisions.

The paper is organized as follows: Section 2 is a brief review of relevant studies. In Section 3, a Stackelberg supply chain game model is established and the equilibrium points and stability of the system are analyzed. In Section 4, the influences of relevant factors on complexity and expected utilities are given and compared by numerical simulation. In Section 5, system chaos control based on the parameter control method is investigated. Section 6 gives the conclusions of the paper.

2. Literature Review

Our research is mainly related to three streams of literature: cooperative advertising, risk aversion, and complexity in the supply chain. To highlight the contributions of the article, some representative literature examples are reviewed in these three streams.

2.1. Cooperative Advertising in Supply Chain

Cooperative advertising was extended by many authors under different settings [25,26]. Karray and Zaccour [27] demonstrated that cooperative advertising could also have harmful impacts on the retailers considering a bilateral duopoly. Xie and Wei [28] discussed coordinating advertising and pricing in a supply chain channel. Compared with the non-cooperative mode, the cooperative mode achieved better coordination by generating higher channel profits. The retailer price was lower to consumers and the advertising efforts were higher for all channel participants. Cooperative advertising was applied to the collaboration policy in supply chain management under uncertain conditions by Sarkar and Omair [29]. Through these research studies, it was concluded that the optimal decision of cooperative advertising made the revenue of the supply chain grow rapidly. Xiao and Zhou [30] established a mixed manufacturer–retailers advertising cooperation model, in which the manufacturer acted as the leading participant and played the Stackelberg game with the retailer coalition.

In recent years, more researchers have applied cooperative advertising to the supply chain and obtained a large amount of research results. Further research has shown that the investment of cooperative advertising affects the stability of the supply chain and the revenue of each participant.

2.2. Risk Aversion in Supply Chain

The crucial issue of supply chain risk management was risk measurement. The risk preference of participants had an important impact on the profit of each node enterprise and the whole supply chain. Therefore, more and more scholars began to study the influence of the risk preferences on production, price and profit, as well as on the coordination mechanism of conflicts among supply chain participants. Agrawal and Seshadri [31] demonstrated a single-cycle supply chain involving one supplier and multiple risk-averse retailers using a mean-variance approach. The research found that retailers could increase the number of orders to the expected optimal level and thus share the risk by offering reciprocal risk-sharing contracts. Xu and Dan [32] focused on the manufacturer's dual-channel pricing strategy in a risk-averse supply chain. He found the price with risk aversion was lower than that with risk-neutral in a complete information environment of the dual-channel supply chain, and then proposed a two-way coordination mechanism of a profit-sharing contract. Yan and Jin [33] concentrated on the optimal decision-making problem of a supply chain with risk aversion under the condition of demand disturbance. Zhou and Li [34] built cooperative advertising and ordering in a two-level supply chain with risk aversion that considered manufacturers selling through retailers.

Based on the analysis of the literature, supply chain management with risk aversion has become a trending issue in the fields of economics and management. Researchers pay more attention to how participants effectively avoid risks in the competition so as to obtain maximum revenue.

2.3. Complexity in Supply Chain Game

In recent years, more researchers have applied game theory to study the system of the supply chain. However, the decision-making stage of the participants in the supply chain is a dynamic process. Based on this fact, nonlinear dynamics theory was introduced into related problems about supply chain management.

Karray and Amin [35] built a cooperative advertising model in a channel with competing retailers. The study discussed the equilibrium solution of the game model under centralized and decentralized decision making, and gave the corresponding coordination mechanism. The conclusion was that cooperative advertising stimulated retailers' spending.

Based on game theory and chaotic dynamics theory, Ma and Ren [36] established a closed-loop supply chain model with dual recycling channels. The results show that a precipitous speeding of recycling price adjustment led the system into a chaotic state and caused the entropy of the system to increase. Ma and Hong [37] explored the dynamic game with risk attitudes and free ride. The pricing and service strategy were studied in a dual-channel supply chain. The study indicated that the system fell to chaos when the risk preference of the retailer was too high.

Stackelberg game theory was widely used in various fields of supply chain research as a conventional processing method of an economic model. Khademi and Ferrara [38] studied a two-level Stackelberg game model with two followers and a leader, and applied control theory and algorithm to obtain the optimal solution of the game model. This study was applicable to a long-term investment game model. Ma and Tian [39] constructed a long-term dynamic Stackelberg game model in three situations: without cap-and-trade regulation, grandfathering approach, and benchmarking approach. The work focused on the dynamic complexity study of a multi-channel supply chain which contained a manufacturer, an e-retailer, and a traditional retailer under cap-and-trade regulation.

Combing the previous literature, researchers find that most scholars established Stackelberg game models based on single risk aversion or cooperative advertising. Few scholars paid attention to research on the dynamic behavior of the Stackelberg game model in a cooperative advertising supply chain considering risk aversion. The influences of relevant factors on complexity and expected utilities of the supply chain were not studied and compared by numerical simulation.

3. Supply Chain Model

3.1. Model Description

In this research, researchers consider a supply chain which consists of a manufacturer and a retailer as shown in Figure 1. The manufacturer produces products and sells them to customers through the retailer. To improve sales volume, the manufacturer usually invests in national advertising and the retailer invests in local advertising. Considering the manufacturer as the leader deciding on national advertising investment and the retailer as the follower deciding on local advertising investment, researchers establish a Stackelberg game model of a manufacturer-led supply chain with risk-averse behavior.

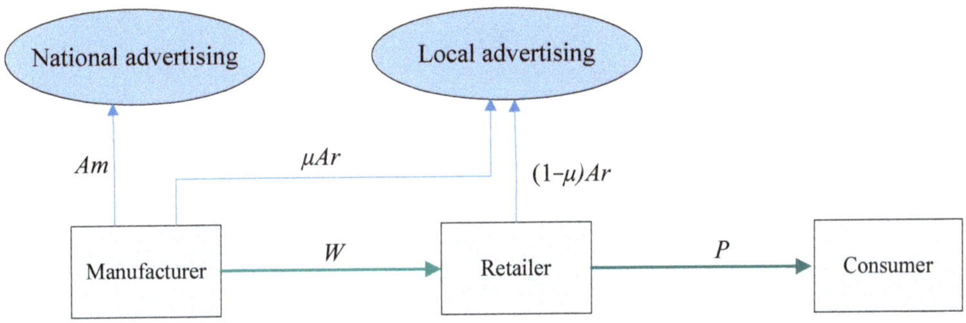

Figure 1. The model framework.

Notations used throughout this paper are listed in Table 1.

Table 1. Notations.

Notation	Explanation
p	Price of retailer, $p > w > 0$
v	Baseline demand, $v > 0$
A_m	National advertising expenditure of manufacturer, $A_m \geq 0$
A_r	Local advertising expenditure of retailer, $A_r \geq 0$
k_m	Effect coefficient of national advertising expenditure, $k_m > 0$
k_r	Effect coefficient of local advertising expenditure, $k_r > 0$
w	Wholesale price (manufacturer's price to the retailers)
μ	Manufacturer participation rate in percentage, $0 \leq \mu \leq 1$
α_1	Adjustment speed of manufacturer advertising expenditure, $\alpha_1 > 0$
α_2	Adjustment speed of retailer advertising expenditure, $\alpha_2 > 0$
λ_1	Risk tolerance level of manufacturer, $\lambda_1 > 0$
λ_2	Risk tolerance level of retailer, $\lambda_2 > 0$

3.2. Model Assumption

The following assumptions are used for the game model of a manufacturer–retailer supply chain:

(1) The unit production cost of the manufacturer and unit handling cost of the retailer incurred in addition to purchasing cost are assumed to be equal to zero.

(2) Cooperative advertising is divided into the manufacturer's national advertising expenditure A_m and the retailer's local advertising Due to different formula editors, some formula formats are not uniform. We will adjust the format of this part in later stage of production. A_r, which are decision variables for the supply chain participants.

(3) In this model, the manufacturer and retailer are risk-averse, which revenues are represented by exponential utility functions and follow a normal distribution.

3.3. Model Construction

The consumer demand function depends on the retail price p and the advertising expenditures A_m and A_r, which is linear in prices and concave in advertising expenditure to account for decreasing marginal returns. This specification was widely used in previous research. The demand decreases in price and increases with advertising expenditures of the supply chain participants. The consumer demand function can be expressed as follows [35,40]:

$$V(p, A_m, A_r) = v - p + k_m\sqrt{A_m} + k_r\sqrt{A_r} \tag{1}$$

The effect coefficients of advertising expenditure are captured by the positive parameters k_m and k_r. Parameter μ is the participation rate by the manufacturer, the percentage that the manufacturer agrees to pay the retailer to subsidize the local advertising expenditure. The profits of the manufacturer, the retailer, and the system are as follows, respectively [28]:

$$\Pi_m = w(v - p + k_m\sqrt{A_m} + k_r\sqrt{A_r}) - \mu A_r - A_m \tag{2}$$

$$\Pi_r = (p - w)(v - p + k_m\sqrt{A_m} + k_r\sqrt{A_r}) - (1 - \mu)A_r \tag{3}$$

$$\Pi_{m+r} = p(v - p + k_m\sqrt{A_m} + k_r\sqrt{A_r}) - A_r - A_m \tag{4}$$

Parameter v is a random variable, as follows $v = \overline{v} + \varepsilon$. Here, $\overline{v}$ is the mean of the baseline demand and ε follows a normal distribution, such that $E(\varepsilon) = 0$, $Var(\varepsilon) = \sigma^2$, and $E(v) = E(\overline{v})$. The exponential utility function is one form of the utility function, which dominates in both theoretical and applied study in the field of managerial decision making. Here, the manufacturer and retailer are risk-averse participants. The profit Π follows a normal distribution. Expected utility can been expressed by the following equation [37,41–43]:

$$E(U) = E(\Pi) - \lambda\sqrt{Var(\Pi)} \tag{5}$$

This paper investigates the risk attitudes of the manufacturer and retailer, mainly referring to Xie and Yue [41] and Choi and Li [44]. Parameter λ is the risk tolerance level. The value of $\lambda > 0$ implies risk-averse behavior. When the value of λ is higher, the risk-averse degree is greater. When λ approaches 0, risk-neutral behavior is implied. Researchers can obtain the expected value $E(\Pi)$ and the variance value $Var(\Pi)$ of the manufacturer and retailer as follows:

$$E(\Pi_m) = w(v - p + k_m\sqrt{A_m} + k_r\sqrt{A_r}) - \mu A_r - A_m \tag{6}$$

$$Var(\Pi_m) = w^2\sigma^2 \tag{7}$$

$$E(\Pi_r) = (p - w)(v - p + k_m\sqrt{A_m} + k_r\sqrt{A_r}) - (1 - \mu)A_r \tag{8}$$

$$Var(\Pi_r) = (p - w)^2\sigma^2 \tag{9}$$

According to the Formulas (5)–(9), the expected utility function of the manufacturer and retailer can be expressed by the following equation:

$$E(U_m) = w(v - p + k_m\sqrt{A_m} + k_r\sqrt{A_r}) - \mu A_r - A_m - \lambda_1 w\sigma \tag{10}$$

$$E(U_r) = (p - w)(v - p + k_m\sqrt{A_m} + k_r\sqrt{A_r}) - (1 - \mu)A_r - \lambda_2(p - w)\sigma \tag{11}$$

From Formulas (10) and (11), researchers can obtain the revenue functions of the manufacturer and retailer which are more in accordance with the actual situation using the utility function. The prospects of the manufacturer and the retailer are to maximize respective expected utilities.

The manufacturer firstly determines advertising expenditure A_m and the wholesale price w of products. Based on the manufacturer's decision information, the retailer determines advertising expenditure A_r and the retail price p of products. To solve the Stackelberg equilibrium, first consider the retailer's optimal decisions [45].

The solutions can be solved by first order equations $\partial E(U_r)/\partial A_r = 0$ and $\partial E(U_r)/\partial p = 0$.

$$\begin{cases} \frac{\partial E(U_r)}{\partial A_r} = \frac{(p-w)k_r}{2\sqrt{A_r}} - 1 + \mu = 0 \\ \frac{\partial E(U_r)}{\partial p} = v - 2p + k_m\sqrt{A_m} + k_r\sqrt{A_r} + w - \lambda_2\sigma = 0 \end{cases} \tag{12}$$

Obtain the optimal values A_r^* and p^*:

$$A_r^* = \frac{{k_r}^2(-v - k_m\sqrt{A_m} + w + \lambda_2\sigma)^2}{(-4 + 4\mu + {k_r}^2)^2} \tag{13}$$

$$p^* = \frac{1}{2}\left(v + w + k_m\sqrt{A_m} - \lambda_2\sigma + k_r\sqrt{\frac{{k_r}^2(-v - k_m\sqrt{A_m} + w + \lambda_2\sigma)^2}{({k_r}^2 + 4\mu - 4)^2}}\right) \tag{14}$$

Then, the expected utility function of the manufacturer $E(U_m)$ has the form

$$E(U_m) = w(v - p^* + k_m\sqrt{A_m} + k_r\sqrt{{A_r}^*}) - \mu {A_r}^* - A_m - \lambda_1 w\sigma \tag{15}$$

Researchers can obtain $\partial E(U_m)/\partial A_m$, where $F = (-4 + 4u + {k_r}^2)^2$ and $G = -v + w + \lambda_2\sigma$.

$$\frac{\partial E(U_m)}{\partial A_m} = \frac{wk_m}{4\sqrt{A_m}}\left(1 - \frac{{k_r}^3(G - k_m\sqrt{A_m})}{F\sqrt{\frac{{k_r}^2(G-k_m\sqrt{A_m})^2}{F}}}\right) + \frac{u{k_r}^2k_m(G - k_m\sqrt{A_m})}{F\sqrt{A_m}} - 1 \tag{16}$$

Then, $\partial E(U_r)/\partial A_r$ can be obtained by Formulas (12) and (14):

$$\frac{\partial E(U_r)}{\partial A_r} = \left(k_m\sqrt{A_m} + k_r\sqrt{\frac{{k_r}^2(G - k_m\sqrt{A_m})^2}{F}} - G\right)\frac{k_r}{4\sqrt{A_r}} - 1 + u \tag{17}$$

In this paper, researchers develop a dynamic Stackelberg game model. Researchers assume that they adjust their advertising expenditures adaptively based on limited information and each enterprise can obtain its marginal utility information by means of market experiments [45]. In fact, when making advertising expenditure decisions, the manufacturer cannot obtain the whole market information, as they are restricted by the objective conditions such as the decision-making ability. Suppose that the decision in next period can be adjusted with bounded rationality based on partial estimation of marginal utility of the current period [46]. In each period, adjustments occur as follows: Firstly, manufacturer estimates marginal utility with respect to national advertising expenditure and makes decisions based on the bounded rational expectation. After the manufacturer carries out this decision, the retailer estimates the marginal utility with respect to local advertising expenditure according to the advertising expenditure of the manufacturer [45]. The model can be constructed as follows:

$$\begin{cases} A_m(t+1) = A_m(t) + \alpha_1 A_m(t)\left\{\frac{wk_m}{4\sqrt{A_m(t)}}\left(1 - \frac{{k_r}^3(G-k_m\sqrt{A_m(t)})}{F\sqrt{\frac{{k_r}^2(G-k_m\sqrt{A_m(t)})^2}{F}}}\right) + \frac{u{k_r}^2k_m(G-k_m\sqrt{A_m(t)})}{F\sqrt{A_m(t)}} - 1\right\} \\ A_r(t+1) = A_r(t) + \alpha_2 A_r(t)\left\{\left(k_m\sqrt{A_m(t+1)} + k_r\sqrt{\frac{{k_r}^2(G-k_m\sqrt{A_m(t+1)})^2}{F}} - G\right)\frac{k_r}{4\sqrt{A_r(t)}} - 1 + u\right\} \end{cases} \tag{18}$$

Here, α_1 and α_2 represent the advertising expenditure adjustment speed of the manufacturer and retailer, respectively.

3.4. The Stability of the System

In order to intuitively describe the dynamic competition process and complex system dynamics of enterprises, the numerical simulation is carried out in this section to investigate the influence of relevant parameters on system stability. Referring to related literature [5,41,47], the researchers used similar parameter settings for the numerical simulation. Based on satisfying constraints, we consider the assignment of relevant parameters.

The baseline demand should be large enough, so $v = 400$ was assigned to be larger than other parameters. The researchers set $\mu = 0.3$ as μ should be within $[0,1]$. Other parameters were assigned as $w = 200$, $k_r = 0.2$, $k_m = 0.4$, $\lambda_1 = 3$, $\lambda_2 = 4$, and $\sigma = 8$.

In system (18), let $A_m(t+1) = A_m(t)$ and $A_r(t+1) = A_r(t)$. Then, researchers can obtain four groups of system equilibrium solutions as follows:

$$E_0 = (0,0), E_1 = (407.2,0), E_2 = (0,148.2), E_3 = (407.2,162.8). \tag{19}$$

Obviously, E_0, E_1, and E_2 are partly or entirely zero. They are boundary unstable equilibrium points. E_3 is the unique non-negative equilibrium point of system (18). It is very significant to analyze the unique positive equilibrium point for participants. To study the local stability of equilibrium points, the Jacobian matrix at E_3 has the form

$$J(E_3) = \begin{bmatrix} 1 - 0.5001\alpha_1 & 0 \\ 0.6414\mathrm{e}^{-2}\alpha_2 & 1 - 0.3500\alpha_2 \end{bmatrix} \tag{20}$$

Its characteristic equation is $F(\lambda) = \lambda^2 - Tr\lambda + Det = 0$, where Tr is the trace and Det is the determinant of matrix $J(E_3)$, which are given by

$$Det = (1 - 0.5001\alpha_1)(1 - 0.3500\,\alpha_2) \tag{21}$$

$$Tr = 2 - 0.5001\alpha_1 - 0.3500\alpha_2 \tag{22}$$

Researchers can obtain $Tr^2 - 4Det > 0$. Thus, the eigenvalues of the Nash equilibrium are real. According to the Jury condition, the necessary and sufficient condition for the local stability of the Nash equilibrium can be obtained:

$$\begin{cases} 1 - Tr + Det > 0 \\ 1 + Tr + Det > 0 \\ 1 - Det > 0 \end{cases} \tag{23}$$

The long-term strategies of the manufacturer and retailer can remain stable when constraint (24) is satisfied. The above conditions are defined as the stability region of the adjustment parameters (α_1,α_2).

To study the influence of key parameters on the stability region of system (18), researchers examined a two-dimensional stable region with adjustment parameters (α_1,α_2) to examine the changing trend of the stable region when other parameters were fixed. If the values of (α_1,α_2) are in the stable region, the system will be stable at the Nash equilibrium $E_3^*(407.2, 162.8)$ after a number of games. The two-dimensional stable region of the system is presented in Figure 2 according to Equations (22)–(24).

The blue areas indicate the stable region of the system and show the interaction effects between the adjustment parameters α_1 and α_2. Obviously, the stable region $\alpha_2 \in [0, 5.71]$ is larger than $\alpha_1 \in [0, 3.99]$. From the range of the stability region, researchers can see that the stability of the system is more sensitive to the manufacturer. If α_1 is relatively lower, even if α_2 is relatively higher, the system is stable, which also will be proved by the following figures. If the enterprises keep adjustment speeds (α_1,α_2) within the stability region, no matter what initial advertising expenditures are chosen by the enterprises, the market will eventually reach the Nash equilibrium after a series of games. As a result, the enterprises can obtain optimal strategies. Otherwise, complex phenomena such as bifurcation and chaos will appear.

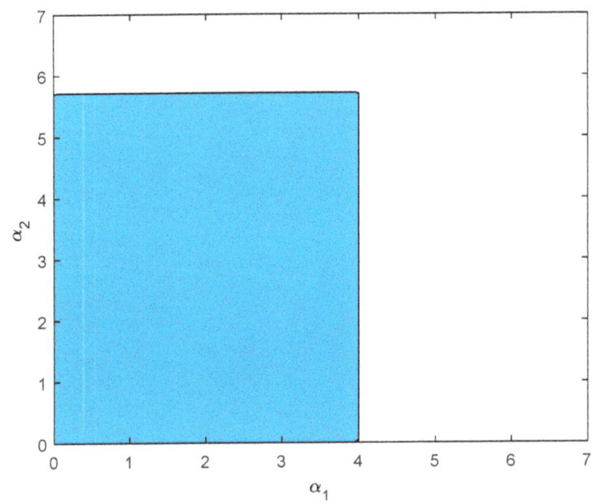

Figure 2. The stable region of Nash Equilibrium in (α_1, α_2) plane.

The economic meaning of the stable region is that whatever initial values are chosen by the manufacturer and retailer in the local stable region, they will eventually achieve Nash equilibrium $E_3^*(407.2, 162.8)$ after finite games. Stability is beneficial for enterprises to make long-term strategies. Therefore, enterprises pay more attention to the stable region. The enterprises can forecast the development of the market but have to change their advertising expenditures frequently [48]. If the parameters exceed the stable region, the system will eventually be out of control and fall into chaos.

Based on the above reasons, researchers investigated the dynamics of system (18) and the impacts of relevant factors by means of dynamics analysis and numerical simulation, as described in the following sections. The complex dynamical behaviors of the system including the stable region, bifurcation, and chaos will be studied in detail.

4. Dynamics Analysis and Numerical Simulation

4.1. The Dynamics with Respect to Adjustment Speeds α_1 and α_2

Dynamic characteristics can be expressed in the 2D bifurcation diagram [39]. In Figure 3, the light blue area stands for the stable region, while the bright blue, the dark blue, and the deep blue stand for the period-2, period-4, and period-8 stable regions. With the parameters α_1 and α_2 increasing continuously, the system will go into a chaos state. The chaos and divergence regions are shown by the gray and the white areas, respectively, which indicates that one enterprise will quit out of the market in economics. Figure 3 shows that α_2 has better stability than α_1, illustrated by its larger stable region.

Figure 4a,b show the bifurcation diagram of A_m and A_r with respect to α_1. The corresponding largest Lyapunov exponent (LLE) is shown by Figure 4c,d, which is usually seen as characteristic value of chaos. It is shown that when the parameter $\alpha_1 < 3.99$, A_m and A_r become asymptotically stable with $\alpha_2 = 3.5$ and $\alpha_2 = 6$. Researchers can observe that the stable region of A_m is not changed with parameter α_2. For $\alpha_1 = 3.99$, the system goes into period-2 bifurcation, which is improved by LLE from another aspect. After that, the system enters the 2-period state ($3.99 < \alpha_1 < 4.64$), which also can be depicted by its attractor in Figure 5a,b. A further increase of α_1 will lead to the appearance of the period-4 state ($4.64 < \alpha_1 < 4.75$) and chaos ($\alpha_1 > 4.75$). Figure 5c,d show chaotic attractors with $\alpha_1 = 4.95$, which is consistent with Figure 4. Chaos attractors represent the disordered fluctuation of the system.

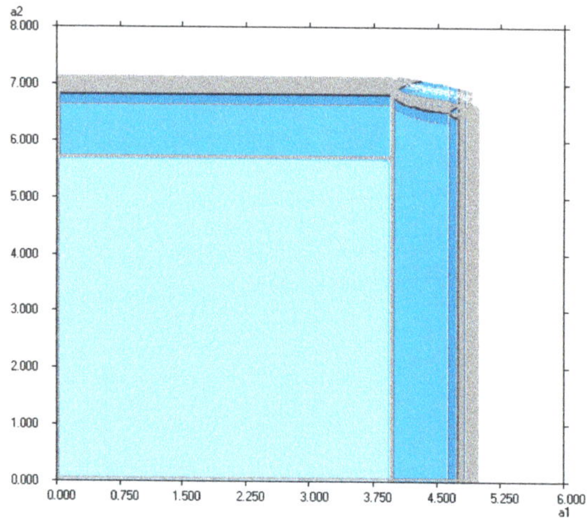

Figure 3. Two-dimensional bifurcation diagram with respect to α_1 and α_2.

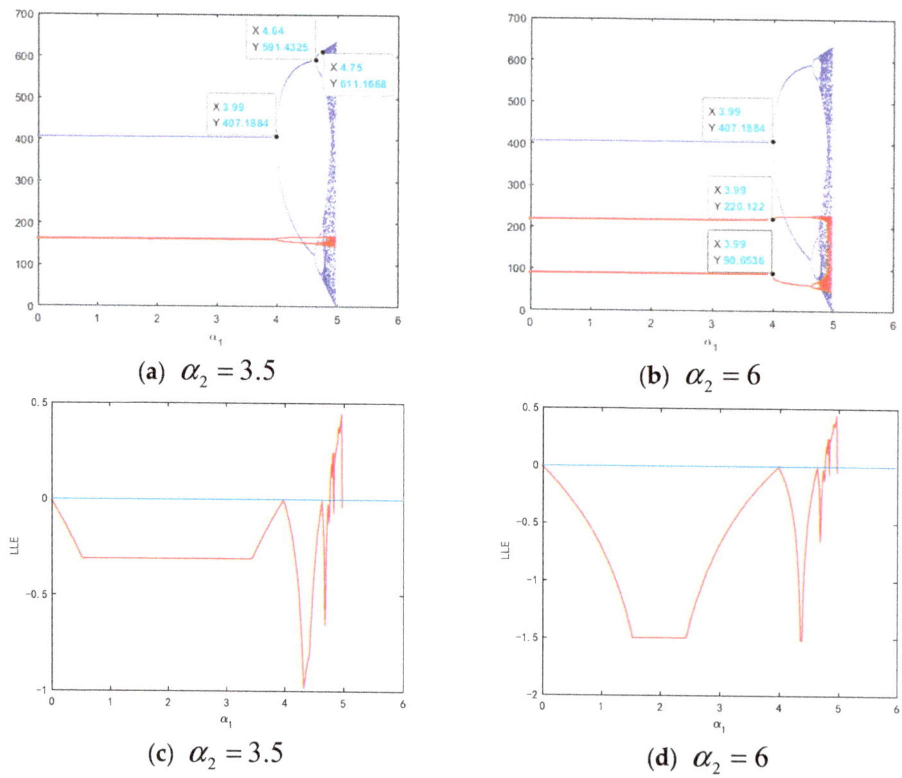

Figure 4. The Bifurcation diagram and largest Lyapunov exponent with respect to α_1.

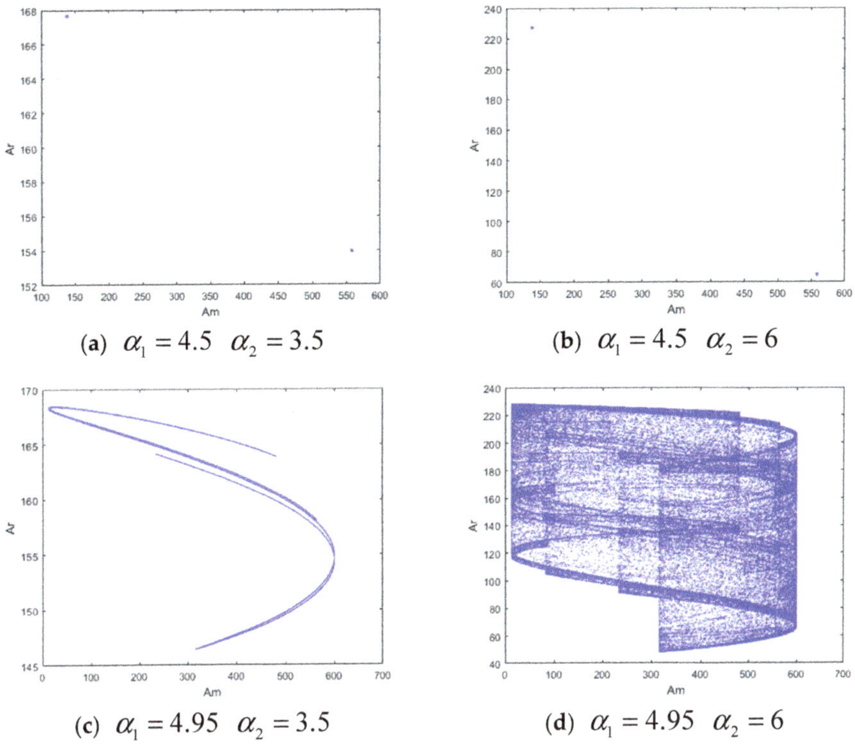

(**a**) $\alpha_1 = 4.5 \quad \alpha_2 = 3.5$ (**b**) $\alpha_1 = 4.5 \quad \alpha_2 = 6$

(**c**) $\alpha_1 = 4.95 \quad \alpha_2 = 3.5$ (**d**) $\alpha_1 = 4.95 \quad \alpha_2 = 6$

Figure 5. The attractors of system with different α_1 and α_2.

The corresponding Double Largest Lyapunov Exponents (DLLE) with respect to α_1 and α_2 are shown in Figure 6a,b. The blue areas are LLE < 0 with a color transition from dark blue to light blue. When the LLE tends to 0, the color becomes light khaki and the system is in period-2 and period-4 stable cycles, etc. The dark khaki areas with the LLE > 0 means the system is in chaos and divergence. Researchers can see the trend of system state from order to disorder by the perspective of LLE.

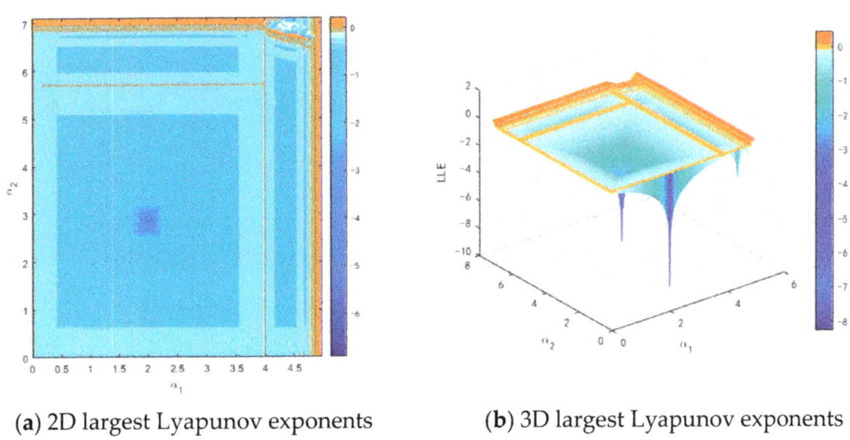

(**a**) 2D largest Lyapunov exponents (**b**) 3D largest Lyapunov exponents

Figure 6. The largest Lyapunov exponents with respect to α_1 and α_2.

In the market, the objectives of the manufacturer and the retailer are to maximize respective expected utility. The effects of the parameters α_1 and α_2 on expected utilities and average expected utilities are shown in Figure 7. According to Figure 7a,b, when α_1 is in the stable region, expected utilities of the manufacturer and the retailer are stable at $(E(U_m), E(U_r)) = (19006, 7863)$. With α_1 increasing, the system enters period-doubling bifurcation and a chaotic state. When $E(U_m)$ and $E(U_r)$ go into chaos, the oscillation of expected utilities intensifies, which has similar bifurcation properties to A_m and A_r.

In Figure 7c,d, it can be found that the average expected utilities of the manufacturer and the retailer are in a downward state with the instability of the system. When the system loses the stability, the average expected utility will decrease, which is not in accord with the revenue of enterprises. It is favorable for participants to stabilize the system to the equilibrium point. Hence, the manufacturer and retailer will control their adjustment speed to maintain the stability of the market in order to obtain the maximum revenue.

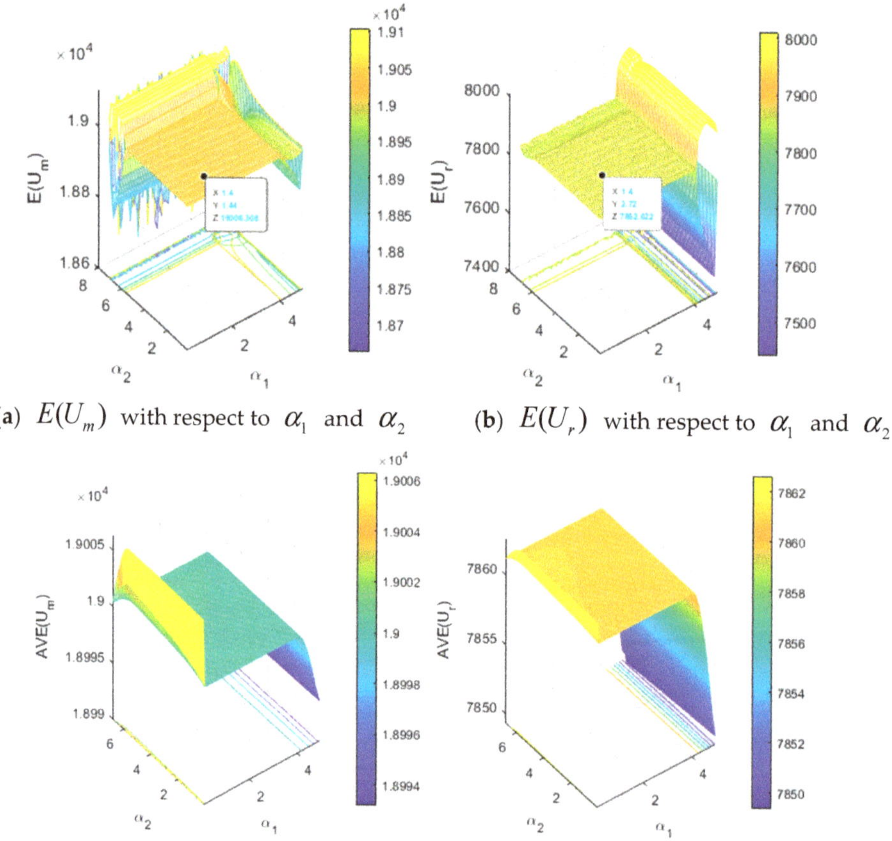

(a) $E(U_m)$ with respect to α_1 and α_2 **(b)** $E(U_r)$ with respect to α_1 and α_2

(c) $AVE(U_m)$ with respect to α_1 and α_2 **(d)** $AVE(U_r)$ with respect to α_1 and α_2

Figure 7. The expected utilities and the average expected utilities with respect to α_1 and α_2.

4.2. The Influence of the Manufacturer Participation Rate μ

In this section, the influence of parameter μ on system stability and dynamic behavior will be studied.

Figure 8 clearly indicates the 2D bifurcation diagrams of the system with $\mu = 0.2, 0.4, 0.6, 0.8$, respectively. From the 2D bifurcation diagrams, researchers can clearly know that different bifurcation curves correspond to different values of parameter μ. Researchers can find that μ rarely has influence on the stable region of α_1. However, μ has a greater influence on the stable region of α_2, which becomes larger with a greater μ. This indicates that the participation rate of local advertising by the manufacturer has a positive impact on the system's stability, that is, the higher the degree of the manufacturer's participation rate, the harder for the Nash equilibrium point to lose its stability.

Figure 9 shows the bifurcation diagram of A_m and A_r with respect to μ when the system is stable, during period-2 bifurcation, period-4 bifurcation, and chaos, respectively, with α_1 and α_2 taking different values. Researchers can see that A_m decreases slightly, but A_r increases dramatically as μ increases. Therefore, the greater the participation rate in local advertising by the manufacturer, the more the retailer spends on the local advertising expenditure. However, the manufacturer's national advertising expenditure A_m only has a slight downward adjustment with the increase of participation rate μ.

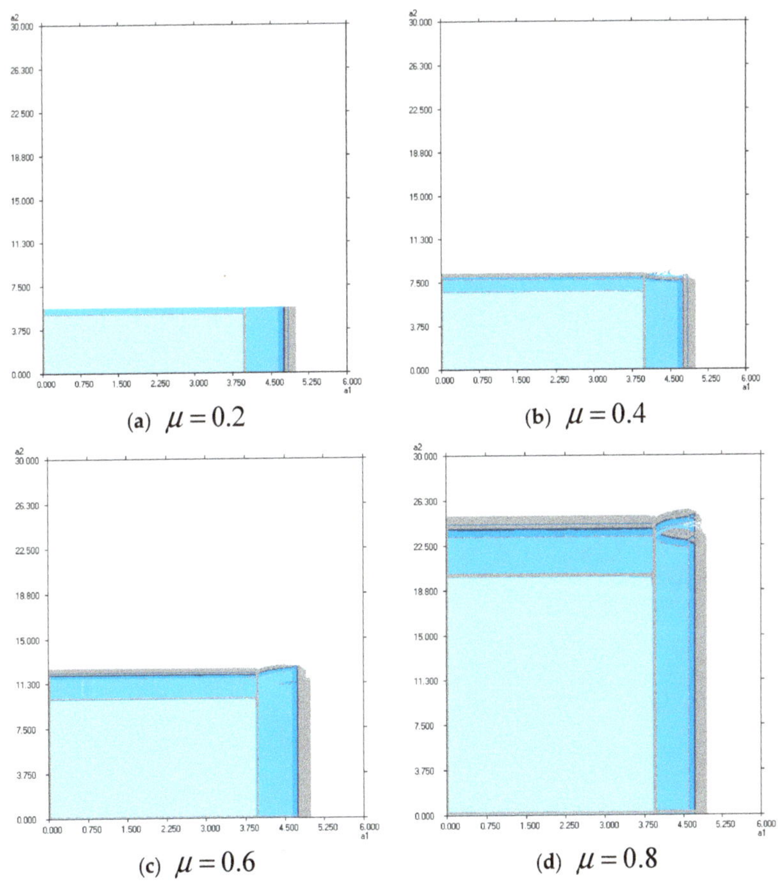

Figure 8. The 2D bifurcation diagrams with respect to α_1 and α_2 with different μ.

The expected utilities of manufacturer and retailer are shown in Figure 10. With the increase in participation rate μ and the change of system state, the manufacturer's expected utility decreases, while the retailer's expected utility increases slightly on the whole. Therefore, in real market competition, the manufacturer tries to avoid taking too large a value of the participation rate for obtaining the maximum expected utility. Although the retailer will obtain more expected utility by increasing μ, it will also increase local advertising expenditure rapidly. In order to maintain market stability and maximize revenue for both participants, as the dominant enterprise in the market, the manufacturer will adjust the participation rate μ appropriately, neither too high nor too low.

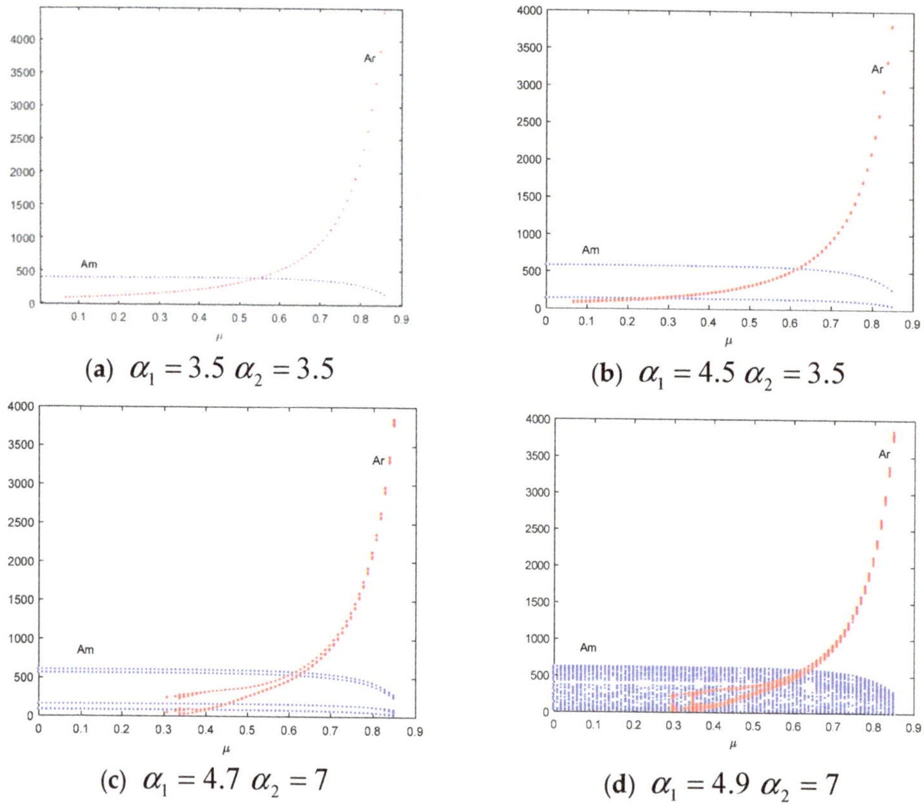

Figure 9. The Bifurcation diagram of A_m and A_r with respect to μ with different α_1 and α_2.

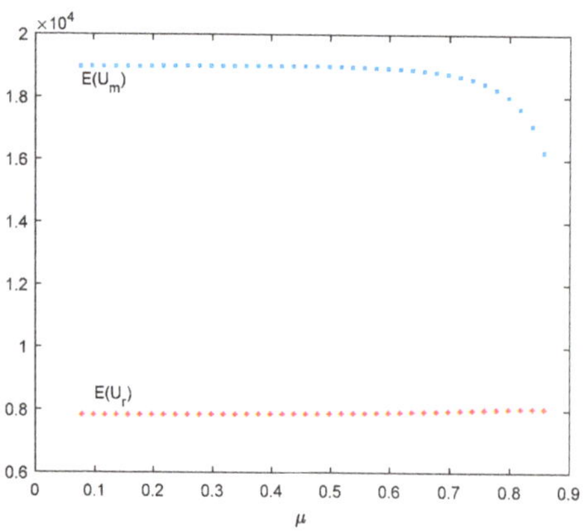

Figure 10. The expected utilities with respect to μ with $\alpha_1 = 3.5$ and $\alpha_2 = 3.5$.

4.3. The Influence of Risk Tolerance Levels λ_1 and λ_2

In this section, the influence of risk tolerance levels λ_1 and λ_2 is to be analyzed with respect to the Nash equilibrium point and expected utilities.

In order to analyze the effects of risk tolerance levels on the stable region, the corresponding 2D bifurcation diagrams are shown in Figure 11. Researchers chose (λ_1, λ_2) as $(2,3)$, $(3,3)$, $(4,4)$, and $(4,5)$, respectively, and other parameters remained unchanged as in system (18). Researchers can find that the stable region has no variation with risk tolerance levels λ_1 and λ_2.

Figure 12a,b are the 3D bifurcation diagrams with respect to the parameters α_1, λ_1, A_m and α_1, λ_1, A_r, respectively. Figure 12c,d are the 3D bifurcation diagrams with the parameters α_1, λ_2, A_m and α_1, λ_2, A_r, respectively. Researchers can find that the Nash equilibrium (A_m, A_r) of the system does not change with risk tolerance levels λ_1. Figure 12c,d show the change of Nash equilibrium A_m is very slight, whereas Nash equilibrium A_r is obvious with risk tolerance levels λ_2.

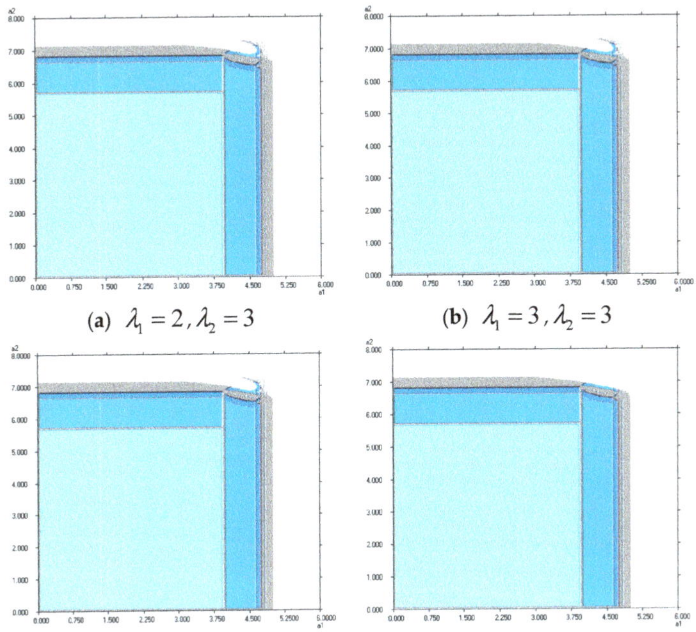

(a) $\lambda_1 = 2, \lambda_2 = 3$ (b) $\lambda_1 = 3, \lambda_2 = 3$ (c) $\lambda_1 = 4, \lambda_2 = 4$ (d) $\lambda_1 = 4, \lambda_2 = 5$

Figure 11. The 2D bifurcation diagram with respect to α_1 and α_2 with different λ_1 and λ_2.

Figure 13 shows the Nash equilibrium points of A_m and A_r with respect to λ_1 and λ_2. Figure 13a,b indicate the influence of risk tolerance level λ_1 and λ_2 on advertising expenditures A_m and A_r, respectively, when the system is in a stable state. The expected utilities of the manufacturer and retailer with respect to λ_1 and λ_2 are described by Figure 13c,d. In order to compare the Nash equilibrium (A_m, A_r) and expected utilities $(E(U_m), E(U_r))$ with different values of λ_1 and λ_2 more clearly, Table 2 has been created with respect to λ_1 and λ_2 with $\alpha_1 = 3.5$ and $\alpha_2 = 3.5$. Researchers can conclude that

(1) The Nash equilibrium A_m rises slightly and A_r reduces with increasing risk tolerance level λ_2, while A_m and A_r remain unchanged with an increase in risk tolerance level λ_1, which corresponds to Figure 13a,b. Therefore, the retailer will try to increase their own risk tolerance level for less local advertising expenditure.

(2) The expected utility $E(U_m)$ decreases and $E(U_r)$ remains unchanged with an increase in risk tolerance level λ_1. The expected utility $E(U_m)$ increases and $E(U_r)$ decreases

with an increase in risk tolerance level λ_2. These correspond to Figure 13c,d. Therefore, in order to obtain more expected utilities, the manufacturer and retailer will try to reduce their own risk tolerance levels. Meanwhile, the manufacturer wants to increase the risk tolerance level of retailer for its own revenue.

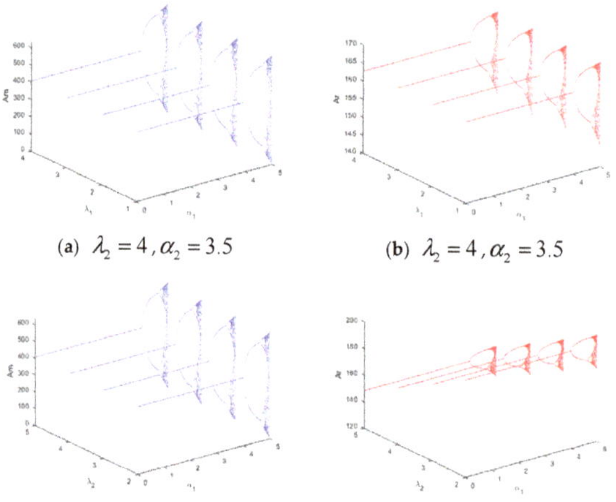

(a) $\lambda_2 = 4, \alpha_2 = 3.5$ (b) $\lambda_2 = 4, \alpha_2 = 3.5$

(c) $\lambda_1 = 3, \alpha_2 = 3.5$ (d) $\lambda_1 = 3, \alpha_2 = 3.5$

Figure 12. The 3D bifurcation diagrams of A_m and A_r with respect to α_1 with (**a**,**b**) $\lambda_1 = 1, 2, 3, 4$ and (**c**,**d**) $\lambda_2 = 2, 3, 4, 5$.

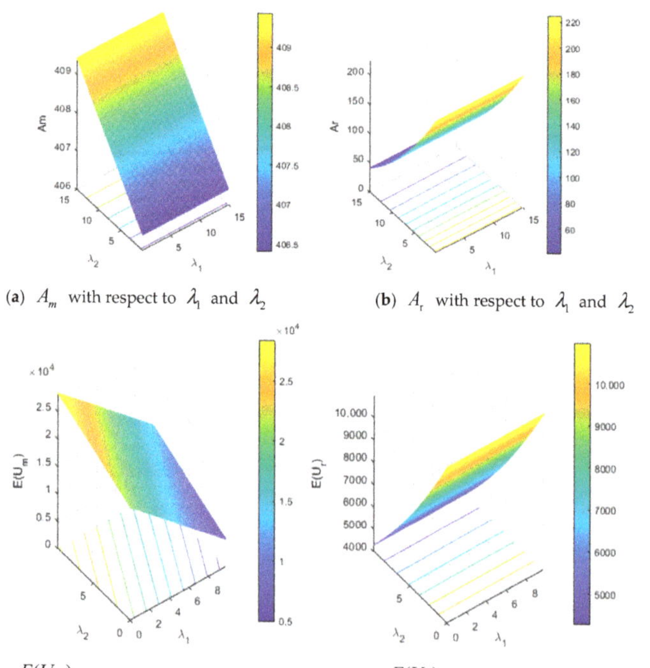

(a) A_m with respect to λ_1 and λ_2 (b) A_r with respect to λ_1 and λ_2

(c) $E(U_m)$ with respect to λ_1 and λ_2 (d) $E(U_r)$ with respect to λ_1 and λ_2

Figure 13. Nash equilibrium points and the expected utilities with $\alpha_1 = 3.5$ $\alpha_2 = 3.5$.

Table 2. Nash equilibrium points (A_m, A_r) and expected utilities $(E(U_m), E(U_r))$ with respect to λ_1 and λ_2 with $\alpha_1 = 3.5$ and $\alpha_2 = 3.5$.

(A_m, A_r) $(E(U_m), E(U_r))$	$\lambda_1 = 2$	$\lambda_1 = 3$	$\lambda_1 = 4$	$\lambda_1 = 5$	$\lambda_1 = 6$
$\lambda_2 = 2$	(406.8, 193.7) (19,020, 9356)	(406.8, 193.7) (17,420, 9356)	(406.8, 193.7) (15,820, 9356)	(406.8, 193.7) (14,220, 9356)	(406.8, 193.7) (12,620, 9356)
$\lambda_2 = 3$	(407, 177.9) (19,810, 8593)	(407, 177.9) (18,210, 8593)	(407, 177.9) (16,610, 8593)	(407, 177.9) (15,010, 8593)	(407, 177.9) (13,410, 8593)
$\lambda_2 = 4$	(407.2, 162.8) (20,610, 7863)	(407.2, 162.8) (19,010, 7863)	(407.2, 162.8) (17,401, 7863)	(407.2, 162.8) (15,810, 7863)	(407.2, 162.8) (14,210, 7863)
$\lambda_2 = 5$	(407.4, 148.3) (21,400, 7165)	(407.4, 148.3) (19,800, 7165)	(407.4, 148.3) (18,200, 7165)	(407.4, 148.3) (16,600, 7165)	(407.4, 148.3) (15,000, 7165)
$\lambda_2 = 6$	(407.6, 134.6) (22,190, 6499)	(407.6, 134.6) (20,590, 6499)	(407.6, 134.6) (18,990, 6499)	(407.6, 134.6) (17,390, 6499)	(407.6, 134.6) (15,790, 6499)

4.4. *The Influence of Advertising Expenditure Effect Coefficients k_m and k_r*

In this section, the influence of effect coefficients k_m and k_r is to be analyzed with respect to the Nash equilibrium point and expected utilities.

The 2D bifurcation diagrams are shown in Figure 14. Researchers can find that the stable region has no variation when effect coefficients (k_m, k_r) are $(0.3, 0.3)$, $(0.5, 0.3)$, $(0.7, 0.5)$, and $(0.7, 0.7)$, respectively. No matter how k_m and k_r change, the system stability remains fixed with $\alpha_1 \in [0, 3.99]$ and $\alpha_2 \in [0, 5.71]$.

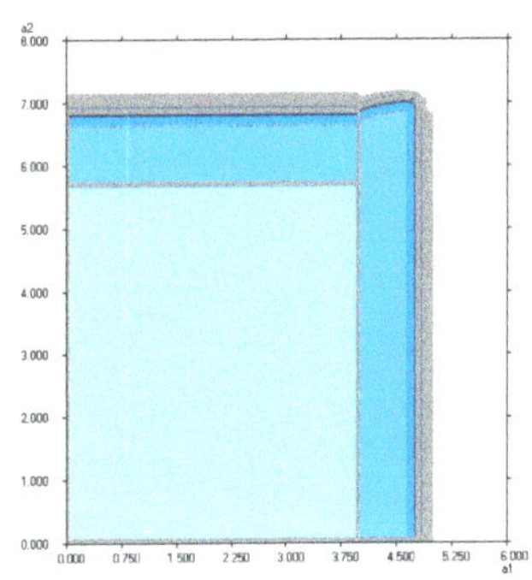

(**a**) $k_m = 0.3, k_r = 0.3$

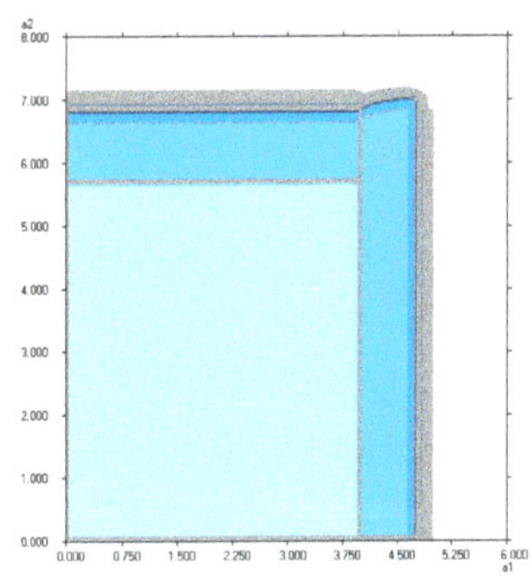

(**b**) $k_m = 0.5, k_r = 0.3$

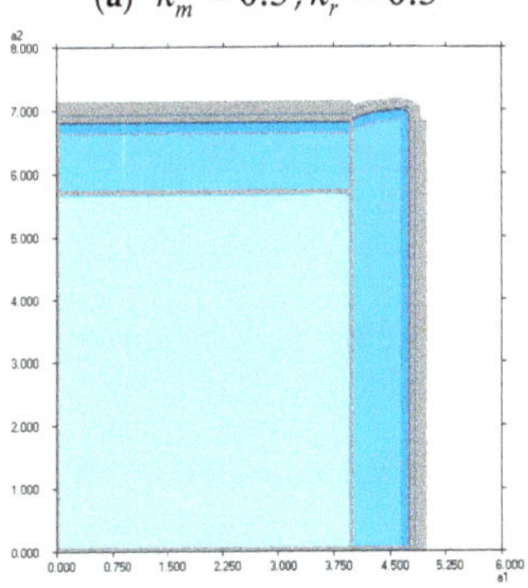

(**c**) $k_m = 0.7, k_r = 0.5$

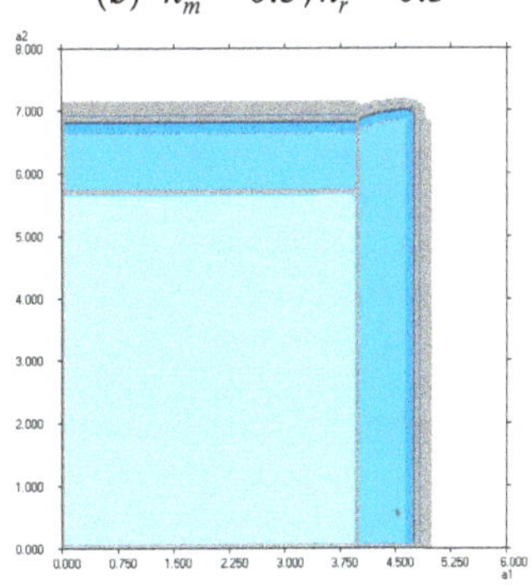

(**d**) $k_m = 0.7, k_r = 0.7$

Figure 14. The 2D bifurcation diagram with respect to α_1 and α_2 with different k_m and k_r.

Figure 15a,b are the 3D bifurcation diagrams with respect to the parameters α_1, k_m, A_m and α_1, k_m, A_r, respectively. Figure 15c,d are the 3D bifurcation diagrams with the parameters α_1, k_r, A_m and α_1, k_r, A_r, respectively. Researchers can find that the Nash equilibrium (A_m, A_r) is changed with effect coefficients k_m and k_r. As shown in Figure 15a,b, the values of the Nash equilibrium (A_m, A_r) increase with the increase in k_m. As shown in Figure 15c,d, the increase in k_r can increase the values of the Nash equilibrium (A_m, A_r). Meanwhile, the bifurcations and the chaos of the system are indicated clearly.

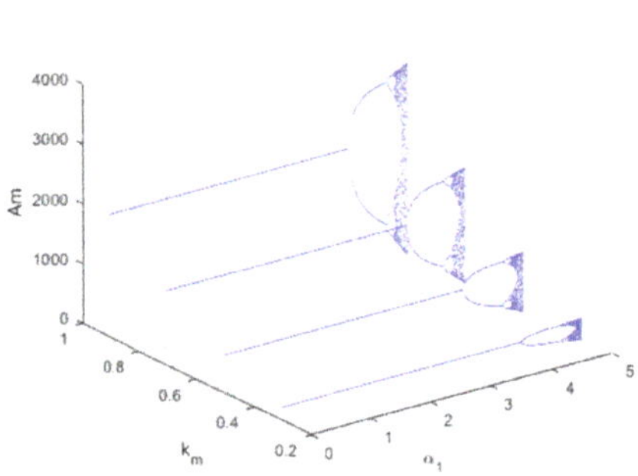

(**a**) $k_r = 0.2$, $k_m = 0.3, 0.5, 0.7, 0.9$

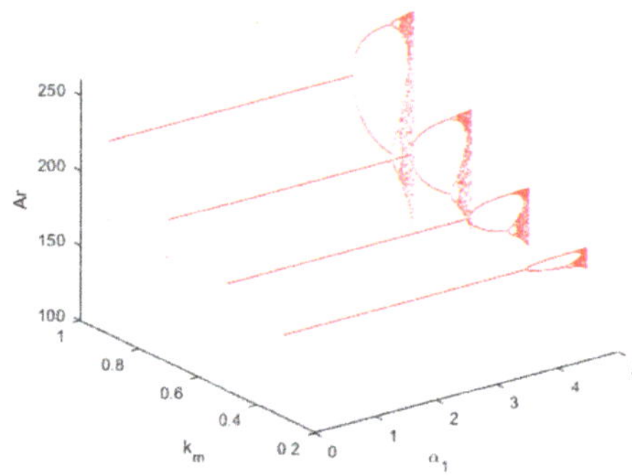

(**b**) $k_r = 0.2$, $k_m = 0.3, 0.5, 0.7, 0.9$

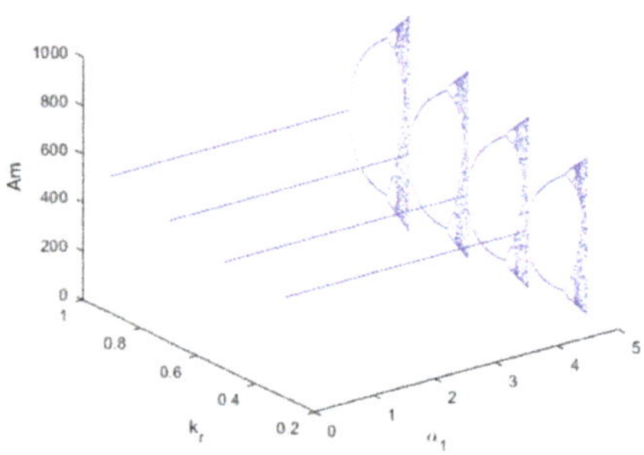

(**c**) $k_m = 0.4$, $k_r = 0.3, 0.5, 0.7, 0.9$

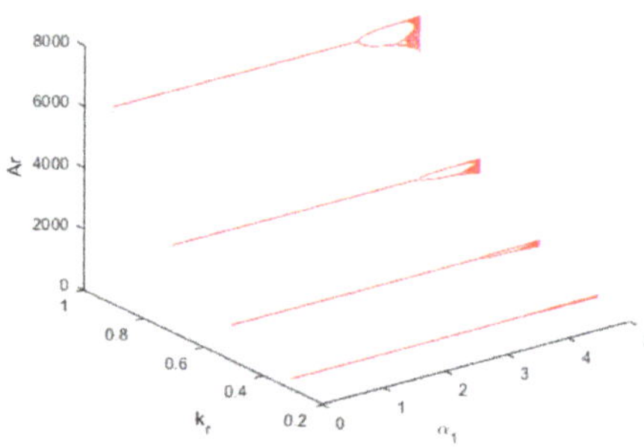

(**d**) $k_m = 0.4$, $k_r = 0.3, 0.5, 0.7, 0.9$

Figure 15. The 3D bifurcation diagrams of A_m and A_r with respect to α_1 with $\alpha_2 = 3.5$.

Figure 16 shows the Nash equilibrium points of A_m and A_r with respect to k_m and k_r. Figure 16a,b indicate the influence of effect coefficients k_m and k_r on advertising expenditures A_m and A_r when the system is stable. Figure 16c,d describe the expected utilities of the manufacturer and retailer with respect to k_m and k_r. In order to compare the Nash equilibrium (A_m, A_r) and expected utilities $(E(U_m), E(U_r))$ more clearly, the following Table 3 was created with respect to k_m and k_r with $\alpha_1 = 3.5$ and $\alpha_2 = 3.5$. By comparison, researchers can find that

(1) With an increase in effect coefficients k_m and k_r, the stable region remains unchanged.

(2) The Nash equilibrium A_m and A_r rise with an increase in effect coefficients k_m and k_r. A_m grows rapidly but A_r slowly and relative to the increase in k_m, meanwhile A_m grows slowly but A_r rapidly relative to the increase in k_r. Researchers can see from the trend of the graph in Figure 16a,b that A_m and A_r are greatly influenced by k_m and k_r, respectively. Therefore, the manufacturer and retailer will try to reduce their own effect coefficients k_m and k_r for less advertising expenditures. More noteworthy is that the increase in their effect coefficient has a greater impact on their own advertising investment.

(3) The expected utilities $E(U_m)$ and $E(U_r)$ rise with an increase in effect coefficients k_m and k_r. The expected utility $E(U_m)$ grows slowly with an increase in k_m; meanwhile, growth is rapid relative to an increase in k_r. Researchers can also see from the trend of the graph in Figure 16c that the expected utility $E(U_r)$ grows slowly with an increase in k_r; meanwhile, growth is rapid relative to an increase in k_m. These correspond to Figure 16d. The effect coefficient of the opponent has a greater impact on its own expected utility. Therefore, in order to obtain more expected utilities and less advertising expenditures, while decreasing their own effect coefficient, the manufacturer and retailer will be willing to increase the opponent's effect coefficient to gain the most revenue.

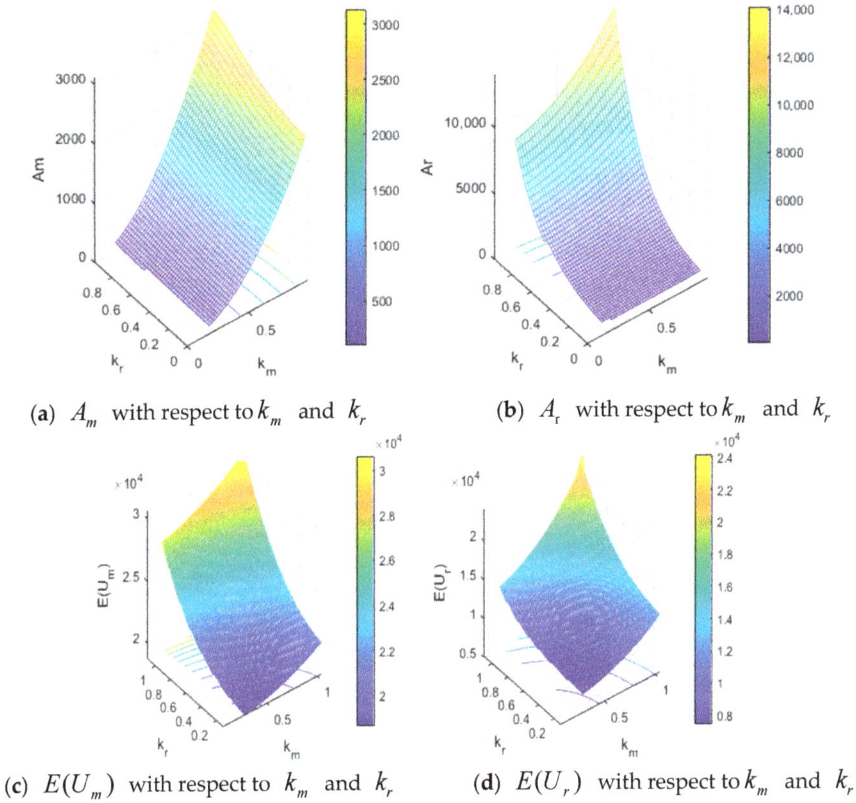

Figure 16. Nash equilibrium points and the expected utilities with $\alpha_1 = 3.5$ $\alpha_2 = 3.5$.

Table 3. Nash equilibrium points (A_m, A_r) and expected utilities $(E(U_m), E(U_r))$ with respect to k_m and k_r with $\alpha_1 = 3.5$ and $\alpha_2 = 3.5$.

(A_m, A_r) $(E(U_m), E(U_r))$	$k_m = 0.3$	$k_m = 0.5$	$k_m = 0.7$	$k_m = 0.9$
$k_r = 0.3$	(234.3, 365) (19,090, 7694)	(650.1, 400.4) (19,500, 8439)	(1272, 456.3) (20,130, 9618)	(2098, 536.4) (20,960, 11,310)
$k_r = 0.5$	(251.9, 1148) (19,970, 8193)	(697.3, 1262) (20,420, 9013)	(1359, 1444) (21,090, 10,310)	(2230, 1704) (21,970, 12,160)
$k_r = 0.7$	(280.6, 2749) (21,470, 9072)	(772.7, 3038) (21,960, 10,030)	(1459, 3494) (22,700, 11,530)	(2429, 4141) (23,670, 13,660)
$k_r = 0.9$	(320.9, 6148) (23,830, 10,570)	(874.3, 6834) (24,390, 11,750)	(1665, 7903) (25,220, 13,590)	(2650, 9396) (26,290, 16,160)

5. Chaos Control

Based on the discussion above, the chaotic state frequently appears against the stability of the supply chain, which is harmful to market returns. The manufacturer and retailer in the supply chain try their best to take measures to prevent the period-doubling bifurcation and chaotic state.

Referring to previous studies, the parameters control method is used to control chaos. Here, set the control parameter m [33,49–51]. The manufacturer and retailer should control their adjustment speed of relevant variables when they formulate a management strategy. The system (18) can be rewritten as:

$$\begin{cases} A_m(t+1) = (1-m)\left\{ A_m(t) + \alpha_1 A_m(t)\left\{ \frac{wk_m}{4\sqrt{A_m(t)}}\left(1 - \frac{k_r{}^3(G-k_m\sqrt{A_m(t)})}{F\sqrt{\frac{k_r{}^2(G-k_m\sqrt{A_m(t)})^2}{F}}}\right) + \frac{uk_r{}^2k_m(G-k_m\sqrt{A_m(t)})}{F\sqrt{A_m(t)}} - 1 \right\}\right\} + mA_m(t) \\ A_r(t+1) = (1-m)\left\{ A_r(t) + \alpha_2 A_r(t)\left\{ \left(k_m\sqrt{A_m(t)} + k_r\sqrt{\frac{k_r{}^2(G-k_m\sqrt{A_m(t)})^2}{F}} - G\right)\frac{k_r}{4\sqrt{A_r(t)}} - 1 + u \right\}\right\} + mA_r(t) \end{cases} \tag{24}$$

Figure 17 shows the bifurcation and LLE diagram with respect to parameter m with $\alpha_1 = 4.9$ and $\alpha_2 = 7$. The system changes from chaos to stability when parameter is $m = 0.185$. (A_m, A_r) returns to Nash equilibrium $(407.2, 162.8)$ with an increase in parameter m.

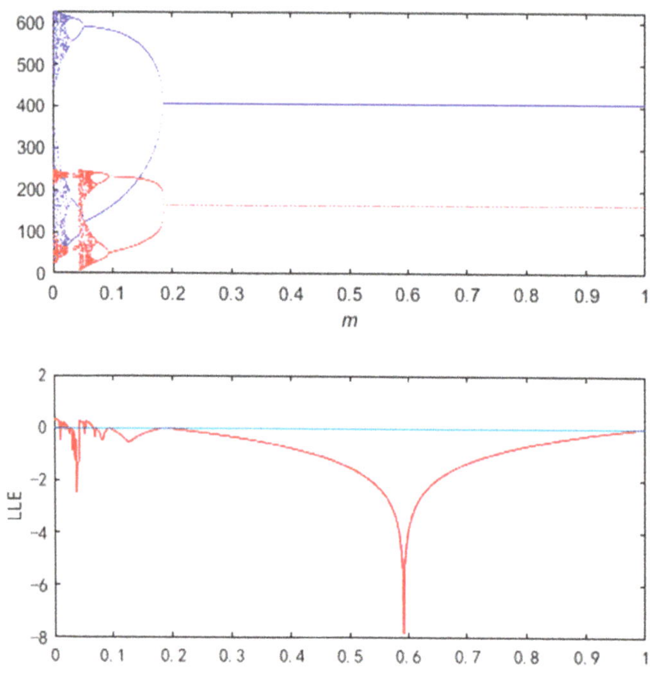

Figure 17. The bifurcation diagram of A_m and A_r and the largest Lyapunov exponent with respect to m.

Through comparison between the stable regions in Figure 18, researchers can find that the range of the stable region is extended compared with the one without chaos control in Figure 3. By $m = 0.1, 0.2, 0.3, 0.4$, the stable region expands with an increase in parameter m. Nevertheless, the Nash equilibrium (A_m, A_r) does not change with parameter m and stays at $(407.2, 162.8)$.

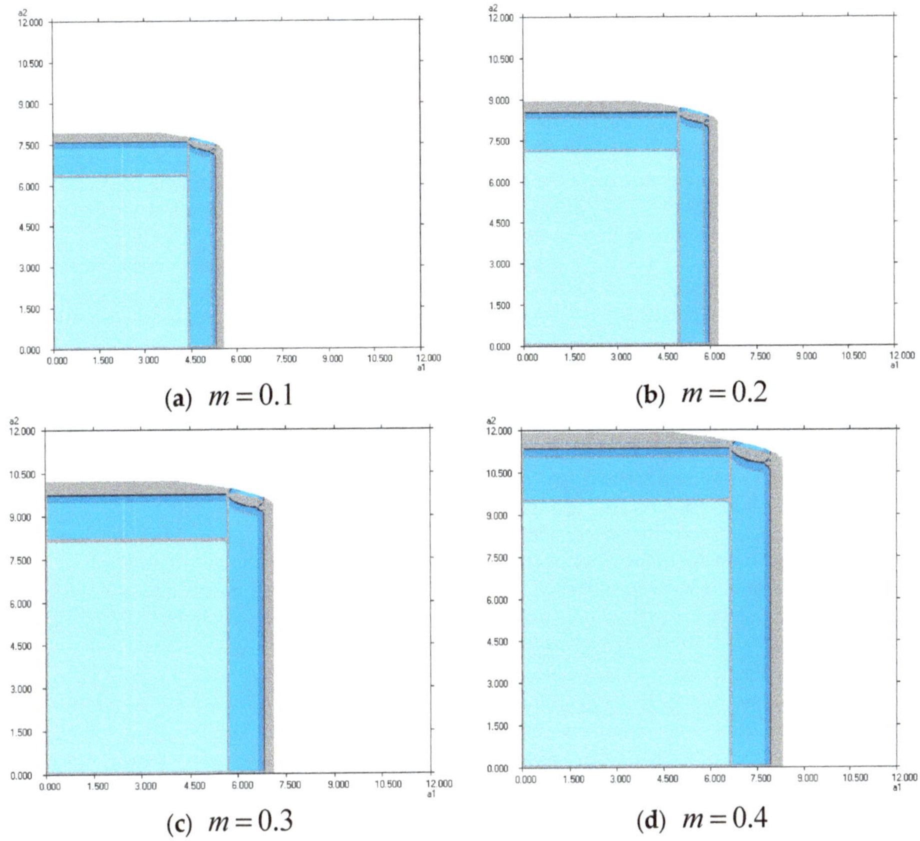

Figure 18. The 2D bifurcation diagrams with respect to α_1 and α_2 with different m.

6. Contribution and Implication

The major findings obtained in this paper can be employed by managers of supply chains as follows:

(1) If the enterprises keep adjustment speeds within the stability region, no matter what initial advertising expenditures are chosen by the enterprises, the market will eventually reach the Nash equilibrium after a series of games. As a result, the enterprises can obtain optimal strategies. Otherwise, if the adjustment speeds of advertising expenditure are too fast, the system loses stability and enters chaotic states, which lead to a rapid decrease in expected utilities. The manufacturer and retailer will control their adjustment speeds to maintain the stability of the supply chain market for the maximum revenue.

(2) The stability range of the system expands with an increase in the participation rate of local advertising expenditure by the manufacturer. As the manufacturer increases the participation rate of local advertising, the retailer spends more on local advertising expenditures. However, the national advertising expenditure of the manufacturer only has a slight downward adjustment with the increase in the participation rate. The expected utility of the manufacturer decreases, while the retailer increases slightly on the whole by an increase in the participation rate. For market stability and maximizing revenue, the manufacturer should adjust the participation rate appropriately, avoiding too high or too low values.

(3) The stable region has no variation with risk tolerance levels of the manufacturer and retailer. The risk tolerance level of the manufacturer has no effect on the national advertising expenditure by the manufacturer and local advertising expenditure by the retailer. However, the manufacturer's national advertising expenditure rises and the retailer's local advertising expenditure reduces with an increase in the risk tolerance level of the retailer when the system is in a stable state. The expected utility of the manufacturer decreases and the expected utility of retailer remains unchanged with an increase in the risk tolerance level of the manufacturer. The expected utility of the manufacturer rises and the expected utility of the retailer decreases with an increase in the risk tolerance level of the retailer. Hence, the manufacturer will try to reduce their own risk tolerance level for economic revenue. In order to make neither too much advertising expenditure nor too little expected utility, the retailer will appropriately adjust the risk tolerance level to adapt to their own development according to their own enterprise strategy in the supply chain market competition.
(4) With increasing effect coefficients of advertising expenditure, the stable region remains unchanged. The advertising expenditures and expected utilities rise with an increase in effect coefficients of advertising expenditure. The increase in its effect coefficient has a greater impact on its own advertising investment. In order to obtain more expected utilities and less advertising expenditures, both the manufacturer and retailer will reduce their own effect coefficients of advertising expenditure; meanwhile, they will attempt to increase their opponent's effect coefficients to gain the most revenue.
(5) It is profitable to keep the relevant factors in a certain range for manufacturer and retailer. Once the system falls into chaos, the chaos phenomenon can be delayed or eliminated effectively by the parameters control method.

Some valuable managerial implications for managers are summarized as follows: First, the appropriate participation rate of local advertising expenditure not only is good for the revenue of participants, but also is conducive to the stability of the supply chain, which needs to be constantly adjusted by the manufacturer in the competitive supply chain market. Second, the manufacturer and retailer will try to reduce their own risk tolerance levels for economic revenue, which makes it impossible for enterprises to avoid risks. However, in the realistic market competition, enterprises need to avoid risks as much as possible to ensure the stability of the market, avoiding market turbulence. Therefore, enterprises cannot develop steadily for a long time in the fierce supply chain market competition. Finally, once the market is in a chaotic state, the control method is applied by the manager effectively to avoid some loss in revenue.

7. Conclusions

In this paper, considering cooperative advertising, researchers have established a Stackelberg dynamic game model of a manufacturer–retailer supply chain with risk-averse behavior. Based on game and dynamics theory, the influence of relevant parameters on the stability of the model has been researched. The impact of adjustment speeds, risk tolerance levels, participation rate of local advertising expenditure by the manufacturer, and effect coefficients of advertising expenditure on system complexity and expected utilities have been analyzed by the numerical simulation method. Based on the parameters control method, the chaotic state of the system is effectively controlled. The research results of this study can provide important reference for advertising expenditure decisions and revenue maximization of participants in the context of risk aversion.

There are still some limitations in the paper. The analysis of the Stackelberg dynamic game model of the manufacturer and retailer might be simple in practice. The future work can consider a complicated multi-channel supply chain game model with a combination of traditional and online factors, and take into account the impact of more factors on the multi-channel supply chain stability, including but not limited to risk aversion and cooperative advertising.

Author Contributions: Writing—original draft, J.L.; Funding acquisition, C.L. All authors have read and agreed to the published version of the manuscript.

Funding: This work is supported by the National Natural Science Foundation of China (No. 71673042).

Data Availability Statement: Not applicable.

Acknowledgments: The authors are grateful to the reviewers and associate editor for their extremely useful comments.

Conflicts of Interest: The authors declare no conflict of interest regarding the publication of this paper.

References

1. Berger, P.D. Vertical Cooperative Advertising Ventures. *J. Mark. Res.* **1972**, *9*, 309–312. [CrossRef]
2. Fulop, C. The Role of Advertising in the Retail Marketing Mix. *Int. J. Advert.* **1988**, *7*, 99–117. [CrossRef]
3. Bergen, M.; John, G. Understanding Cooperative Advertising Participation Rates in Conventional Channels. *J. Mark. Res.* **1997**, *34*, 357–369. [CrossRef]
4. Zhang, J.; Gou, Q.; Liang, L.; Huang, Z. Supply chain coordination through cooperative advertising with reference price effect. *Omega* **2013**, *41*, 345–353. [CrossRef]
5. Giri, B.C.; Sharma, S. Manufacturer's pricing strategy in a two-level supply chain with competing retailers and advertising cost dependent demand. *Econ. Model.* **2014**, *38*, 102–111. [CrossRef]
6. Stuart, A. Portfolio Selection: Efficient Diversification of Investment. *OR* **1959**, *10*, 253–254. [CrossRef]
7. Markowitz, H.M. *Portfolio Selection: Efficient Diversification of Investments*; Wiley: New York, NY, USA, 1959.
8. Eeckhoudt, L.; Gollier, C.; Schlesinger, H. The Risk-Averse (and Prudent) Newsboy. *Manag. Sci.* **1995**, *41*, 786–794. [CrossRef]
9. Martínez-de-Albéniz, V.; Simchi-Levi, D. Mean-variance trade-offs in supply contracts. *Nav. Res. Logist.* **2006**, *53*, 603–616. [CrossRef]
10. Kahneman, D.; Tversky, A. Prospect Theory: An Analysis of Decision under Risk. *Econometrica* **1979**, *47*, 263–291. [CrossRef]
11. Zhao, H.; Wang, H.; Liu, W.; Song, S.; Liao, Y. Supply Chain Coordination with a Risk-Averse Retailer and the Call Option Contract in the Presence of a Service Requirement. *Mathematics* **2021**, *9*, 787. [CrossRef]
12. Marshall Fisher, A.R. Reducing the Cost of Demand Uncertainty Through Accurate Response to Early Sales. *Oper. Res.* **1996**, *44*, 87–99. [CrossRef]
13. Anupindi, R.; Bassok, Y. Centralization of stocks: Retailers vs. manufacturer. *Manag. Sci.* **1999**, *45*, 178–191. [CrossRef]
14. Gan, X.; Sethi, S.P.; Yan, H. Channel Coordination with a Risk-Neutral Supplier and a Downside-Risk-Averse Retailer. *Prod. Oper. Manag.* **2005**, *14*, 80–89. [CrossRef]
15. Rockafellar, R.T.; Uryasev, S. Conditional value-at-risk for general loss distributions. *J. Bank. Financ.* **2002**, *26*, 1443–1471. [CrossRef]
16. Gao, T.; Wang, K.; Mei, Y.; He, S.; Wang, Y. Supply Chain Pricing Models Considering Risk Attitudes under Free-Riding Behavior. *Mathematics* **2022**, *10*, 1723. [CrossRef]
17. Choi, T.-M.; Ma, C.; Shen, B.; Sun, Q. Optimal pricing in mass customization supply chains with risk-averse agents and retail competition. *Omega* **2019**, *88*, 150–161. [CrossRef]
18. Xiao, T.; Yang, D. Price and service competition of supply chains with risk-averse retailers under demand uncertainty. *Int. J. Prod. Econ.* **2008**, *114*, 187–200. [CrossRef]
19. Zhang, L.; Wang, J. Coordination of the traditional and the online channels for a short-life-cycle product. *Eur. J. Oper. Res.* **2017**, *258*, 639–651. [CrossRef]
20. Ma, J.; Wang, H. Complexity analysis of dynamic noncooperative game models for closed-loop supply chain with product recovery. *Appl. Math. Model.* **2014**, *38*, 5562–5572. [CrossRef]
21. Liu, J.; Liu, G.; Li, N.; Xu, H. Dynamics Analysis of Game and Chaotic Control in the Chinese Fixed Broadband Telecom Market. *Discret. Dyn. Nat. Soc.* **2014**, *2014*, 275123. [CrossRef]
22. Askar, S.S.; Al-Khedhairi, A. Local and Global Dynamics of a Constraint Profit Maximization for Bischi–Naimzada Competition Duopoly Game. *Mathematics* **2020**, *8*, 1458. [CrossRef]
23. Askar, S.S. The Influences of Asymmetric Market Information on the Dynamics of Duopoly Game. *Mathematics* **2020**, *8*, 1132. [CrossRef]
24. Sun, L.; Ma, J. Study and Simulation on Dynamics of a Risk-Averse Supply Chain Pricing Model with Dual-Channel and Incomplete Information. *Int. J. Bifurc. Chaos* **2016**, *26*, 1650146. [CrossRef]
25. Berger, P.D.; Magliozzi, T. Optimal Co-Operative Advertising Decisions in Direct-Mail Operations. *J. Oper. Res. Soc.* **1992**, *43*, 1079–1086. [CrossRef]
26. Dant, R.P.; Berger, P.D. Modelling Cooperative Advertising Decisions in Franchising. *J. Oper. Res. Soc.* **1996**, *47*, 1120–1136. [CrossRef]
27. Karray, S.; Zaccour, G. Could co-op advertising be a manufacturer's counterstrategy to store brands? *J. Bus. Res.* **2006**, *59*, 1008–1015. [CrossRef]

28. Xie, J.; Wei, J.C. Coordinating advertising and pricing in a manufacturer–retailer channel. *Eur. J. Oper. Res.* **2009**, *197*, 785–791. [CrossRef]
29. Sarkar, B.; Omair, M.; Kim, N. A cooperative advertising collaboration policy in supply chain management under uncertain conditions. *Appl. Soft Comput.* **2020**, *88*, 105948. [CrossRef]
30. Xiao, D.; Zhou, Y.-W.; Zhong, Y.; Xie, W. Optimal cooperative advertising and ordering policies for a two-echelon supply chain. *Comput. Ind. Eng.* **2019**, *127*, 511–519. [CrossRef]
31. Agrawal, V.; Seshadri, S. Impact of Uncertainty and Risk Aversion on Price and Order Quantity in the Newsvendor Problem. *Manuf. Serv. Oper. Manag.* **2000**, *2*, 410–423. [CrossRef]
32. Xu, G.; Dan, B.; Zhang, X.; Liu, C. Coordinating a dual-channel supply chain with risk-averse under a two-way revenue sharing contract. *Int. J. Prod. Econ.* **2014**, *147*, 171–179. [CrossRef]
33. Yan, B.; Jin, Z.; Liu, Y.; Yang, J. Decision on risk-averse dual-channel supply chain under demand disruption. *Commun. Nonlinear Sci. Numer. Simul.* **2018**, *55*, 206–224. [CrossRef]
34. Zhou, Y.-W.; Li, J.; Zhong, Y. Cooperative advertising and ordering policies in a two-echelon supply chain with risk-averse agents. *Omega* **2018**, *75*, 97–117. [CrossRef]
35. Karray, S.; Amin, S.H. Cooperative advertising in a supply chain with retail competition. *Int. J. Prod. Res.* **2014**, *53*, 88–105. [CrossRef]
36. Ma, J.; Ren, H.; Yu, M.; Zhu, M. Research on the Complexity and Chaos Control about a Closed-Loop Supply Chain with Dual-Channel Recycling and Uncertain Consumer Perception. *Complexity* **2018**, *2018*, 9853635. [CrossRef]
37. Ma, J.; Hong, Y. Dynamic game analysis on pricing and service strategy in a retailer-led supply chain with risk attitudes and free-ride effect. *Kybernetes* **2021**, *51*, 1199–1230. [CrossRef]
38. Khademi, M.; Ferrara, M.; Pansera, B.; Salimi, M. A dynamic game on Green Supply Chain Management. *arXiv* **2015**, arXiv:1503.04772.
39. Ma, J.; Tian, Y.; Xu, T.; Koivumäki, T.; Xu, Y. Dynamic game study of multi-channel supply chain under cap-and-trade regulation. *Chaos Solitons Fractals* **2022**, *160*, 112131. [CrossRef]
40. Karray, S. Periodicity of pricing and marketing efforts in a distribution channel. *Eur. J. Oper. Res.* **2013**, *228*, 635–647. [CrossRef]
41. Xie, G.; Yue, W.; Wang, S.; Lai, K.K. Quality investment and price decision in a risk-averse supply chain. *Eur. J. Oper. Res.* **2011**, *214*, 403–410. [CrossRef]
42. Walls, M.R. Combining decision analysis and portfolio management to improve project selection in the exploration and production firm. *J. Pet. Sci. Eng.* **2004**, *44*, 55–65. [CrossRef]
43. Avinadav, T.; Chernonog, T.; Perlman, Y. Mergers and acquisitions between risk-averse parties. *Eur. J. Oper. Res.* **2017**, *259*, 926–934. [CrossRef]
44. Choi, T.-M.; Li, D.; Yan, H. Mean–variance analysis of a single supplier and retailer supply chain under a returns policy. *Eur. J. Oper. Res.* **2008**, *184*, 356–376. [CrossRef]
45. Guo, Z.; Ma, J. Dynamics and implications on a cooperative advertising model in the supply chain. *Commun. Nonlinear Sci. Numer. Simul.* **2018**, *64*, 198–212. [CrossRef]
46. Li, T.; Ma, J. Complexity analysis of dual-channel game model with different managers' business objectives. *Commun. Nonlinear Sci. Numer. Simul.* **2015**, *20*, 199–208. [CrossRef]
47. Ma, J.; Li, Q.; Bao, B. Study on Complex Advertising and Price Competition Dual-Channel Supply Chain Models Considering the Overconfidence Manufacturer. *Math. Probl. Eng.* **2016**, *2016*, 2027146. [CrossRef]
48. Ma, J.; Guo, Z. The influence of information on the stability of a dynamic Bertrand game. *Commun. Nonlinear Sci. Numer. Simul.* **2016**, *30*, 32–44. [CrossRef]
49. Xi, X.; Zhang, Y. Complexity analysis of production, fertilizer-saving level, and emission reduction efforts decisions in a two-parallel agricultural product supply chain. *Chaos Solitons Fractals* **2021**, *152*, 111358. [CrossRef]
50. Tian, Y.; Ma, J.; Xie, L.; Koivumäki, T.; Seppänen, V. Coordination and control of multi-channel supply chain driven by consumers' channel preference and sales effort. *Chaos Solitons Fractals* **2020**, *132*, 109576. [CrossRef]
51. Chen, X.; Zhou, J. The complexity analysis and chaos control in omni-channel supply chain with consumer migration and advertising cost sharing. *Chaos Solitons Fractals* **2021**, *146*, 110884. [CrossRef]

mathematics

MDPI

Article

Implications of CSR Practices for a Development Supply Chain in Alleviating Farmers' Poverty

Qingyu Zhang * and Tianlong Luo *

Research Institute of Business Analytics and Supply Chain Management, College of Management, Shenzhen University, Shenzhen 518060, China
* Correspondence: q.yu.zhang@gmail.com (Q.Z.); luo_tl@126.com (T.L.)

Abstract: To alleviate farmer poverty, this paper investigates the effect of a retailer's different socially responsible practices on a two-echelon supply chain consisting of one rural (poor) farmer, one suburban farmer, and one common retailer. Different from a commercial supply chain (whose members' objectives are to maximize their profits) and a humanitarian supply chain (whose objective is to save more people, rather than to prioritize profits), the paper aims to study a development supply chain where the CSR-conscious retailer aims to lift the poor farmer out of poverty through cost sharing, altruistic practices, or fairness practices. Can the CSR-conscious retailer (and the development supply chain) do well by doing good? To answer the above question, four models of potential CSR investment are established and analyzed. Considering the different influences of the retailer's CSR practices, this paper uses a Stackelberg game to analyze the decisions and profits of the farmers and the retailer in these four models. Our study finds that, first, the retailer's CSR practices can improve the whole supply chain's performance, which means that the supply chain has the potential to achieve the Pareto improvement for both the farmers and the retailer. Second, the retailer's CSR practices yield benefits while implementing cost-sharing or fairness practices. Third, the rural farmer always benefits from the retailer's CSR practices and may prefer the altruistic practice from which they can benefit the most. In addition, to benefit their profit more, the rural farmer should grow high- or low-value-added crops rather than medium-value-added ones. Fourth, from the suburban farmer's perspective, the retailer's CSR practices are not beneficial for their performance. However, the extent to which the suburban farmer's performance decreases is much lower than the extent to which the rural farmer's performance improves. The results of this paper might be used by stakeholders to alleviate poverty.

Keywords: socially responsible practices; development supply chain; emerging markets; operational improvements; poverty alleviation; game theory

MSC: 91C99; 91A80; 90B50

1. Introduction

In emerging markets, the agricultural sector accounts for a significant portion of economic activities. Most farmers in developing economies such as India or China remain poor, partly because most farmers have limited access to advanced farming practices, low-cost logistics systems, efficient sale channels, and reliable market information, in addition to other obstacles such as access to traffic, credit, loans, land, electricity, and water [1]. It is an obstacle for farmers in developing areas to make a living since they have a cost disadvantage over farmers in developed markets [2,3]. In contrast, many large, modern firms are expanding rapidly in developed regions. For example, with the reform and opening up in China, many firms have grown, e.g., the Yonghui retail superstore. Facing these challenges, many companies try to help poor farmers out of poverty [1]. The question of how to help people in impoverished areas to escape poverty through the efforts of firms from developed areas is a vital topic in economic imbalance, poverty alleviation, or corporate social responsibility (CSR) research.

Many companies take poverty alleviation as social responsibility in developing countries. However, since industrial poverty alleviation requires overcoming the harsh business environment, companies have to invest considerable money. This will bring tremendous financial pressure to CSR-conscious companies, which means that implementing corporate social responsibility does not necessarily increase profits in the short term. They believe that conducting CSR is strategic philanthropy. They hope that, in the long run, it may be helpful to increase their product quality, reputation, or marketing effectiveness, and even increase profits [4]. For example, the Yonghui superstore invested 50 million RMB in building value-added food industrial parks and organizing product deep-processing workshops in Min county of China to support local people to establish their businesses, which helped at least 5000 households to escape poverty in 2019. As a leading domestic retail company, Yonghui actively fought against poverty with its abundant supply chain resources and perfect sales channels through altruism, charity, and fairness. Yonghui also created the "charity supermarket" model, aiming to target poverty alleviation by various means of profit surrender [5]. Purchasing and using more products and services from poor areas is an effective strategy to help poor people to increase their incomes and eliminate poverty. The Chinese government advocates for supporting some role-model firms, which take the lead in poverty alleviation through purchasing and reselling products from poor areas under the same conditions as those in developed regions [6]. With the government's pro-poor initiative as an endorsement and the proper profit surrender from CSR-conscious firms, farmers can provide quality products and sell them at reasonable prices rather than accept the market price passively. For example, by the end of August 2020, Hubei province had included 15,235 consumer poverty-relief products, with a total sale volume of 21.25 billion yuan, effectively offsetting the impact of the price drop on the income of poor households due to the epidemic. In addition to low-value-added (i.e., low-profit-margin) products such as vegetables and shiitake mushrooms, farmers can plant more high-value-added products such as tea and crayfish [7].

A commercial supply chain consists of individual players whose objective is to maximize their profits. Whereas a humanitarian supply chain does not maximize its profit, it always aims to minimize the loss of life and alleviate suffering after a disaster. It does not matter if it ends up with a higher cost. Unlike a humanitarian or commercial supply chain, a development supply chain aims for the economic development of the poor [8]. Specifically, a development supply chain refers to the integration of farmers from poor areas into the supply chain system as suppliers or distributors [9], in which the core firms implement corresponding strategies according to the characteristics of poor farmers, such as providing technologies, information, and funds, to improve the performance of poor farmers and/or the supply chain. CSR-conscious core firms help poor farmers through cost sharing, altruistic preference, or fairness concerns during this process. These firms might sacrifice their profits to lift poor farmers out of poverty, and, by doing so, the whole supply chain's performance might improve. Such supply chain poverty alleviation practice is of great help in developing poor areas around the world [10]. All transactions in the development supply chain are commercial. Unlike commercial supply chains, they require core firms to implement CSR practices to support farmers struggling to make a living. Such supply chains can play an essential role in developing areas to bridge the period in which humanitarian assistance has ended and the economy has grown enough to allow commercial supply chains to function. Moreover, eventually, farmers from rural areas can make a living under the competition of suburban farmers, which is a key feature of the development supply chain.

Typically, rural farmers incur higher costs than suburban farmers in planting the same types of crops due to the previously mentioned technical and infrastructure disadvantages and obstacles. A CSR-conscious retailer can help the rural farmer to survive via different practices, e.g., cost sharing, profit surrendering, or selling more rural products. Doing so might cause competition between rural farmers and suburban farmers. Our research purports to study the impact of the retailer's different CSR practices on the development supply chain in the presence of competition. Specifically, given a 2-to-1 development supply chain with one rural farmer (poor farmer), one suburban farmer (modern farmer), and one common retailer, our model is intended to examine the following research questions: (1) What are the effects of the retailer's CSR practices on the development (i.e., anti-poverty) supply chain? (2) How can the poor farmer benefit from the retailer's CSR practices? (3) Are the retailer's CSR practices financially sustainable (i.e., can the retailer and the development supply chain do well by doing good)? (4) What are the determining factors that affect the performance of the anti-poverty actions and their sustainability?

To examine the research questions, we use a game-theoretic framework to capture the underlying strategic interactions among three parties: a poor farmer, a suburban farmer, and a retailer. Our core contribution is to study a development supply chain where the retailer may sacrifice their profit to lift the poor farmer out of poverty through cost sharing, altruistic preference, or fairness concerns. Our analysis shows that, with the retailer's CSR practices, the performance of poor farmers and even the supply chain has been improved significantly, providing theoretical support for the development supply chain and poverty alleviation in impoverished areas. Specifically, our key findings are as follows. (1) The retailer's CSR practices improve the whole supply chain's performance, which means that the supply chain has the potential to achieve the Pareto improvement for both farmers and the retailer. (2) The retailer's CSR practices also yield benefits while implementing cost-sharing or fairness practices. (3) The rural farmer always benefits from the retailer's CSR practices. The effect of altruistic preference practice and fairness concern practice on the rural farmer's profit is better than the cost-sharing practice. In addition, to benefit their profit more, the rural farmer should grow high- or low-value-added crops rather than medium-value-added ones. (4) From the suburban farmer's perspective, the retailer's CSR practices are not good for their performance. However, the extent to which the suburban farmer's performance decreases is much less than the rural farmer's performance gains.

The rest of this study is organized as follows. In Section 2, we review the relevant literature. Section 3 describes the model formulations. In Sections 4 and 5, we carry out the equilibrium analysis and compare four models to articulate the management implications. Section 6 presents numerical results and insights. Finally, Section 7 concludes the paper.

2. Literature Review

This paper is related to two streams in the literature. The first studies the development supply chain and the second research stream focuses on CSR practices in the supply chain.

2.1. Development Supply Chain

Agricultural production is easily affected by the weather. Thus, the output is random, and the demand volatility is high. Coupled with the perishability factor, farmers in an unstable environment are faced with tremendous pressure and many risks [11]. For areas with better infrastructure, contract farming is an essential aspect of agricultural institutions that aid in the transition to modern agriculture [12]. However, farmers in poverty areas usually face higher production costs and lower profit margins because they are isolated from main production areas and markets and lack the infrastructure, knowledge, technology, and information [13]. Although there are cases in which individual enterprises have led the implementation of initiatives for poverty alleviation, this type of initiative typically requires resources that an average enterprise would not possess [14]. Thus, an enterprise without a sense of CSR will tend not to participate in poverty alleviation.

In the development supply chain literature, it can be seen that CSR practices have attracted considerable attention. To alleviate farmer poverty, enterprises, governments, and nongovernmental organizations are developing different mechanisms for aiding farmers in developing countries. Specifically, Zhou et al. [15] examine whether the broader dissemination of information will always benefit farmers. Providing information to only one farmer is optimal, but providing information to all farmers can be detrimental. Chen et al. [16] investigate a new business model in which I.T.C. Limited developed the "e-Choupals" for the rural areas of India, and they find that the implicit agreement behaves as a formal contract, regardless of the price elasticity of the local market. Upon reaching an agreement with I.T.C., the farmers always give priority to delivering directly to I.T.C. Since most consumers in rural areas of developing countries cannot afford to purchase certain livelihood improvement products such as home appliances, Yu et al. [17] develop a parsimonious model to determine the optimal subsidy program in different settings to gain a better understanding about the conditions under which it is optimal for the government to subsidize consumers only, manufacturers only, or both, and they find that the structure of the optimal subsidy program depends on (a) whether there is a well-established market selling price for the products; and (b) the relative emphasis that the government places on consumer welfare versus manufacturer profit.

Although these studies examine the issues of economic value creation for farmers in different ways, they all lack the investigation of the CSR practices of supply chain enterprises. They do not study the issue of whether CSR contributes to improving the economic performance of the supply chain.

2.2. CSR Practices in Supply Chains

There is limited modeling literature on socially responsible operations because this topic is an emerging research area in operations management [18,19]. Accordingly, most of the relevant articles are recent. They are divided into three categories: cost-sharing practice, altruistic preference practice, and fairness concern practice.

Cost-sharing practice. Zhou et al. [20] investigate a two-echelon supply chain where the retailer provides customers with some pre-sales services, which positively impacts the market demand. Moreover, the manufacturer's online channel benefits from the retailer's pre-sales services by sharing the retailer's sales effort cost. The findings show that a service-cost-sharing contract can effectively stimulate the retailer to improve their service level, the manufacturer would share a larger proportion of the service cost, and the retailer would set a higher service level in the differential pricing scenario than in the non-differential pricing scenario if the degree of free-riding is not very high. Xie et al. [21] investigate a dual-channel closed-loop supply chain combining the revenue-sharing contract in the forward channel with the channel investment cost-sharing contract and introduce the Stackelberg game to explore the contract coordination mechanism. The results show that the proposed contract can increase the profits of supply chain members. Li et al. [22] investigate the impact of revenue-sharing and cost-sharing contracts offered by a retailer on emission reduction efforts and firms' profitability. The findings suggest that supply chain coordination and the range of sharing rates depend critically on parameters. Hosseini-Motlagh et al. [23] propose a cost-sharing contract to explore the coordination issue in a manufacturer–retailer supply chain where the manufacturer is socially responsible and invests in CSR activities.

Altruistic preference practice. Wang et al. [24] investigate the influence of a government subsidy and a remanufacturer's altruistic preference on the decision making in a low-carbon e-commerce closed-loop supply chain, and find that the altruistic preference behavior increases the revenue of the e-commerce platform and improves the efficiency of the supply chain, but is not advantageous to the remanufacturer. In addition, the effects of altruistic behavior on promoting the recycling of waste products are inferior to the impact of government subsidies of the same strength. Wan et al. [25] construct a dual-channel supply chain network equilibrium model comprising hotels, online travel agent platforms, and demand markets to study the influence of the decision-maker's altruistic preference and the consumers' low-carbon preference on decision-making behavior in all layers of a dual-channel environmental hotel supply chain, and they find that the altruistic preferences have different impacts on the supply chain under the merchant mode and the agency mode. Xu and Wang [26] propose a competitive–cooperative strategy based on altruistic behavior for the dual-channel supply chain by applying the theory of the co-competition game, and prove the existence of the equilibrium strategy and the stability of the proposed model through mathematical deduction. In the same context, Wang et al. [27,28] study the decision making, coordination contracts, and altruistic preferences of an e-commerce supply chain composed of a manufacturer and a third-party e-commerce platform, and find that altruistic preferences help to improve sales prices and sales service levels.

Fairness concerns practice. Guan et al. [29] examine channel coordination under fairness concerns in a supply chain. They determine the optimal strategies by differential game models. A dominant channel member's sensitivity to fairness is more significant in the decision-making process and channel efficiency. Caliskan-Demirag et al. [30] extend the model developed by Cui et al. [31] to a nonlinear demand setting. They find that compared to the linear demand, the conditions to coordinate the decentralized channel are relaxed, in which only the retailer has fairness concerns. Yang et al. [32] investigate the effect of the retailer's fairness concerns on a distribution channel composed of a manufacturer and a retailer. They find that there exists a Pareto improvement for the supply chain member's profit when the retailer has changed from having no fairness concerns to being fair-minded. Zheng et al. [33] propose three coordination mechanisms to allocate surplus profit and examine how the retailer's fairness concerns practice affects profit allocation in a three-echelon closed-loop supply chain (CLSC). Zhou et al. [34] consider a low-carbon supply chain channel with a manufacturer and a retailer and show how the optimal decision and coordination change when a retailer has fairness concerns. Shen et al. [35] construct decentralized decision models with and without fairness concerns to study the impact of network externalities and fairness concerns on platform supply chain advertising strategies.

There is literature that considers CSR in other contexts. Arslan and Turkay [36] investigate the EOQ model with sustainability considerations that include environmental and social responsibility criteria and conventional economic considerations. Modak et al. [37] develop a socially responsible closed-loop supply chain (CLSC) model that considers donation as a CSR activity and recycling of the used products for environmental sustainability to investigate the optimal channel structures. Considering the main reasons behind the lack of social responsibility of multinational corporations, Liu et al. [38] develop a model with punishment to analyze whether the penalty rate will impact corporate social responsibility's input and how it will affect the self-interest of supply chain members.

However, the papers above do not address the comparison of different CSR practices in a development supply chain with two representative farmers and a retailer. In other words, they do not give reasonable explanations of how different CSR practices influence the supply chain equilibrium in different ways. In sum, the literature positioning of this article is shown in Table 1 by incorporating the supply chain, poor farmer, CSR practices (cost-sharing practice; altruistic practice; fairness concerns practice), competition, and game theory. Most existing studies have not considered poor farmers in developing areas. In addition, although some study the poor farmer in the supply chain (e.g., Sodhi and Tang [9,39]), they adopt the conceptual approach and do not consider the competition context, which is one of the main factors that influences pricing decisions in the supply chain. On the other hand, the researchers including CSR practices in a supply chain usually focus on a noncompetitive context and ignore the poor farmer's characteristics and their impact on the decisions with game theory, which is an essential phenomenon in a development supply chain. Thus, we will fill the research gap by introducing the poor farmer's characteristics, the competition, and the CSR practices in the context of the supply chain structure.

This study extends the literature related to the development supply chain and CSR practices and contributes to the literature in the following ways. First, we assume a positive relationship between CSR practices and consumers' purchasing intention and propose the corresponding analytical models to study the optimal decisions of farmers and retailers. Second, we deploy three retailers' CSR practices, i.e., cost-sharing practice, altruistic preference practice, and fairness concern practice, and investigate whether these CSR practices have the same or different implications for the development supply chain.

Table 1. Positioning of this article.

	Poor Farmer	Supply Chain	Cost-Sharing Practice	Altruistic Preference Practice	Fairness Concern Practice	Competition	Game Theory
Sodhi & Tang [9]	√	√					
Sodhi & Tang [39]	√	√					
Zhou et al. [15]	√					√	√
Chen et al. [16]	√		√				√
Zhou et al. [20]		√	√			√	√
Xie et al. [21]		√	√				√
Wu [40]		√	√				√
Yu et al. [17]			√			√	√
Wang et al. [24]		√		√			√
Wan et al. [25]		√		√		√	√
Xu & Wang [26]		√		√			√
Guan et al. [29]		√			√		√
Zheng et al. [33]		√			√	√	√
Zhou et al. [34]		√			√		√
This paper	√	√	√	√	√	√	√

3. Model Assumptions

We analyze a two-level supply chain that has two representative farmers ("she") who produce and sell similar agricultural products to a common retailer ("he"), as depicted in Figure 1. The farmer (farmer 1) from a poor rural area can label their crops as anti-poverty products, which are certified by the government (e.g., in China) so that customers know they are purchasing an anti-poverty product at the point of purchase. The crops produced by the farmer (farmer 2) from a suburban area are considered a general product. The farmers offer a reasonable wholesale price for their agricultural products before the sale season. The retailer signs a purchase contract that specifies the agreed price and volume with the farmers to meet the demand of the retail market. We assume that the farmers engage in the Bertrand competition, which is similar to those used in prior literature [16,18,41]. Farmer 1 incurs a unit production cost $c + \mu$, and farmer 2 incurs a unit production cost c. Without loss of generality, let $c = 0$ and the unit production cost to the retailer is 0. In this development supply chain, the retailer implements social responsibility practices through cost sharing, altruistic preference, and fairness concerns to support the farmers from rural areas. Consumers have become more sensitive to CSR practices over the past few decades, which strongly influences corporate strategies and decisions [40]. Existing literature suggests that CSR practices have often positively influenced consumers' purchase intentions and company profitability [42,43]. According to a survey conducted by Yuen et al. [44], CSR practices positively affect consumer satisfaction and consumers are willing to pay extra premiums for CSR. For example, Chongqing municipality created a public poverty-alleviation brand, which can be used for all identified anti-poverty products and enhance added value through the brand effect. In 2020, the sales of anti-poverty products in Chongqing amounted to 2 billion yuan, increasing by 7.1 percent over 2019.

We assume that consumers are willing to pay extra money for each anti-poverty-labeled product 1 ($\kappa\tau$, $\kappa\theta$, $\kappa\phi$), where (τ, θ, ϕ) represents the degree of retailer's different types of CSR practices on farmer 1, as depicted in Table 2. κ represents the extent to which consumers perceive the retailer's CSR practices. $\kappa = 0$ means that even if the retailer implements CSR practices, the practices are not perceived by consumers, which is equivalent to the effect when there is no anti-poverty label. $\kappa > 0$ means that consumers can perceive the retailer's CSR practices, and the larger κ, the more pronounced the degree of perception. We consider a one-period model. The two farmers set the wholesale price w_1 and w_2 first, and the retailer sets the order quantities q_1 and q_2 after observing the wholesale prices. The market prices p_1 and p_2 without the retailer's CSR practices follow a downward-sloping inverse demand function:

$$p_1 = a_1 - bq_1 - \gamma q_2 \quad (1)$$

$$p_2 = a_2 - bq_2 - \gamma q_1 \quad (2)$$

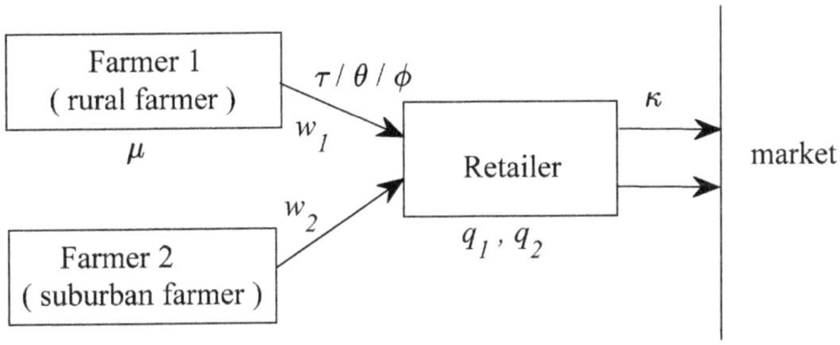

Figure 1. Two-echelon two-to-one development supply chain structure.

Table 2. Explanation of CSR practices.

CSR Practices	Explanation
Cost-sharing practice	Due to technical and infrastructure disadvantages, the rural farmer pays more than the suburban farmer to grow the same agricultural products. As a CSR practice, the CSR-conscious retailer shares the rural farmer's cost, but not the suburban farmer's.
Altruistic preference practice	The CSR-conscious retailer has altruistic behavior toward the rural farmer but not the suburban farmer.
Fairness concern practice	Typically, the retailer will order more products from the suburban farmer due to the cost advantage. As a fairness concerns practice, the CSR-conscious retailer tries to order a similar amount from both farmers.

This type of inverse demand function is widely used in alternative products [45,46]. a_1 and a_2 correspond to the market potential. Since these two products come from different regions and brands (anti-poverty products, general products), a_1 and a_2 are different and relatively independent. For example, Yangshan, a poor county in China, is known for chicken farming, and the selling price in the Guangzhou wholesale market is very competitive ($a_1 > a_2$). On the other hand, the two products from the farmers are popular commodities whose quantities sensitivity and cross-quantities sensitivity are very close. For the convenience of processing, we assume that the quantities sensitivity and cross-quantities sensitivity are the same in these two inverse demand functions. b represents quantities sensitivity, and cross-quantities sensitivity γ reflects the degree to which the products of the two farmers are substitutes ($b > \gamma$). All parameters are positive. Table 3 lists the definitions of the symbols used in the paper.

Four models are established and analyzed based on three CSR practices. For expositional simplicity, the basic model without any CSR practice, the model with the CSR practice being the retailer sharing the cost proactively, the model with the CSR practice being the retailer's altruistic preference, and the model with the CSR practice being the retailer's fairness concerns strategy are denoted as Model N, Model S, Model A, and Model F, respectively.

Table 3. Summary of notations.

Notation	Definition
a_i	Market potential of farmers' products
b	The quantities sensitivity parameter of two products
γ	The competition intensity of two farmers' products
κ	The elasticity of market price with regard to CSR performance
μ	The cost difference between two farmers' products
τ	The degree of retailer's cost-sharing practice
θ	The degree of retailer's altruistic preference practice
ϕ	The degree of retailer's fairness concern practice
p_i^j	Market prices of the two products
w_i^j	The wholesale price of two products (farmers' decision variable)
q_i^j	The order quantities of two products (retailer's decision variable)
π_i^j	Profit of two farmers
π_R^j	Profit of the retailer
π_{SC}^j	Profit of the supply chain

Note: $i = \{1, 2\}$ represent farmer 1 and farmer 2, respectively. $j = \{N, S, A, F\}$ represent model N, model S, model A, and model F, respectively.

4. Model Analysis

4.1. Model N: The Base Model

In Model N, the supply chain members are rational economic individuals who make decisions to maximize their profits. This model is a benchmark to analyze the effect of retailers' CSR practices.

Farmer 1's profit is determined by

$$\pi_1 = (w_1 - \mu)q_1 \tag{3}$$

Farmer 2's profit is determined by

$$\pi_2 = w_2 q_2 \tag{4}$$

The retailer's profit is determined by

$$\pi_R = (p_1 - w_1)q_1 + (p_2 - w_2)q_2 \tag{5}$$

The following proposition details the equilibrium results for the two-stage game, which can be solved analytically using the backward induction method, and we obtain Lemma 1.

Lemma 1. *In model N, the farmers' equilibrium wholesale prices are*

$$w_1^N = \frac{2a_1b^2 - a_1\gamma^2 + 2b^2\mu - a_2b\gamma}{4b^2 - \gamma^2} \tag{6}$$

$$w_2^N = \frac{2a_2b^2 - a_2\gamma^2 - a_1b\gamma + b\gamma\mu}{4b^2 - \gamma^2} \tag{7}$$

The retailer's equilibrium order quantities are

$$q_1^N = \frac{b(2a_1b^2 - a_1\gamma^2 - 2b^2\mu + \gamma^2\mu - a_2b\gamma)}{2(4b^4 - 5b^2\gamma^2 + \gamma^4)} \tag{8}$$

$$q_2^N = \frac{b(2a_2b^2 - a_2\gamma^2 - a_1b\gamma + b\gamma\mu)}{2(4b^4 - 5b^2\gamma^2 + \gamma^4)} \tag{9}$$

All the proofs are given in Appendix A.

We next examine the effects of the model parameters on the farmers' wholesale prices, the retailer's order quantities, the market prices, and the supply chain member's profits in sequence. Specifically, we study the effects of the cost difference μ.

Proposition 1. *(i) w_1^N, w_2^N, p_1^N, p_2^N, q_2^N increase with μ, q_1^N decreases with μ; (ii) π_1^N decreases with μ, and π_2^N increases with μ; (iii) π_R^N increases with μ if λ_1 is positive; otherwise, it decreases with μ, where $\lambda_1 = 2a_2\gamma^3 - (a_1 - \mu)(8b^3 - 6b\gamma^2)$.*

Proposition 1 indicates that, on the one hand, the higher the cost difference μ, the higher the wholesale price w_1 will be established to maintain farmer 1's optimal profit. It will force the retailer to increase the market price p_1, which leads to the lower demand q_1 for product 1, and then affects the profit of product 1; that is, farmer 1 obtains a lower profit. On the other hand, farmer 2 has a competitive advantage over farmer 1 for the increasing cost difference, which always benefits farmer 2. Since the demand of product 1 can be transferred to product 2 through competition, thus, not only the wholesale price w_2 and the market price p_2 will be set higher, but also the demand q_2 for product 2 increases, which will increase farmer 2's profit. As the retailer's profit is affected by product 1 and product 2, only if the maximum unit profit of product 1 $(a_1 - \mu)$ is greater than that of product 2 (a_2) will the retailer's profit decrease with the increasing cost difference μ.

4.2. Model S: Model with a Cost-Sharing Retailer

Next, we explore a scenario in which the retailer implements social responsibility through cost sharing with farmers from rural areas. Since the consumers are willing to pay more for an anti-poverty product labeled with the retailer's cost-sharing practices, the demand functions of product 1 become $p_1 = a_1 - bq_1 - \gamma q_2 + \kappa\tau$ [42,43], and the demand functions of product 2 remain the same as 2. Thus, farmer 1's profit is determined by

$$\pi_1 = [w_1 - \mu(1 - \tau)]q_1 \tag{10}$$

Farmer 2's profit is determined by

$$\pi_2 = w_2 q_2 \tag{11}$$

The retailer's profit is determined by

$$\pi_R = (p_1 - w_1 - \tau\mu)q_1 + (p_2 - w_2)q_2 \tag{12}$$

Lemma 2. *In Model S, (i) the farmers' equilibrium wholesale prices are*

$$w_1^S = \frac{2a_1b^2 - a_1\gamma^2 - a_2b\gamma + 2b^2\mu - 4b^2\mu\tau + \gamma^2\mu\tau - \gamma^2\kappa\tau + 2b^2\kappa\tau}{4b^2 - \gamma^2} \tag{13}$$

$$w_2^S = \frac{2a_2b^2 - a_2\gamma^2 - a_1b\gamma + b\gamma\mu - b\gamma\kappa\tau}{4b^2 - \gamma^2} \tag{14}$$

The retailer's equilibrium order quantities are

$$q_1^S = \frac{b(2a_1b^2 - a_1\gamma^2 - a_2b\gamma - 2b^2\mu + \gamma^2\mu + 2b^2\kappa\tau - \gamma^2\kappa\tau)}{2(4b^4 - 5b^2\gamma^2 + \gamma^4)} \tag{15}$$

$$q_2^S = \frac{b(2a_2b^2 - a_2\gamma^2 - a_1b\gamma + b\gamma\mu - b\gamma\kappa\tau)}{2(4b^4 - 5b^2\gamma^2 + \gamma^4)} \tag{16}$$

(ii) The retailer's equilibrium profit is π_R^S, and farmer 1's equilibrium profit and farmer 2's equilibrium profit are as follows:

$$\pi_1^S = \frac{b\left(a_1\gamma^2 - 2a_1b^2 + 2b^2\mu - \gamma^2\mu - 2b^2\kappa\tau + \gamma^2\kappa\tau + a_2b\gamma\right)^2}{2(b^2-\gamma^2)(4b^2-\gamma^2)^2} \tag{17}$$

$$\pi_2^S = \frac{b\left(a_2\gamma^2 - 2a_2b^2 + a_1b\gamma - b\gamma\mu + b\gamma\kappa\tau\right)^2}{2(b^2-\gamma^2)(4b^2-\gamma^2)^2} \tag{18}$$

We next examine the effects of the model parameters on the farmers' wholesale prices, the retailer's order quantities, the market prices, and the supply chain member's profits in sequence. Specifically, we study the effects of the cost difference μ and the degree of the retailer's cost sharing τ.

Proposition 2. *(i) w_1^S increases with μ if $\tau < 2b^2/\left(4b^2-\gamma^2\right)$, and decreases with μ if $2b^2/\left(4b^2-\gamma^2\right) < \tau < 1$; (ii) w_2^S, p_1^S, p_2^S, q_2^S increase with μ, and q_1^S decreases with μ; (iii) π_1^S, π_R^S decrease with μ and π_2^S increase with μ; (iv) when $0 < \mu < \mu_0$, π_{SC}^S decreases with μ; when $\mu > \mu_0$, π_{SC}^S increases with μ, where $\mu_0 = (a_1 + \kappa\tau) - a_2\left(8b^3\gamma - 3b\gamma^3\right)/\left(12b^4 - 9b^2\gamma^2 + 2\gamma^4\right)$.*

Proposition 2 indicates that when the degree of the retailer's CSR practice is low ($\tau < 2b^2/\left(4b^2-\gamma^2\right)$), the retailer shares a small proportion of the cost difference, and farmer 1 still bears most of the cost difference. The higher the cost difference μ, the higher the wholesale price w_1 will be established to maintain farmer 1's optimal profit. It will force the retailer to increase the market price (p_1), which leads to the lower demand for product 1 (q_1), and then affects the profit of product 1; that is, farmer 1 obtains a lower profit. When the degree of the retailer's CSR practice is high ($2b^2/\left(4b^2-\gamma^2\right) < \tau < 1$), most of the cost difference will be transferred to the retailer. Farmer 1 will set a lower wholesale price despite facing a higher cost difference μ. However, it will induce the retailer to set a higher market price for product 1, leading to a lower demand for product 1. Thus, farmer 1 still obtains a lower profit; the cost difference always hurts farmer 1. As we know from Proposition 1, farmer 2 has a competitive advantage over farmer 1 for the increasing cost difference, which always benefits farmer 2. Since the demand of product 1 can be transferred to product 2 through competition, thus, not only the wholesale price w_2 and the market price p_2 will be set higher, but also the demand q_2 for product 2 increases, which will increase farmer 2's profit. As for the retailer, even though his profit is affected by product 1 and product 2, he takes on a certain proportion of the cost difference μ as implementing social responsibility. Thus, his profit will decrease with increasing cost differences.

However, the impact of cost difference on supply chain profit falls into two cases, which are affected by the relationship between the elasticity of the sale price κ and the potential markets of product 1 (a_1) and product 2 (a_2). When the cost difference is small ($0 < \mu < \mu_0$), the supply chain profit will decrease with μ. The reason is that a smaller cost difference ($\mu < \mu_0$) means that the consumers are willing to pay more for the retailer's CSR practice (a larger κ) and prefer product 1 over product 2 ($a_1 > a_2$). There may be little room for the retailer's CSR practice to be effective; that is, the increase in farmer 1's profit is less than the retailer's giveaway. As for case two, when the cost difference is slightly higher than the threshold (μ_0), the supply chain profit may benefit from μ. The probable reason is that a high cost difference may undermine the advantage of product 1 over product 2, which means that the retailer's CSR practice is effective, allowing farmer 1 to obtain more profit than what the retailer provides. Therefore, in poverty alleviation, on the one hand, firms and governments should improve the business infrastructure to minimize cost differences for those anti-poverty products that will help the supply chain to obtain a better profit (case one). On the other hand, for those anti-poverty products with a relatively high cost difference, firms should seek methods to improve the effectiveness of CSR practices, which may help to improve the supply chain profit, e.g., taking measures of publicity to strengthen consumers' recognition of CSR. Otherwise, it may need support from the government.

Proposition 3. *(i) w_1^S increases with τ if $(2b^2-\gamma^2)\kappa-(4b^2-\gamma^2)\mu>0$; otherwise, it decreases; (ii) p_1^S, q_1^S increase with τ, and w_2^S, p_2^S, q_2^S decrease with τ; (iii) π_1^S, π_R^S increase with τ, and π_2^S decreases with τ; (iv) when $0<\tau<\min(\tau_0,1)$, π_{SC}^S decreases with τ, and when $\max(0,\tau_0)<\tau<1$, π_{SC}^S increases with τ, where $\tau_0=\frac{1}{\kappa}\left[\frac{a_2(8b^3\gamma-3b\gamma^3)}{(12b^4-9b^2\gamma^2+2\gamma^4)}-(a_1-\mu)\right]$.*

Proposition 3 indicates that, relative to the cost difference, when the elasticity of the sale price concerning the degree of cost sharing is large ($\kappa>\mu(4b^2-\gamma^2)/(2b^2-\gamma^2)$), consumers are more willing to pay extra money for product 1 after observing the retailer's CSR practice. With a higher degree of cost sharing τ, not only the wholesale price w_1 and the market price p_1 will be set higher, but also the demand of product 1 increases, which will increase farmer 1's profit. Furthermore, relative to the cost difference, when the elasticity of the sale price concerning the degree of cost sharing is small ($0<\kappa<\mu(4b^2-\gamma^2)/(2b^2-\gamma^2)$), farmer 1 will set a lower wholesale price even though the retailer implements a higher degree of cost sharing τ. However, because of the consumer purchasing intention, the retailer still can set a higher market price p_1 and obtain a higher demand for product 1 at the same time, which means that farmer 1 obtains a higher profit. Thus, the retailer's cost-sharing practices always benefit farmer 1.

Farmer 2 has a competitive disadvantage over farmer 1 for the increasing degree of the retailer's cost sharing, which always hurts farmer 2. Since the demand of product 2 can be transferred to product 1 through competition, thus, not only the wholesale price w_2 and the market price p_2 will be set lower, but also the demand q_2 of product 2 will decrease, which will decrease farmer 2's profit.

Since the retailer's profit is affected by product 1 and product 2, the retailer's profit will increase with an increasing degree of cost sharing τ, which means that the retailer's cost-sharing practice benefits not only farmer 1 but also himself. It proves that the consumer market recognizes the retailer's CSR practice to some extent. Thus, higher cost sharing may help the retailer to obtain better profits.

As for the supply chain, the effect of the cost sharing on its profit depends on the relationship between the elasticity of the sale price κ, the potential markets of product 1 (a_1), and the cost difference μ. For example, a higher degree of cost sharing τ may induce a higher supply chain profit because consumers are willing to pay more for the retailer's CSR practice (a larger κ) and prefer product 1 with a low cost difference than product 2 ($a_1>a_2$). Thus, the management implication is similar to Proposition 2. In poverty alleviation, on the one hand, firms and governments should build brands for poverty-alleviation products and pay attention to quality, packaging, advertising, etc., to improve their market potential. On the other hand, firms can organize various activities to promote CSR, which may help to strengthen consumers' recognition of their CSR behaviors.

Let $\Delta=p_1^N-\mu$ to capture the added value of the anti-poverty product. $\pi_{sc}^S=\pi_1^S+\pi_2^S+\pi_R^S$ is the supply chain profit in Model S. Substituting $\mu=p_1^N-\Delta$ into π_{sc}^S, we obtain Proposition 4.

Proposition 4. *(i) When $0<\tau<\min(\tau_{\Delta S},1)$, π_{sc}^S decreases first and then increases with Δ; when $\min(\tau_{\Delta S},1)<\tau<1$, it increases with Δ; (ii) when $0<\Delta<\Delta_{\tau S}$, π_{sc}^S decreases first and then increases with τ; when $\Delta_{\tau S}<\Delta$, π_{sc}^S increases with τ, where $\tau_{\Delta S}=\frac{a_2b\gamma(36b^4-25b^3\gamma^2+4\gamma^4)}{\kappa(72b^6-78b^4\gamma^2+30b^2\gamma^4-4\gamma^6)}$, $\Delta_{\tau S}=\frac{a_2b\gamma(9b^2-4\gamma^2)}{24b^4-18b^2\gamma^2+4\gamma^4}$.*

Proposition 4 shows the interaction effect of the added value Δ and the degree of cost sharing τ on the supply chain profit. To obtain a higher supply chain profit, on the one hand, from the retailer's perspective, if farmer 1 has produced high-value-added products, e.g., medicinal materials, then the retailer should implement a higher degree of CSR practice—that is, to share a greater proportion of the cost difference. If farmer 1 has produced low-value-added products, e.g., rice or soybeans, the retailer should implement a slightly lower CSR practice to avoid reducing the supply chain profit. When farmer 1 has planted low-value-added products, a higher retailer's CSR practice may make competition between farmers fiercer, which will not help to increase the supply chain profit.

On the other hand, from farmer 1's perspective, if the retailer has a high degree of CSR practice, farmer 1 should plant products with high added value. If the retailer performs a low degree of CSR practice, farmer 1 should grow low-value-added products. Farmer 1 should not grow medium-value-added products, which do not help to increase the supply chain profit. The possible reason may be that the threshold for farmer 1 to plant high-value-added products is relatively high. The added value can compensate for the threshold cost, causing the supply chain to profit at a higher level. For example, it requires farmer 1 to have increased operational capacity and investment to produce rare medicinal materials, but the added value of these products is much higher than the cost. However, the value of medium-value-added products cannot cover the threshold cost for farmer 1. For instance, farmers from poor areas in China invest in such medium-value-added products as fruits, e.g., mandarin oranges. Even if such a product does not require a significant investment, it cannot help them to obtain a better profit since it cannot be sold at a higher price.

4.3. Model A: Model with an Altruistic Preference Retailer

Altruistic preference behavior is another type of retailer's CSR practice in which the retailer takes farmers' profit into account when making decisions. Let the degree of the retailer's altruistic preference for farmer 1 be θ $(0 < \theta < 1)$, which means that the retailer favors farmer 1. The consumers are willing to pay more for an anti-poverty product labeled with the retailer's altruistic preference practice. Thus, the demand function of product 1 becomes $p_1 = a_1 - bq_1 - \gamma q_2 + \kappa\theta$ [42,43], and the demand function of product 2 remains the same as $p_2 = a_2 - bq_2 - \gamma q_1$.

Farmer 1's profit is determined by

$$\pi_1 = (w_1 - \mu)q_1 \tag{19}$$

Farmer 2's profit is determined by

$$\pi_2 = w_2 q_2 \tag{20}$$

The retailer's utility is determined by

$$U_R = (p_1 - w_1)q_1 + (p_2 - w_2)q_2 + \theta\pi_1 \tag{21}$$

Lemma 3. *In Model A, (i) the farmers' equilibrium wholesale prices are*

$$w_1^A = \frac{2a_1b^2 - a_1\gamma^2 + 2b^2\mu + 2b^2\kappa\theta - 4b^2\mu\theta - \gamma^2\kappa\theta + \gamma^2\mu\theta - a_2b\gamma}{(4b^2 - \gamma^2)(1 - \theta)} \tag{22}$$

$$w_2^A = -\frac{a_2\gamma^2 - 2a_2b^2 + a_1b\gamma - b\gamma\mu + b\gamma\kappa\theta}{4b^2 - \gamma^2} \tag{23}$$

The retailer's equilibrium order quantities are

$$q_1^A = -\frac{b(a_1\gamma^2 - 2a_1b^2 + 2b^2\mu - \gamma^2\mu - 2b^2\kappa\theta + \gamma^2\kappa\theta + a_2b\gamma)}{2(4b^4 - 5b^2\gamma^2 + \gamma^4)} \tag{24}$$

$$q_2^A = -\frac{b(a_1\gamma^2 - 2a_1b^2 + 2b^2\mu - \gamma^2\mu - 2b^2\kappa\theta + \gamma^2\kappa\theta + a_2b\gamma)}{2(4b^4 - 5b^2\gamma^2 + \gamma^4)} \tag{25}$$

(ii) The retailer's equilibrium profit is $\pi_R^A = (p_1^A - w_1^A)q_1^A + (p_2^A - w_2^A)q_2^A$, and farmers 1's equilibrium profit and farmer 2's equilibrium profit are as follows:

$$\pi_1^A = \frac{b\left(a_1\gamma^2 - 2a_1b^2 + 2b^2\mu - \gamma^2\mu - 2b^2\kappa\theta + \gamma^2\kappa\theta + a_2b\gamma\right)^2}{2(b^2-\gamma^2)(4b^2-\gamma^2)^2(1-\theta)} \quad (26)$$

$$\pi_2^A = \frac{b\left(a_2\gamma^2 - 2a_2b^2 + a_1b\gamma - b\gamma\mu + b\gamma\kappa\theta\right)^2}{2(b^2-\gamma^2)(4b^2-\gamma^2)^2} \quad (27)$$

We next examine the effects of the model parameters on the farmers' wholesale prices, the retailer's order quantities, the market prices, and the supply chain member's profits in sequence. Specifically, we study the effects of the cost difference μ and the degree of the retailer's altruistic preference θ for farmer 1.

Proposition 5. *(i) w_1^A increases with μ if $0 < \theta < \theta_0$, and decreases with θ if $\theta_0 < \theta < 1$, where $\theta_0 = \frac{2b^2}{4b^2-\gamma^2}$; (ii) w_2^A, p_1^A, p_2^A, q_2^A increase with μ; and q_1^A decreases with μ; (iii) π_1^A decreases with μ, and π_2^A increases with μ; (iv) If $0 < \theta < \theta_1$ ($\theta_1 < \theta < 1$), then when $0 < \mu < \mu_1$, π_R^A decreases (increases) with μ; and when $\mu > \mu_1$, π_R^A increases (decreases) with μ, where $\theta_1 = \frac{4b^4-3b^2\gamma^2}{12b^4-11b^2\gamma^2+2\gamma^4}$, $\mu_1 = a_1 + \theta\kappa - \frac{a_2(4b^3\gamma\theta - b\gamma^3\theta - b\gamma^3)}{(12b^4-11b^2\gamma^2+2\gamma^4)\theta - 4b^4 + 3b^2\gamma^2}$; (v) when $0 < \mu < \mu_2$, π_{SC}^A decreases with μ; and when $\mu > \mu_2$, π_{SC}^A increases with μ, where $\mu_2 = (a_1 + \kappa\theta) - \frac{a_2(8b^3\gamma - 3b\gamma^3)}{12b^4-9b^2\gamma^2+2\gamma^4}$.*

Proposition 5 indicates that when the degree of the retailer's CSR practice is low ($0 < \theta < \theta_0$)—that is, the retailer takes a small proportion of farmers' profit into account when making decisions—the higher the cost difference μ, the higher the wholesale price w_1 will be established to maintain farmer 1's optimal profit. It will force the retailer to increase the market price (p_1), which leads to the lower demand q_1 of product 1, and then affects the profit of product 1; that is, farmer 1 obtains a lower profit. When the degree of the retailer's CSR practice is high ($\theta_0 < \theta < 1$), it means that the retailer is willing to transfer a higher part of the profit to farmer 1 when making a decision. Farmer 1 will set a lower wholesale price despite facing a higher cost difference. However, this will induce the retailer to set a higher market price for product 1, which leads to a lower demand of product 1. Thus, farmer 1 still obtains a lower profit; that is, the cost difference always hurts farmer 1. As we know from Propositions 1 and 2, farmer 2 has a competitive advantage over farmer 1 for the increasing cost difference, which always benefits farmer 2. Since the demand of product 1 can be transferred to product 2 through competition, thus, not only the wholesale price w_2 and the market price p_2 will be set higher, but also the demand q_2 of product 2 increases, which will increase farmer 2's profit.

As we know, the retailer's profit is affected by product 1 and product 2. There are two ways for the retailer to increase his profit when facing the increasing cost difference. The first one is that when the cost difference is low ($0 < \mu < \mu_1$), the retailer can implement a high degree of CSR practice ($\theta_1 < \theta < 1$). The reason is that a high degree of CSR practice helps to mitigate the double marginal effect between farmer 1 and the retailer, which results in a better profit. The second one is that when the cost difference is high ($\mu > \mu_1$), the retailer can implement a low degree of CSR practice ($0 < \theta < \theta_1$). It means that farmer 1 bears most of the increasing cost difference, which may cause the retailer to take part of the profit from farmer 1.

From the perspective of the supply chain, the impact of the cost difference on the supply chain profit is affected by the relationship between the elasticity of the sale price κ and the potential markets of product 1 (a_1) and product 2 (a_2), which is similar to Proposition 2(iv). For example, when the cost difference is slightly higher than the threshold (μ_2), the supply chain profit may benefit from μ. The reason may be that with a slightly higher cost difference, the marginal revenue of the retailer's CSR practice for farmer 1 is higher than the marginal cost of CSR for the retailer, which means that the retailer's CSR practice is effective.

Proposition 6. *(i) w_1^A increases (decreases) with θ if $(a_1-\mu+\kappa) > (<) \frac{a_2 b\gamma}{2b^2-\gamma^2}$; (ii) p_1^A, q_1^A increase with θ; and w_2^A, p_2^A, q_2^A decrease with θ; (iii) π_2^A decreases with θ; (iv) when $(a_1+\kappa\theta-\mu) < \frac{a_2 b\gamma}{2b^2-\gamma^2} < [a_1+\kappa(2-\theta)-\mu]$, π_1^A decreases with θ; when $\frac{a_2 b\gamma}{2b^2-\gamma^2} < (a_1+\kappa\theta-\mu)$ or $\frac{a_2 b\gamma}{2b^2-\gamma^2} > [a_1+\kappa(2-\theta)-\mu]$, π_1^A increases with θ; (v) when $0<\theta<\min(\theta_2,1)$, π_{SC}^A decreases with θ; when $\max(\theta_2,0)<\theta<1$, π_{SC}^A increases with θ, where $\theta_2 = \frac{1}{\kappa}\left[\frac{a_2\left(8b^3\gamma-3b\gamma^3\right)}{\left(12b^4-9b^2\gamma^2+2\gamma^4\right)}-(a_1-\mu)\right]$.*

Proposition 6 indicates that if the maximum unit profit of product 1 is relatively greater than that of product 2 $(a_1-\mu+\kappa) > a_2 b\gamma/\left(2b^2-\gamma^2\right)$, farmer 1 will set a higher wholesale price even though the retailer implements a higher degree of altruistic preference. On the other hand, if the maximum unit profit of product 1 is relatively smaller than that of product 2 $(a_1-\mu+\kappa) < a_2 b\gamma/\left(2b^2-\gamma^2\right)$, the higher degree of the retailer's altruistic preference for farmer 1 leads to a lower wholesale price w_1. However, because of consumers' purchasing intentions, the retailer can still set a higher market price p_1 and obtain a higher demand q_1 of product 1. Thus, farmer 1 may obtain a higher profit when the retailer performs a higher level of altruistic preference. Since the retailer's altruistic preference practice is motivated by helping farmer 1, farmer 2 has a competitive disadvantage over farmer 1, which is always not beneficial for farmer 2. The higher the degree of altruistic preference, the more demand of product 2 can be transferred to product 1 through competition. Thus, not only the wholesale price w_2 and the market price p_2 will be set lower, but also the demand q_2 of product 2 will decrease, which will decrease farmer 2's profit. The effect of the retailer's altruistic preference practice on the supply chain profit depends on the relationship between the elasticity of the sale price κ, the potential markets a_1 of product 1, and the cost difference μ, which is similar to Proposition 3(iv). For example, a higher degree of altruistic preference θ may induce a higher supply chain profit because the consumers are willing to pay more for the retailer's altruistic preference practice (a larger κ) and prefer product 1 with a low cost difference than product 2 $(a_1 > a_2)$. Thus, in addition to the measures proposed in Proposition 3, firms and governments can work together to improve the business infrastructure to reduce cost differentials, which may help in poverty alleviation.

Let $\Delta = p_1^N - \mu$ to capture the added value of a poverty-alleviation-labeled product. $\pi_{sc}^S = \pi_1^S + \pi_2^S + \pi_R^S$ is the supply chain profit. Substituting $\mu = p_1^N - \Delta$ into π_{sc}^S, we have Proposition 7.

Proposition 7. *(i) When $0<\theta<\min(\theta_{\Delta A},1)$, π_{sc}^A decreases first and then increases with Δ; when $\min(\theta_{\Delta A},1)<\theta<1$, π_{sc}^A increases with Δ; (ii) when $0<\Delta<\Delta_{\theta A}$, π_{sc}^A decreases first and then increases with θ; when $\Delta_{\theta A}<\Delta$, π_{sc}^A increases with θ, where $\theta_{\Delta A} = \frac{a_2 b\gamma\left(36b^4-25b^3\gamma^2+4\gamma^4\right)}{\kappa\left(72b^6-78b^4\gamma^2+30b^2\gamma^4-4\gamma^6\right)} \equiv \tau_{\Delta S}$; $\Delta_{\theta A} = \frac{a_2 b\gamma\left(36b^4+4\gamma^4-25b^2\gamma^2\right)}{96b^6-96b^4\gamma^2+34b^2\gamma^4-4\gamma^6}$.*

Proposition 7 shows the interaction effect of the added value Δ and the degree of the retailer's altruistic preference θ on the supply chain profit. From the perspective of farmer 1, she should produce crops with high added value after observing the retailer implementing a high level of altruistic preference. Otherwise, farmer 1 should produce crops with low added value. From the perspective of the retailer, if farmer 1 has produced high-value-added crops, the retailer should perform a high degree of altruistic preference. If farmer 1 has produced low-value-added products, the retailer should implement a slightly lower degree of CSR practice to avoid reducing the supply chain profit. Thus, only the retailer and farmer 1 cooperate well, which can maximize the supply chain profit.

4.4. Model F: Model with a Fair-Minded Retailer

Fairness concerns practice is a type of retailer's indirect CSR practice for farmer 1, and will affect the distribution of supply chain profits in the supply chain [47]. The fair-minded retailer takes the order quantities of product 1 and product 2 into account when making decisions. We model fairness concerns practice as quantities inequity aversion such that the fair-minded retailer is willing to "give up some monetary payoff to move in the direction of more equitable outcomes" [31,48]. Let the degree of the retailer's fairness concerns be ϕ. Since farmer 2 has an advantage over farmer 1 because of the cost difference, the retailer's fairness concerns practice means caring for farmer 1. Since the consumers are willing to pay more for an anti-poverty product labeled with the retailer's fairness concern practice, the demand function of product 1 becomes $p_1 = a_1 - bq_1 - \gamma q_2 + \kappa\phi$ [42,43] and the demand function of product 2 remains the same as $p_2 = a_2 - bq_2 - \gamma q_1$.

Farmer 1's profit is determined by

$$\pi_1 = (w_1 - \mu)q_1 \tag{28}$$

Farmer 2's profit is determined by

$$\pi_2 = w_2 q_2 \tag{29}$$

The retailer's profit is determined by

$$\pi_R = (p_1 - w_1)q_1 + (p_2 - w_2)q_2 \tag{30}$$

The retailer's utility is determined by

$$U_R = (p_1 - w_1)q_1 + (p_2 - w_2)q_2 - \phi(q_2 - q_1) \tag{31}$$

Lemma 4. *In Model F, (i) the farmers' equilibrium wholesale prices are*

$$w_1^F = \frac{2a_1b^2 - a_1\gamma^2 + 2b^2\mu + 2b^2\phi - \gamma^2\phi - \gamma^2\kappa\phi - a_2b\gamma + b\gamma\phi + 2b^2\kappa\phi}{4b^2 - \gamma^2} \tag{32}$$

$$w_2^F = -\frac{a_2\gamma^2 - 2a_2b^2 + 2b^2\phi - \gamma^2\phi + a_1b\gamma - b\gamma\mu + b\gamma\phi + b\gamma\kappa\phi}{4b^2 - \gamma^2} \tag{33}$$

The retailer's equilibrium order quantities are

$$q_1^F = \frac{b\left(2a_1b^2 - a_1\gamma^2 - 2b^2\mu + 2b^2\phi + \gamma^2\mu - \gamma^2\phi - \gamma^2\kappa\phi - a_2b\gamma + b\gamma\phi + 2b^2\kappa\phi\right)}{2(4b^4 - 5b^2\gamma^2 + \gamma^4)} \tag{34}$$

$$q_2^F = -\frac{b\left(a_2\gamma^2 - 2a_2b^2 + 2b^2\phi - \gamma^2\phi + a_1b\gamma - b\gamma\mu + b\gamma\phi + b\gamma\kappa\phi\right)}{2(4b^4 - 5b^2\gamma^2 + \gamma^4)} \tag{35}$$

(ii) The retailer's equilibrium profit is $\pi_R^F = (p_1^F - w_1^F)q_1^F + (p_2^F - w_2^F)q_2^F$, and farmer 1's equilibrium profit and farmer 2's equilibrium profit are as follows:

$$\pi_1^F = \frac{b\left(2a_1b^2 - a_1\gamma^2 - 2b^2\mu + 2b^2\phi + \gamma^2\mu - \gamma^2\phi - \gamma^2\kappa\phi - a_2b\gamma + b\gamma\phi + 2b^2\kappa\phi\right)^2}{2(b^2 - \gamma^2)(4b^2 - \gamma^2)^2} \tag{36}$$

$$\pi_2^F = \frac{b\left(a_2\gamma^2 - 2a_2b^2 + 2b^2\phi - \gamma^2\phi + a_1b\gamma - b\gamma\mu + b\gamma\phi + b\gamma\kappa\phi\right)^2}{2(b^2 - \gamma^2)(4b^2 - \gamma^2)^2} \tag{37}$$

We next examine the effects of the model parameters on the farmers' wholesale prices, the retailer's order quantities, the market prices, and the supply chain member's profits in sequence. Specifically, we study the effects of the cost difference μ and the degree of retailer's fairness concerns ϕ here.

Proposition 8. *(i) w_1^F, w_2^F, p_1^F, p_2^F, q_2^F, π_2^F increase with μ; (ii) q_1^F, π_1^F decrease with μ; (iii) π_R^F decreases with μ in $0 < \mu < \mu_3$, and increases with μ in $\mu > \mu_3$, where $\mu_3 = (a_1 + \phi\kappa) - \frac{a_2 b\gamma^3 + \phi(b+\gamma)^2(2b-\gamma)^2}{b^2(4b^2-3\gamma^2)}$; (iv) π_{SC}^F decreases with μ in $0 < \mu < \mu_4$, and increases with μ in $\mu > \mu_4$, where $\mu_4 = (a_1 + \phi\kappa) - \frac{a_2(8b^3\gamma - 3b\gamma^3) - \phi(2b-\gamma)^2(b+\gamma)^2}{(12b^4 - 9b^2\gamma^2 + 2\gamma^4)}$.*

Proposition 8 indicates that, on the one hand, the higher the cost difference μ, the higher the wholesale price w_1 will be established to maintain farmer 1's optimal profit. The higher wholesale price w_1 will force the retailer to increase the market price p_1, which leads to the lower demand q_1 of product 1, affecting the profit of product 1; that is, farmer 1 obtains a lower profit. On the other hand, farmer 2 has a competitive advantage over farmer 1 for the increasing cost difference, which always benefits farmer 2. Since the demand of product 1 can be transferred to product 2 through competition, thus, not only the wholesale price w_2 and the market price p_2 will be set higher, but also the demand q_2 of product 2 increases, which will increase farmer 2's profit.

However, the effect of the cost difference on the retailer's profit and supply chain profit are similar. When the cost difference is small ($0 < \mu < \mu_3$, $0 < \mu < \mu_4$), both the retailer's profit and supply chain profit will decrease. A small cost difference means that the competitive advantages of product 1 and product 2 are similar. Thus, there may be little room for the retailer's CSR practice to be effective; the increase in farmer 1's profit is less than the retailer's giveaway.

When the cost difference is slightly higher than the threshold (μ_3, μ_4), both the retailer's profit and supply chain profit may benefit from μ. In this situation, farmer 1 has a competitive disadvantage over farmer 2, which means that the marginal revenue of the retailer's CSR practice for farmer 1 is high. To some extent, it demonstrates the retailer's CSR practice's effectiveness.

Proposition 9. *(i) w_1^F, q_1^F increase with ϕ; (ii) w_2^F, q_2^F, π_2^F decrease with ϕ; (iii) when $\kappa > (<)(2b^2 - b\gamma)/(6b^2 - 2\gamma^2)$, p_1^F increases (decreases) with ϕ; when $\kappa > (<)(2b - \gamma)/\gamma$, p_2^F increases (decreases) with ϕ; when $\kappa > (2b^2 - b\gamma)/(2b^2 - \gamma^2)$, π_1^F increases with ϕ.*

Proposition 9 shows that a retailer's greater fairness concern can help farmer 1 to obtain a higher profit if consumers have relatively high recognition of the retailer's CSR practice ($\kappa > (2b^2 - b\gamma)/(2b^2 - \gamma^2)$). On the one hand, if consumers have relatively high recognition of the retailer's CSR practice ($\kappa > (2b^2 - b\gamma)/(6b^2 - 2\gamma^2)$), the market price of product 1 will be higher. On the other hand, a fair-minded retailer is more willing to purchase from farmer 1 (larger q_1^F), which may help farmer 1 to increase slightly the wholesale price w_1^F. Thus, farmer 1 will obtain a higher profit. As for farmer 2, since the retailer's purchases are skewed towards farmer 1, farmer 2 has a competitive disadvantage over farmer 1, which means that the order quantity of product 2 will decrease. Furthermore, the market price of product 2 will also decrease due to consumers preferring product 1, which can lead to a lower wholesale price w_2^F of product 2. Thus, farmer 2 will also obtain a lower profit.

Let $\Delta = p_1^N - \mu$ to capture the added value of the anti-poverty product. $\pi_{sc}^F = \pi_1^F + \pi_2^F + \pi_R^F$ is the supply chain profit. Substituting $\mu = p_1^N - \Delta$ into π_{sc}^F, we obtain Proposition 10.

Proposition 10. *(i) When $0 < \phi < \min(\phi_{\Delta F}, 1)$, π_{sc}^F decreases first and then increases with Δ; when $\min(\phi_{\Delta F}, 1) < \phi < 1$, π_{sc}^F increases with Δ; (ii) when $0 < \Delta < \Delta_{\phi F}$, π_{sc}^F decreases first and then increases with ϕ; when $\Delta_{\phi F} < \Delta$, π_{sc}^F increases with ϕ, where $\Delta_{\phi F} =$*

$$\frac{\begin{bmatrix} a_2(4b\gamma^5+36b^5\gamma-25b^3\gamma^3)-\kappa\varphi(72b^6-78b^4\gamma^2-4\gamma^6+30b^2\gamma^4) \\ -\varphi(24b^6-2\gamma^6+12b^2\gamma^4-20b^3\gamma^3-26b^4\gamma^2+4b\gamma^5+24b^5\gamma) \end{bmatrix}}{(96b^6-96b^4\gamma^2+34b^2\gamma^4-4\gamma^6)} \quad and \quad \phi_{\Delta F} = \frac{(\gamma^2-4b^2)\begin{pmatrix} 6a_2b^2\gamma^2-6a_2b^4-2a_2\gamma^4+3a_2b\gamma^3-5a_2b^3\gamma+4a_2b\gamma^3\kappa-9a_2b^3\gamma\kappa \\ +24\Delta b^4\kappa+4\Delta\gamma^4\kappa-18\Delta b^2\gamma^2\kappa+8\Delta b^4+2\Delta\gamma^4-6\Delta b^2\gamma^2-4\Delta b\gamma^3+8\Delta b^3\gamma \end{pmatrix}}{2(3b^2-\gamma^2)\begin{pmatrix} 12b^4\kappa^2+8b^4\kappa+8b^3\gamma\kappa-9b^2\gamma^2\kappa^2-6b^2\gamma^2\kappa \\ -4b\gamma^3\kappa+2\gamma^4\kappa^2+2\gamma^4\kappa-8b^4+6b^2\gamma^2-2b\gamma^3 \end{pmatrix}}.$$

Proposition 10 shows the interaction effect of the added value Δ and the degree of retailer's fairness concerns ϕ on supply chain profit. There is a game between farmer 1 and the retailer to maximize the supply chain profit; each one can adopt a strategy based on the other's action. From the perspective of farmer 1, she should produce crops with high value-added after observing a retailer with high fairness concerns. Otherwise, farmer 1 should produce crops with low added value. From the retailer's perspective, if farmer 1 has produced high-value-added crops, the retailer should perform a high degree of fairness concern. If farmer 1 has produced low-value-added products, the retailer should implement a slightly lower degree of CSR practice to avoid reducing the supply chain profit. Thus, the retailer and farmer 1 can work together to maximize the supply chain profit.

5. Comparison of Four Models

This section discusses the comparison of Model N, Model S, Model A, and Model F. Since the two farmers seek competitive pricing policies while the retailer is making optimal quantities decisions, each player considers interaction with the competitor to maximize their individual profit. We compare the equilibria of each player in the same degree of the retailer's CSR practice and let $\theta = \phi = \tau$.

Proposition 11. *(i)* $w_2^A = w_2^S, q_1^A = q_1^S, q_2^A = q_2^S, p_1^A = p_1^S, p_2^A = p_2^S$; *when* $2a_1b^2 - a_1\gamma^2 + 2b^2\mu - a_2b\gamma + \tau(2b^2\kappa - \gamma^2\kappa - 4b^2\mu + \gamma^2\mu) > (<)0$, $w_1^A > (<)w_1^S$. *(ii)* $w_1^F > w_1^S, w_2^F < w_2^S, q_1^F > q_1^S, q_2^F < q_2^S, p_1^F < p_1^S, p_2^F > p_2^S$. *(iii)* $w_2^F < w_2^A, q_1^F > q_1^A, q_2^F < q_2^A, p_1^F < p_1^A, p_2^F > p_2^S$; *when* $\tau < (>)\frac{b\gamma-2a_1b^2+a_1\gamma^2+2b^2\mu-\gamma^2\mu+2b^2-\gamma^2+a_2b\gamma}{(2b^2-\gamma^2+b\gamma+2b^2\kappa-\gamma^2\kappa)}$, $w_1^F > (<)w_1^A$.

Proposition 11 shows the comparison of decision equilibria between the four models. In Models A and S, the decision equilibria of product 2 are the same ($w_2^A = w_2^S, q_2^A = q_2^S, p_2^A = p_2^S$). Furthermore, the order quantities q_1 and market price p_1 of product 1 remain the same. It means that the form of the retailer's CSR practices will not affect the overall profit of the supply chain, but will affect the profit distribution among the supply chain in Model A and Model S. As for Model F, compared with Models S and A, the retailer will order more product 1 ($q_1^F > q_1^S, q_1^F < q_1^A$), but has a lower market price ($p_1^F < p_1^S, p_1^F < p_1^A$). However, the retailer will order product 2 in a lower quantity ($q_2^F < q_2^S, q_2^F < q_2^A$), but has a higher market price ($p_2^F > p_2^S, p_2^F > p_2^A$). Thus, the retailer implementing different CSR practices can affect the strategic equilibria of supply chain members.

Proposition 12. *(i)* $\pi_1^S > \pi_1^N, \pi_1^A > \pi_1^N, \pi_1^F > \pi_1^N, \pi_1^A > \pi_1^S, \pi_1^F > \pi_1^S$; *(ii) there exists a* τ_1 *whereby, when* $0 < \tau \le \tau_1$, $\pi_1^F \ge \pi_1^A$, *and when* $\tau_1 < \tau < 1$, $\pi_1^F < \pi_1^A$.

Proposition 12 indicates that the retailer's CSR practices are beneficial for farmer 1, which means that the retailer's CSR practices are conducive to poverty alleviation ($\pi_1^S > \pi_1^N$, $\pi_1^A > \pi_1^N$, $\pi_1^F > \pi_1^N$). Furthermore, from farmer 1's perspective, both the retailer's altruistic preference practice and fairness concerns practice are better than the cost-sharing practice ($\pi_1^A > \pi_1^S$, $\pi_1^F > \pi_1^S$). However, the question of which CSR practice is the best for farmer 1 depends on the extent to which the retailer implements it. When the retailer implements a lower degree of CSR practice ($0 < \tau \le \tau_1$), the retailer's fairness concerns practice is better than the altruistic preference practice ($\pi_1^F \ge \pi_1^A$). Otherwise, the retailer's altruistic preference practice is better ($\pi_1^F < \pi_1^A$). Therefore, the retailer should adopt corresponding CSR practices according to the effort he wishes to make.

Proposition 13. $\pi_2^S < \pi_2^N$, $\pi_2^A < \pi_2^N$, $\pi_2^F < \pi_2^N$, $\pi_2^A = \pi_2^S$, $\pi_2^F < \pi_2^S$, $\pi_2^F < \pi_2^A$.

Proposition 13 indicates that the retailer's CSR practices are not always beneficial for farmer 2 ($\pi_2^S < \pi_2^N$, $\pi_2^A < \pi_2^N$, $\pi_2^F < \pi_2^N$). As we know from Proposition 11(i), the decision equilibria of product 2 are the same in Models A and S, which means that farmer 2's profit will be the same in these two models ($\pi_2^A = \pi_2^S$). Compared with Models A and S, farmer 2 will achieve the lowest profit if the retailer implements a fairness concerns practice ($\pi_2^F < \pi_2^S$, $\pi_2^F < \pi_2^A$). It means that the CSR practices implemented by the retailer always benefit farmer 1, but inevitably hurt farmer 2. Therefore, the retailer has to consider a tradeoff to decide which CSR practice is the most appropriate.

Proposition 14. *(i)* $\pi_R^S > \pi_R^N$, $\pi_R^S > \pi_R^A$; *(ii) When* $a_1 - \mu > a_2$, $\pi_R^S > \pi_R^F$.

Proposition 14 indicates that, compared with Model N and Model A, the retailer's profit has increased in Model S ($\pi_R^S > \pi_R^N$, $\pi_R^S > \pi_R^A$), which is consistent with what we find in Proposition 3. However, compared with the fairness concern practice, whether the retailer's cost-sharing practice is better for his profit depends on the competitive advantage between product 1 and product 2. When the potential market of product 1 has a competitive advantage over that of product 2 ($a_1 - \mu > a_2$), the cost-sharing practice is the retailer's best strategy ($\pi_R^S > \pi_R^F$), which also helps the retailer to achieve a win–win result at the same time; that is, the cost-sharing practice benefits not only farmer 1 but also the retailer himself.

Proposition 15. *(i) When* $(a_1 - \mu - a_2 + \kappa\tau)(b + \gamma) - b\tau > (<)0$, $\pi_{sc}^F > (<)\pi_{sc}^S$ *and* $\pi_{sc}^F > (<)\pi_{sc}^A$; *(ii)* $\pi_{sc}^A = \pi_{sc}^S$.

Proposition 15 indicates that the effect of the retailer's CSR practices on the supply chain profit depends on the relationship between the elasticity of the sale price κ, the potential markets of products (a_1, a_2), and the cost difference μ, which is consistent with the previous discussion. For example, the retailer's fairness concern practice may be better than the cost-sharing practice and the altruistic preference practice for the supply chain profit ($\pi_{sc}^F > \pi_{sc}^S$, $\pi_{sc}^F > \pi_{sc}^A$) under the condition that consumers are willing to pay more for the retailer's CSR practice (a larger κ) and prefer product 1 with the low cost difference than product 2 ($a_1 > a_2$). As we know from Proposition 11(i), the form of the retailer's CSR practices will not affect the supply chain profit in Models A and S; the retailer's altruistic preference practice and cost-sharing practice are equally effective for the supply chain profit ($\pi_{sc}^A = \pi_{sc}^S$). Thus, since it can change the effectiveness of CSR practices for the supply chain by adjusting a_1, κ, a_2, and μ, this may be the optimal solution for management to implement the retailer's best CSR practices.

6. Numerical Analysis

This section verifies the above results and further analyzes the impact of μ and the retailer's CSR practices on the four models' pricing, demand, and profit. A series of numerical analyses are carried out using MATLAB software. According to the allowed range of the parameters, let us set $a_1 = 10$, $a_2 = 13$, $b = 1.5$, $\gamma = 0.4$, $\kappa = 1$, $\mu = 3$, and then change the values of parameters τ, θ, and ϕ.

Figure 2a shows that without the retailer's CSR practices, the wholesale price w_1 of product 1 is lower than w_2 of product 2. Moreover, the influence of the retailer's different CSR practices on the wholesale prices is different. Firstly, with the increasing degree of cost-sharing practice, both wholesale prices w_1 and w_2 decrease. Secondly, with the growing degree of the retailer's altruistic preference and fairness concerns, the wholesale price w_1 of product 1 increases while the wholesale price w_2 of product 2 decreases. In addition, the wholesale price w_1 of product 1 with the altruistic preference practice is the highest, while that with cost-sharing practice is the lowest. Moreover, the wholesale price w_2 of product 2 with the altruistic preference practice is equal to that with the cost-sharing practice, while that with the retailer's fairness concern practice is the lowest. We also observe that there are interactions between w_1 and w_2. With a lower degree of the retailer's altruistic preference and fairness concern, w_1 is lower than w_2, while, with a higher degree of the retailer's altruistic preference and fairness concern, w_1 becomes higher than w_2. This may give the retailer an opportunity to manage farmers' decisions w_1 and w_2 by implementing different CSR practices in different degrees.

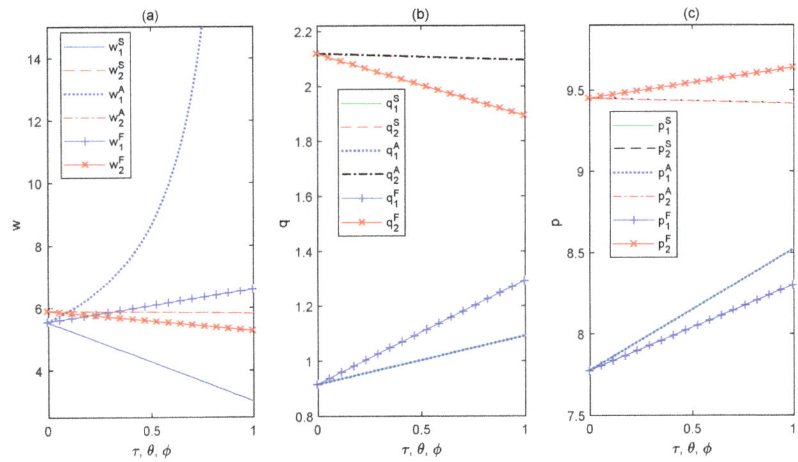

Figure 2. (**a**) Effects of the retailer's CSR practices on w_1 and w_2. (**b**) Effects of the retailer's CSR practices on q_1 and q_2. (**c**) Effects of the retailer's CSR practices on p_1 and p_2.

Figure 2b shows that without the retailer's CSR practices, the order quantity q_1 of product 1 is lower than q_2 of product 2. Moreover, the effect of the retailer's different CSR practices on order quantities is similar. With the increasing degree of the retailer's CSR practices, the order quantity of product 1 increases while that of product 2 decreases. We can also observe that the order quantity of products 1 and 2 in Model A is equal to that in Model S. In addition, the order quantity q_1 of product 1 with the cost-sharing practice is higher than that with the retailer's altruistic preference practice and fairness concern practice, and the order quantity q_2 of product 2 with the fairness concern practice is lower than that with the altruistic preference practice and cost-sharing practice.

Figure 2c shows that without the retailer's CSR practices, the market price p_1 of product 1 is lower than p_2 of product 2. Moreover, the effect of the retailer's different CSR practices on market prices is different. Firstly, with the increasing degree of the cost-sharing practice and altruistic preference practice, the market price of product 1 increases while that of product 2 decreases. Secondly, with the increasing degree of the retailer's fairness concern, both the market prices of product 1 and 2 increase. We can also observe that both the market prices of product 1 and 2 in Model A are equal to those in Model S. In addition, the market price p_1 of product 1 with the retailer's cost-sharing practice and altruistic preference practice is higher than that with the fairness concern practice, and the market price p_2 of product 2 with the fairness concern practice is higher than that with the cost-sharing practice and altruistic preference practice.

Figure 3a shows that without the retailer's CSR practices, farmer 1's profit π_1 is lower than farmer 2's profit π_2. Moreover, the effect of the retailer's different CSR practices on farmers' profits is similar. The retailer's CSR practices always benefit farmer 1's profit, while decreasing farmer 2's profit slightly. In addition, farmer 2's profit with the retailer's cost-sharing practice is equal to that with the altruistic preference practice, which is higher than that with the retailer's fairness concerns practice. However, with the retailer's altruistic preference practice, farmer 1's profit π_1 is higher than that with the retailer's fairness concern practice, which is higher than that with the cost-sharing practice.

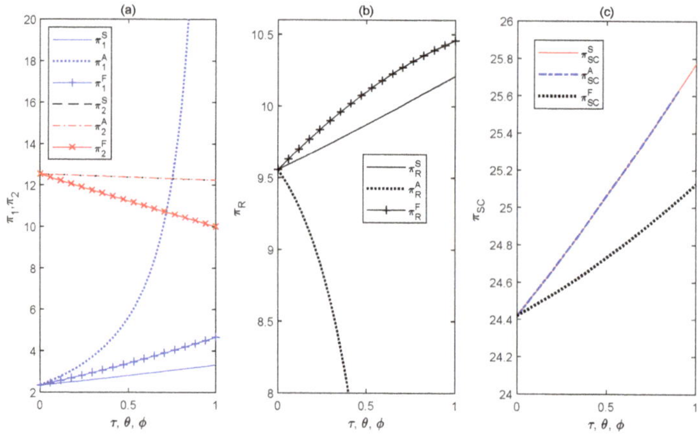

Figure 3. (**a**) Effect of retailer's CSR practice on π_1 and π_2. (**b**) Effect of retailer's CSR practice on π_R. (**c**) Effect of retailer's CSR practice on π_{SC}.

Figure 3b shows that the retailer's profit π_R will increase when he implements the cost-sharing practice and fairness concern practice, and his profit will decrease if he implements an altruistic preference practice. In addition, the retailer will obtain the highest profit when he implements fairness concern practice, and the lowest profit when he implements altruistic preference practice.

Figure 3c shows that the supply chain profit π_{SC} will increase when the retailer implements CSR practices, and the effect of cost-sharing practice and altruistic preference practice on supply chain profit are the same. In addition, with the retailer's cost-sharing practice and altruistic preference practice, the supply chain profit is higher than that with the retailer's fairness concerns practice.

Figure 4 demonstrates the effect of τ, θ, ϕ, and Δ on supply chain profit under different retailer CSR practices. Through the calculation, (1) we obtain $\Delta_{\tau S} = 1.329 > 0$. It can be observed from Figure 4a that when $\tau = 0$, π_{SC}^S decreases with $0 < \Delta < 1.329$ and increases with $\Delta > 1.329$. Furthermore, when $\Delta = 0$, π_{SC}^S decreases with τ. When Δ is large, e.g., $\Delta = 4$, π_{SC}^S increases with τ. Thus, Figure 4a can verify Proposition 4.

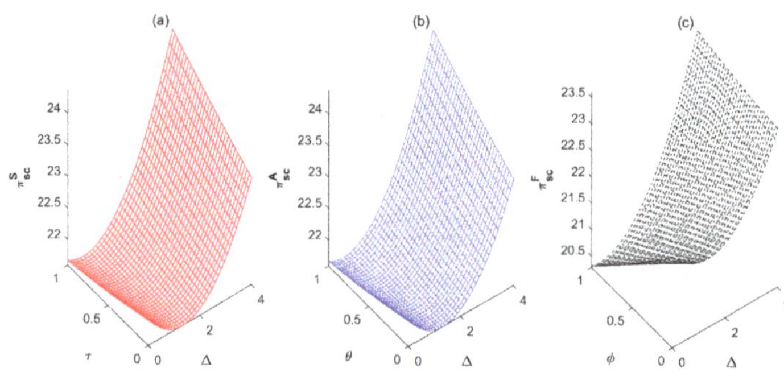

Figure 4. (**a**) Effect of τ and Δ on supply chain profit under retailer's cost-sharing practice. (**b**) Effect of θ and Δ on supply chain profit under retailer's altruistic preference practice. (**c**) Effect of ϕ and Δ on supply chain profit under retailer's fairness concern practice.

(2) Since $\Delta_{\theta A} = 1.329 > 0$, we can observe that the property of Figure 4b is similar to Figure 4a. On one hand, when $\theta = 0$, π_{SC}^{S} decreases with $0 < \Delta < 1.329$ and increases with $\Delta > 1.329$; when $\theta = 1$, π_{SC}^{S} increases with Δ. On the other hand, when $\Delta = 0$, π_{SC}^{S} decreases with Δ. When Δ is large, e.g., $\Delta = 4$, π_{SC}^{S} increases with θ. Thus, Figure 4b can verify Proposition 7.

(3) It also can be observed from Figure 4c that, when Δ is small, e.g., $\Delta = 0$, π_{SC}^{F} decreases with ϕ. When Δ is large, e.g., $\Delta = 4$, π_{SC}^{F} increases with ϕ.

Figure 5 shows the comparison of supply chain profit under the retailer's CSR practices. The upper plane of Figure 5 shows the supply chain profit under the retailer's altruistic preference practice and cost-sharing practice, and the bottom plane shows the supply chain profit under the retailer's fairness concern practice. The three planes intersect when the retailer does not implement any CSR practice. We can find that, in this situation, the retailer's altruistic preference practice and cost-sharing practice are better for supply chain profit than the fairness concern practice.

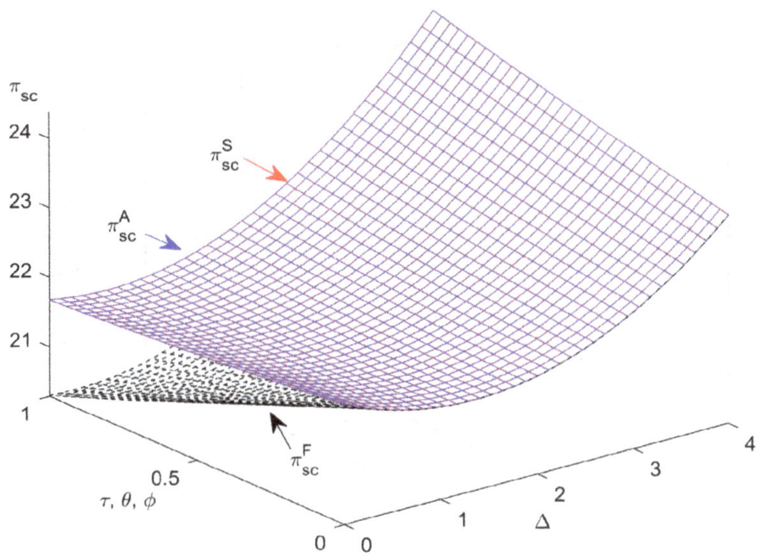

Figure 5. Comparison of the supply chain profit under the retailer's different CSR practices.

7. Conclusions

This paper has investigated a development supply chain where two representative farmers produce and sell a similar crop through a common retailer. The farmers from poor rural areas can label their crops as anti-poverty products, which are certified by the government, and the crops produced by the farmers from suburban areas are considered as general products. By an analytical modeling framework, we have examined the impacts of the retailer's CSR practices (i.e., cost-sharing practice, altruistic preference practice, and fairness concerns practice) on the decision equilibria and profits of the farmers and retailer. Findings are derived and summarized as follows.

First, compared with Model N, each of the retailer's three CSR practices can increase the whole supply chain's profit, which means that the supply chain has the potential to achieve the Pareto improvement for both the farmers and the retailer. From a supply chain perspective, the retailer's altruistic preference and cost-sharing practice are equally effective and better than the fairness concern practice. The results show that the development supply chain can do well by doing good, i.e., the whole development supply chain's performance is improved with the retailer's consideration of the poor farmer rather than their own shareholder value. Therefore, poverty alleviation initiatives that promote the implementation of different CSR practices by core enterprises in the supply chain (e.g., retailers) can generate both economic and social value.

Second, the retailer's CSR practices also benefit them while implementing cost-sharing practice or fairness concern practice. Both CSR practices can create a win–win result for both farmer 1 and the retailer. When implementing the altruistic preference practice, the retailer's situation deteriorates; however, the whole supply chain's performance improves. In this case, a contract can be designed to coordinate the three parties to improve the retailer's performance. The traditional supply chain coordination contracts, such as two-part tariff contracts, can be used to achieve the Pareto improvement among supply chain members. As the core enterprise in the poverty alleviation supply chain, the retailer can realize the incentive conditions and participation constraints of implementing altruistic preference practices among members through contract design. In addition, when the potential market of product 1 has a competitive advantage over that of product 2, the cost-sharing practice is the retailer's dominant strategy. When the retailer implements cost-sharing practice, the wholesale prices of both products will increase if the elasticity of the sale price is large, and will decrease if the elasticity of the sale price is small. In sum, since the development supply chain system becomes better and the retailer can also benefit from CSR behavior, such CSR practices are viable and sustainable, and thus the farmer's poverty can be alleviated.

Third, the rural farmer always benefits from the retailer's CSR practices, and the effect of the altruistic preference practice and fairness concern practice on the rural farmer's profit are better than those of the cost-sharing practice. The cost difference between farmer 1 and 2 always hurts farmer 1, which means that poverty is an obstacle for farmer 1 in making a living. However, the retailer's CSR practices always benefit farmer 1, which means that the retailer's CSR practices are highly effective in alleviating poverty. In addition, farmer 1 should not produce medium-value-added products, which may not help significantly in her profit (shown in Figure 5), while she should develop high-value-added products (e.g., medicinal materials) or low-value-added products (e.g., rice and soybeans), which will benefit her profit more. This form of management may indicate that, from a short-term perspective, the retailer's CSR practices can effectively improve the rural farmer's profits and help them to eliminate poverty. However, in the long run, improving the production efficiency of rural farmers and eliminating the factors leading to poverty, such as upgrading skills, upgrading equipment, and improving infrastructure, are stable strategies to prevent the return to poverty and achieve shared prosperity.

Fourth, from the suburban farmer's perspective, the retailer's CSR practices are not beneficial for their performance. However, the extent to which the suburban farmer's performance decreases is much less than the rural farmer's performance gains. Furthermore, the whole supply chain's performance has been improved significantly with the retailer's CSR practices. Thus, there is always a means to transfer the revenue to the suburban farmer through a contract design to enable the suburban farmer to improve their profit and thus participate in the transactions. In addition, the effect of the cost-sharing practice and altruistic preference practice on farmer 2's profit are the same, being better than those of the fairness concerns practice.

Poverty is an economic and structural problem that is difficult to reduce by using only one approach. It is therefore worth studying the combinations of different CSR practices that enterprises can adopt to help in the reduction of poverty. As we focus on the impacts of the retailer's three CSR practices on the equilibrium decisions of the players in a development supply chain, there are a few limitations in the current model setup and we offer some guidelines for future research. First, in our model, we assume that all consumers value the retailer's three CSR practices equally. However, in reality, there are different forms of retailers' CSR practices, and consumers may be heterogeneous in recognizing different retailers' CSR practices. Second, we assume that farmers incur linear production costs. In reality, there are also nonlinear production cost functions in which farmers face diseconomies of scale due to constraints of land, technology, and other resources. Third, we focus on the most popular commodities and assume that the quantities and cross-quantities sensitivity are the same in the demand function. However, in reality, due to channel and quality differences, there may also be large differences in quantities sensitivity and cross-quantities sensitivity between farmers, which can be further studied. Fourth, our model does not take into account supply chain incentives and coordination issues. For example, a contract can be designed to coordinate the development supply chain that achieves the Pareto improvement for both the farmers and retailer. In addition, knowing the retailer's CSR practices to favor farmer 1, farmer 2 may ally themselves with farmer 1. Fifth, we investigate the impact of the retailer's three CSR practices on the decisions and performance of the supply chain members. In further research, CSR can be treated as an endogenous decision variable instead of a choice from three options exogenously. Sixth, with technological development, the product cycle is becoming shorter and shorter, and consumers' cognition of products is also changing rapidly, which leads to a greater influence of demand uncertainty on supply chain decisions. In this way, demand uncertainty can be incorporated into the development supply chain as a stochastic model. Thus, it would be interesting to propose alternative models to include these aspects in future research.

Author Contributions: Conceptualization, Q.Z. and T.L.; Formal analysis, T.L.; Funding acquisition, Q.Z.; Investigation, Q.Z. and T.L.; Methodology, Q.Z. and T.L.; Software, T.L.; Supervision, Q.Z.; Writing—original draft, Q.Z. and T.L.; Writing—review & editing, Q.Z. and T.L. All authors have read and agreed to the published version of the manuscript.

Funding: This research was supported by Key Project of National Social Science Foundation of China (21AGL014).

Data Availability Statement: All the needed data are within the paper.

Acknowledgments: We'd like to thank the editor and reviewers for their comments to improve the paper.

Conflicts of Interest: There is no conflict of interest to declare.

Appendix A

Proof of Lemma 1. We adopt backward induction to find the equilibrium decisions of the farmers and the retailer. The retailer's profit function is denoted by $\max \pi_R = q_1(a_1 - w_1 - bq_1 - \gamma q_2) + q_2(a_2 - w_2 - bq_2 - \gamma q_1)$.

Given that $\frac{\partial^2 \pi_R}{\partial q_1^2} = -2b < 0$, $\frac{\partial^2 \pi_R}{\partial q_1 \partial q_2} = -2\gamma$, $\frac{\partial^2 \pi_R}{\partial q_2^2} = -2b < 0$, and the Hessian matrix,

$$\begin{vmatrix} \frac{\partial^2 \pi_R}{\partial q_1^2} & \frac{\partial^2 \pi_R}{\partial q_1 \partial q_2} \\ \frac{\partial^2 \pi_R}{\partial q_2 \partial q_1} & \frac{\partial^2 \pi_R}{\partial q_2^2} \end{vmatrix} = \begin{vmatrix} -2b & -2\gamma \\ -2\gamma & -2b \end{vmatrix} = 4\left(b^2 - \gamma^2\right) > 0$$

The retailer's profit function is a concave function. Therefore, the retailer has the optimal decision to maximize profits. Let $\frac{\partial \pi_R}{\partial q_1} = 0$ and $\frac{\partial \pi_R}{\partial q_2} = 0$, then $q_1(w_1, w_2) = \frac{a_1 b - a_2\gamma - bw_1 + \gamma w_2}{2b^2 - 2\gamma^2}$ and $q_2(w_1, w_2) = \frac{a_2 b - a_1\gamma - bw_2 + \gamma w_1}{2b^2 - 2\gamma^2}$. Farmer 1's profit function is denoted by $\pi_1(w_1, w_2) = \frac{(w_1 - \mu)(a_1 b - a_2\gamma - bw_1 + \gamma w_2)}{2b^2 - 2\gamma^2}$. Farmer 2's profit function is denoted by $\pi_2(w_1, w_2) = \frac{w_2(a_2 b - a_1\gamma - bw_2 + \gamma w_1)}{2b^2 - 2\gamma^2}$.

By calculation, we obtain $\frac{\partial^2 \pi_1}{\partial w_1^2} = -\frac{b}{b^2 - \gamma^2} < 0$ and $\frac{\partial^2 \pi_2}{\partial w_2^2} = -\frac{b}{b^2 - \gamma^2} < 0$, which shows that both farmer 1's and farmer 2's profit functions are a concave function. Thus, the farmers have the optimal decision to maximize profits. Let $\frac{\partial \pi_1}{\partial w_1} = 0$ and $\frac{\partial \pi_2}{\partial w_2} = 0$, and then we obtain

$$w_1^N = \frac{2a_1b^2 - a_1\gamma^2 + 2b^2\mu - a_2b\gamma}{4b^2 - \gamma^2},$$

$$w_2^N = \frac{2a_2b^2 - a_2\gamma^2 - a_1b\gamma + b\gamma\mu}{4b^2 - \gamma^2}.$$

Substituting w_1^N and w_2^N into $q_1(w_1, w_2)$, $q_2(w_1, w_2)$, $\pi_1(w_1, w_2)$, $\pi_2(w_1, w_2)$, and π_R, we have equilibria q_1^N, q_2^N, π_1^N, π_2^N, and π_R^N. □

Proof of Proposition 1. (i) $\frac{\partial w_1^N}{\partial \mu} = \frac{2b^2}{4b^2-\gamma^2} > 0$, $\frac{\partial w_2^N}{\partial \mu} = \frac{b\gamma}{4b^2-\gamma^2} > 0$, $\frac{\partial p_1^N}{\partial \mu} = \frac{b^2}{4b^2-\gamma^2} > 0$, $\frac{\partial p_2^N}{\partial \mu} = \frac{b\gamma}{2(4b^2-\gamma^2)} > 0$, $\frac{\partial q_2^N}{\partial \mu} = \frac{b^2\gamma}{2(4b^4-5b^2\gamma^2+\gamma^4)} > 0$, $\frac{\partial q_1^N}{\partial \mu} = \frac{-b(2b^2-\gamma^2)}{2(4b^4-5b^2\gamma^2+\gamma^4)} < 0$.

(ii) $\frac{\partial \pi_1}{\partial \mu} = \left(\frac{\partial w_1}{\partial \mu} - 1\right)q_1 + (w_1 - \mu)\frac{\partial q_1}{\partial \mu} = \left(\frac{2b^2}{4b^2-\gamma^2} - 1\right)q_1 + (w_1 - \mu)\frac{\partial q_1}{\partial \mu} < 0$,
$\frac{\partial \pi_2}{\partial \mu} = \frac{\partial w_2}{\partial \mu}q_2 + w_2\frac{\partial q_2}{\partial \mu} > 0$,

(iii) $\frac{\partial \pi_R^N}{\partial \mu} = \frac{b^2[2a_2\gamma^3-(a_1-\mu)(8b^3-6b\gamma^2)]}{4(b^2-\gamma^2)(4b^2-\gamma^2)^2}$, then $\frac{\partial \pi_R^N}{\partial \mu} > 0$ if $\lambda_1 > 0$, $\frac{\partial \pi_R^N}{\partial \mu} < 0$ if $\lambda_1 < 0$, where $\lambda_1 = [2a_2\gamma^3 - (a_1 - \mu)(8b^3 - 6b\gamma^2)]$. □

Proof of Lemma 2. The proof is similar to Lemma 1, so we omit it here. □

Proof of Proposition 2. (i) $\frac{\partial w_1^S}{\partial \mu} = \frac{2b^2}{4b^2-\gamma^2} - \tau$, then when $\tau < \frac{2b^2}{4b^2-\gamma^2}$, $\frac{\partial w_1^S}{\partial \mu} > 0$; when $\frac{2b^2}{4b^2-\gamma^2} < \tau < 1$, $\frac{\partial w_1^S}{\partial \mu} < 0$.

(ii) $\frac{\partial w_2^S}{\partial \mu} = \frac{b\gamma}{4b^2-\gamma^2} > 0$, $\frac{\partial p_1^S}{\partial \mu} = \frac{b^2}{4b^2-\gamma^2} > 0$, $\frac{\partial p_2^S}{\partial \mu} = \frac{b\gamma}{2(4b^2-\gamma^2)} > 0$, $\frac{\partial q_1^S}{\partial \mu} = -\frac{b(2b^2-\gamma^2)}{2(4b^4-5b^2\gamma^2+\gamma^4)} < 0$, $\frac{\partial q_2^S}{\partial \mu} = \frac{b^2\gamma}{2(4b^4-5b^2\gamma^2+\gamma^4)} > 0$.

(iii) $\frac{\partial \pi_1^S}{\partial \mu} = \frac{b(2b^2-\gamma^2)[a_2b\gamma-(a_1-\mu+\kappa\tau)(2b^2-\gamma^2)]}{(b^2-\gamma^2)(4b^2-\gamma^2)^2} = \frac{\partial \pi_1^N}{\partial \mu} - \frac{\kappa\tau b(2b^2-\gamma^2)^2}{(b^2-\gamma^2)(4b^2-\gamma^2)^2} < 0$, $\frac{\partial \pi_2^S}{\partial \mu} = \frac{\partial w_2^S}{\partial \mu}q_2^S + w_2^S\frac{\partial q_2^S}{\partial \mu} > 0$,

In addition, since $q_1^N = \frac{b(2a_1b^2-a_1\gamma^2-2b^2\mu+\gamma^2\mu-a_2b\gamma)}{2(4b^4-5b^2\gamma^2+\gamma^4)} > 0$, which means $0 < a_2 < \frac{2a_1b^2-a_1\gamma^2-2b^2\mu+\gamma^2\mu}{b\gamma}$, then we have $\frac{\partial \pi_R^S}{\partial \mu} = \frac{b^2[2a_2\gamma^3-(a_1-\mu+\kappa\tau)(8b^3-6b\gamma^2)]}{4(b^2-\gamma^2)(4b^2-\gamma^2)^2}$. Thus, $\max\frac{\partial \pi_R^S}{\partial \mu} = \left.\frac{\partial \pi_R^S}{\partial \mu}\right|_{a_2=\frac{2a_1b^2-a_1\gamma^2-2b^2\mu+\gamma^2\mu}{b\gamma}} = -\frac{b[(b^2-\gamma^2)(4b^4-\gamma^2)(a_1-\mu)+\kappa\tau b^2(4b^2-3\gamma^2)]}{2(b^2-\gamma^2)(4b^2-\gamma^2)^2} < 0$.

(iv) $\frac{\partial \pi_{SC}^S}{\partial \mu} = \frac{b[\mu(12b^4-9b^2\gamma^2+2\gamma^4)-(12b^4-9b^2\gamma^2+2\gamma^4)(a_1+\kappa\tau)+a_2(8b^3\gamma-3b\gamma^3)]}{2(b^2-\gamma^2)(4b^2-\gamma^2)^2}$, which means that $\frac{\partial \pi_{SC}^S}{\partial \mu}$ increases with μ. By solving the equation $\frac{\partial \pi_{SC}^S}{\partial \mu} = 0$, we obtain $\mu_0 = (a_1 + \kappa\tau) - a_2\frac{8b^3\gamma-3b\gamma^3}{12b^4-9b^2\gamma^2+2\gamma^4}$. Thus, when $0 < \mu < \mu_0$, $\frac{\partial \pi_{SC}^S}{\partial \mu} < 0$, when $\mu > \mu_0$, $\frac{\partial \pi_{SC}^S}{\partial \mu} > 0$. □

Proof of Proposition 3.

(i) $\frac{\partial w_1^S}{\partial \tau} = \frac{\kappa(2b^2-\gamma^2)-\mu(4b^2-\gamma^2)}{4b^2-\gamma^2}$, then when $\kappa(2b^2 - \gamma^2) - \mu(4b^2 - \gamma^2) > 0$, $\frac{\partial w_1^S}{\partial \tau} > 0$, when $\kappa(2b^2 - \gamma^2) - \mu(4b^2 - \gamma^2) < 0$, $\frac{\partial w_1^S}{\partial \tau} < 0$;

(ii) $\frac{\partial w_2^S}{\partial \tau} = -\frac{b\gamma\kappa}{4b^2-\gamma^2} < 0$, $\frac{\partial p_1^S}{\partial \tau} = \frac{\kappa(3b^2-\gamma^2)}{4b^2-\gamma^2} > 0$, $\frac{\partial p_2^S}{\partial \tau} = -\frac{b\gamma\kappa}{2(4b^2-\gamma^2)} < 0$, $\frac{\partial q_1^S}{\partial \tau} = \frac{b\kappa(2b^2-\gamma^2)}{2(4b^4-5b^2\gamma^2+\gamma^4)} > 0$, $\frac{\partial q_2^S}{\partial \tau} = -\frac{b^2\gamma\kappa}{2(4b^4-5b^2\gamma^2+\gamma^4)} < 0$.

(iii) Since $\frac{\partial \pi_1^S}{\partial \mu} < 0$, then $[a_2 b\gamma - (a_1 - \mu + \kappa\tau)(2b^2 - \gamma^2)] < 0$, we obtain $\frac{\partial \pi_1^S}{\partial \tau} = \frac{b\kappa(2b^2-\gamma^2)[(a_1-\mu+\kappa\tau)(2b^2-\gamma^2)-a_2 b\gamma]}{(b^2-\gamma^2)(4b^2-\gamma^2)^2} > 0$; $\frac{\partial \pi_2^S}{\partial \tau} = \frac{\partial w_2^S}{\partial \tau} q_2^S + w_2^S \frac{\partial q_2^S}{\partial \tau} < 0$; $\frac{\partial \pi_R^S}{\partial \tau} = -\frac{\kappa b^2[a_2\gamma^3-(a_1-\mu+\kappa\tau)(4b^3-3b\gamma^2)]}{2(b^2-\gamma^2)(4b^2-\gamma^2)^2} = -\kappa \frac{\partial \pi_R^S}{\partial \mu}$.

(iv) $\frac{\partial \pi_{sc}^S}{\partial \tau} = \frac{b\kappa[\tau\kappa(12b^4-9b^2\gamma^2+2\gamma^4)+(a_1-\mu)(12b^4-9b^2\gamma^2+2\gamma^4)+a_2(3b\gamma^3-8b^3\gamma)]}{2(b^2-\gamma^2)(4b^2-\gamma^2)^2}$, which means $\frac{\partial \pi_{sc}^S}{\partial \tau}$ increases with τ. By solving the equation $\frac{\partial \pi_{sc}^S}{\partial \tau} = 0$, we obtain $\tau_0 = \frac{1}{\kappa}\left[\frac{a_2(8b^3\gamma-3b\gamma^3)}{12b^4-9b^2\gamma^2+2\gamma^4} - (a_1-\mu)\right]$. Thus, when $0 < \tau < \min(\tau_0, 1)$, $\frac{\partial \pi_{sc}^S}{\partial \tau} < 0$; when $\max(0, \tau_0) < \tau < 1$, $\frac{\partial \pi_{sc}^S}{\partial \tau} > 0$. □

Proof of Proposition 4. Substituting Equation $\Delta = p_1^N - \mu$ into π_{sc}^S can yield $\pi_{sc}^S(\Delta, \tau)$, which means π_{sc}^S will be affected by parameter Δ and τ. Moreover, we obtain

$$\frac{\partial^2 \pi_{sc}^S(\Delta,\tau)}{\partial \Delta^2} = \frac{12b^5 - 9b^3\gamma^2 + 2b\gamma^4}{(2b^2-2\gamma^2)(3b^2-\gamma^2)^2} > 0$$

$$\frac{\partial^2 \pi_{sc}^S(\Delta,\tau)}{\partial \tau^2} = \frac{b\kappa^2(12b^4 - 9b^2\gamma^2 + 2\gamma^4)}{(2b^2-2\gamma^2)(4b^2-\gamma^2)^2} > 0$$

which means $\pi_{sc}^S(\Delta, \tau)$ is convex in parameter Δ and τ, respectively.

We solve the equation $\frac{\partial \pi_{sc}^S(\Delta,\tau)}{\partial \Delta} = 0$, and obtain

$$\Delta_S = \frac{36a_2b^5\gamma - 72\kappa\tau b^6 + 78\kappa\tau b^4\gamma^2 + 4\kappa\tau\gamma^6 - 25a_2b^3\gamma^3 - 30\kappa\tau b^2\gamma^4 + 4a_2b\gamma^5}{96b^6 - 96b^4\gamma^2 + 34b^2\gamma^4 - 4\gamma^6}$$

We observe that $\Delta_S|_{\tau=0} > 0$. By solving $\Delta_S = 0$, we obtain $\tau_{\Delta S} = \frac{a_2b\gamma(36b^4-25b^3\gamma^2+4\gamma^4)}{\kappa(72b^6-78b^4\gamma^2+30b^2\gamma^4-4\gamma^6)}$. It means that when $0 < \tau < \min(\tau_{\Delta S}, 1)$, then $\Delta_S > 0$, which means that π_{sc}^S decreases first with Δ in $(\Delta < \Delta_S)$ and then increases with Δ in $(\Delta > \Delta_S)$; when $\min(\tau_{\Delta S}, 1) < \tau < 1$, then $\Delta_S < 0$, which means that π_{sc}^S increases with Δ in $(\Delta > 0)$. Thus, Proposition 4(i) is proven. The proof of Proposition 4(ii) is similar to Yuen et al. [44], so we omit it here. □

Proof of Lemma 3. The proof is similar to Lemma 1, so we omit it here. □

Proof of Proposition 5. (i) $\frac{\partial w_1^A}{\partial \mu} = \frac{2b^2-\theta(4b^2-\gamma^2)}{(4b^2-\gamma^2)(1-\theta)}$, then when $\theta < \frac{2b^2}{4b^2-\gamma^2}$, $\frac{\partial w_1^A}{\partial \mu} > 0$; when $\frac{2b^2}{4b^2-\gamma^2} < \theta < 1$, $\frac{\partial w_1^A}{\partial \mu} < 0$.

(ii) $\frac{\partial w_2^A}{\partial \mu} = \frac{b\gamma}{(4b^2-\gamma^2)} > 0$, $\frac{\partial p_1^A}{\partial \mu} = \frac{b^2}{4b^2-\gamma^2} > 0$, $\frac{\partial p_2^A}{\partial \mu} = \frac{b\gamma}{2(4b^2-\gamma^2)} > 0$, $\frac{\partial q_1^A}{\partial \mu} = -\frac{b(2b^2-\gamma^2)}{2(4b^4-5b^2\gamma^2+\gamma^4)} < 0$, $\frac{\partial q_2^A}{\partial \mu} = \frac{b^2\gamma}{2(4b^4-5b^2\gamma^2+\gamma^4)} > 0$.

(iii) $\frac{\partial \pi_1^A}{\partial \mu} = \left(\frac{\partial w_1^A}{\partial \mu} - 1\right) q_1^A + (w_1^A - \mu)\frac{\partial q_1^A}{\partial \mu} = -\frac{2b^2-\gamma^2}{(4b^2-\gamma^2)(1-\theta)} q_1^A + (w_1^A - \mu)\frac{\partial q_1^A}{\partial \mu} < 0$, $\frac{\partial \pi_2^A}{\partial \mu} = \frac{\partial w_2^A}{\partial \mu} q_2^A + w_2^A \frac{\partial q_2^A}{\partial \mu} > 0$,

(iv) $\frac{\partial \pi_R^A}{\partial \mu} = \frac{b[4b^4-3b^2\gamma^2-\theta(12b^4-11b^2\gamma^2+2\gamma^4)](\mu-a_1-\theta\kappa)-a_2(4b^3\gamma\theta-b\gamma^3-b\gamma^3\theta)}{2(4b^2-\gamma^2)^2(1-\theta)(b^2-\gamma^2)}$. We can observe that $12b^4 - 11b^2\gamma^2 + 2\gamma^4 > 0$.

By solving the equation $4b^4 - 3b^2\gamma^2 - \theta(12b^4 - 11b^2\gamma^2 + 2\gamma^4) = 0$, we obtain $\theta_1 = \frac{4b^4-3b^2\gamma^2}{12b^4-11b^2\gamma^2+2\gamma^4}$, which means that when $0 < \theta < \theta_1$, $[4b^4 - 3b^2\gamma^2 - \theta(12b^4 - 11b^2\gamma^2 + 2\gamma^4)] > 0$; when $\theta_1 < \theta < 1$, $[4b^4 - 3b^2\gamma^2 - \theta(12b^4 - 11b^2\gamma^2 + 2\gamma^4)] < 0$. By solving the equation $\frac{\partial \pi_R^A}{\partial \mu} = 0$, we obtain $\mu_1 = a_1 + \theta\kappa - \frac{a_2(4b^3\gamma\theta-b\gamma^3\theta-b\gamma^3)}{(12b^4-11b^2\gamma^2+2\gamma^4)\theta-4b^4+3b^2\gamma^2}$. Thus, if $0 < \theta < \theta_1$, then when $0 < \mu < \mu_1$, $\frac{\partial \pi_R^A}{\partial \mu} < 0$; and when $\mu > \mu_1$, $\frac{\partial \pi_R^A}{\partial \mu} > 0$; if $\theta_1 < \theta < 1$, then when $0 < \mu < \mu_1$, $\frac{\partial \pi_R^A}{\partial \mu} > 0$; and when $\mu > \mu_1$, $\frac{\partial \pi_R^A}{\partial \mu} < 0$.

(v) $\frac{\partial \pi_{SC}^A}{\partial \mu} = \frac{\mu b(12b^4-9b^2\gamma^2+2\gamma^4)-b\left[(a_1+\kappa\theta)(12b^4-9b^2\gamma^2+2\gamma^4)-a_2(8b^3\gamma-3b\gamma^3)\right]}{2(b^2-\gamma^2)(4b^2-\gamma^2)^2}$. By solving the equation $\frac{\partial \pi_{SC}^A}{\partial \mu} = 0$, we obtain $\mu_2 = (a_1+\kappa\theta) - \frac{a_2(8b^3\gamma-3b\gamma^3)}{12b^4-9b^2\gamma^2+2\gamma^4}$, which means that when $0 < \mu < \mu_2$, $\frac{\partial \pi_{SC}^A}{\partial \mu} < 0$; when $\mu > \mu_2$, $\frac{\partial \pi_{SC}^A}{\partial \mu} > 0$. □

Proof of Proposition 6.

(i) $\frac{\partial w_1^A}{\partial \theta} = \frac{(a_1-\mu+\kappa)(2b^2-\gamma^2)-b\gamma a_2}{(4b^2-\gamma^2)(1-\theta)^2}$, then when $(a_1-\mu+\kappa)(2b^2-\gamma^2) - b\gamma a_2 > 0$, $\frac{\partial w_1^A}{\partial \theta} > 0$, when $(a_1-\mu+\kappa)(2b^2-\gamma^2) - b\gamma a_2 < 0$, $\frac{\partial w_1^A}{\partial \theta} < 0$;

(ii) $\frac{\partial w_2^A}{\partial \theta} = -\frac{b\gamma\kappa}{(4b^2-\gamma^2)} < 0$, $\frac{\partial p_1^A}{\partial \theta} = \frac{\kappa(3b^2-\gamma^2)}{4b^2-\gamma^2} > 0$, $\frac{\partial p_2^A}{\partial \theta} = -\frac{b\gamma\kappa}{2(4b^2-\gamma^2)} < 0$, $\frac{\partial q_1^A}{\partial \theta} = \frac{b\kappa(2b^2-\gamma^2)}{2(4b^4-5b^2\gamma^2+\gamma^4)} > 0$, $\frac{\partial q_2^A}{\partial \theta} = -\frac{b^2\gamma\kappa}{2(4b^4-5b^2\gamma^2+\gamma^4)} < 0$.

(iii) $\frac{\partial \pi_2^A}{\partial \theta} = \frac{\partial w_2^A}{\partial \theta} q_2^A + w_2^A \frac{\partial q_2^A}{\partial \theta} < 0$;

(iv) $\frac{\partial \pi_1^A}{\partial \theta} = \frac{b\left[a_2b\gamma-(a_1+\kappa\theta-\mu)(2b^2-\gamma^2)\right]\left[a_2b\gamma-(a_1+2\kappa-\theta\kappa-\mu)(2b^2-\gamma^2)\right]}{2(b^2-\gamma^2)(4b^2-\gamma^2)^2(1-\theta_1)^2}$, thus, when $(a_1+\kappa\theta-\mu)$ $< \frac{a_2b\gamma}{2b^2-\gamma^2} < (a_1+2\kappa-\theta\kappa-\mu)$, $\frac{\partial \pi_1^A}{\partial \theta} < 0$; when $\frac{a_2b\gamma}{2b^2-\gamma^2} < (a_1+\kappa\theta-\mu)$ or $\frac{a_2b\gamma}{2b^2-\gamma^2} > (a_1+2\kappa-\theta\kappa-\mu)$, $\frac{\partial \pi_1^A}{\partial \theta} > 0$.

(v) $\frac{\partial \pi_{SC}^A}{\partial \theta} = \frac{b\kappa\left[\theta\kappa(12b^4-9b^2\gamma^2+2\gamma^4)+(a_1-\mu)(12b^4+2\gamma^4-9b^2\gamma^2)+a_2(3b\gamma^3-8b^3\gamma)\right]}{2(b^2-\gamma^2)(4b^2-\gamma^2)^2}$, which increases with θ. By solving the equation $\frac{\partial \pi_{SC}^A}{\partial \theta} = 0$, we obtain $\theta_2 = \frac{1}{\kappa}\left[\frac{a_2(8b^3\gamma-3b\gamma^3)}{12b^4-9b^2\gamma^2+2\gamma^4} - (a_1-\mu)\right]$. Thus, when $0 < \theta < \min(\theta_2, 1)$, $\frac{\partial \pi_{SC}^A}{\partial \theta} > 0$; when $\max(\theta_2, 0) < \theta < 1$, $\frac{\partial \pi_{SC}^A}{\partial \theta} < 0$. □

Proof of Proposition 7. The proof of Proposition 7 is similar to Proposition 4, so we omit it here. □

Proof of Lemma 4. The proof is similar to Lemma 1, so we omit it here. □

Proof of Proposition 8. (i) $\frac{\partial w_1^F}{\partial \mu} = \frac{2b^2}{4b^2-\gamma^2} > 0$, $\frac{\partial w_2^F}{\partial \mu} = \frac{b\gamma}{4b^2-\gamma^2} > 0$, $\frac{\partial p_1^F}{\partial \mu} = \frac{b^2}{4b^2-\gamma^2} > 0$, $\frac{\partial p_2^F}{\partial \mu} = \frac{b\gamma}{8b^2-2\gamma^2} > 0$, $\frac{\partial q_2^F}{\partial \mu} = \frac{b^2\gamma}{2(4b^4-5b^2\gamma^2+\gamma^4)} > 0$, $\frac{\partial \pi_2^F}{\partial \mu} = \frac{\partial w_2^F}{\partial \mu} q_2^F + w_2^F \frac{\partial q_2^F}{\partial \mu} > 0$.

(ii) $\frac{\partial q_1^F}{\partial \mu} = \frac{-b(2b^2-\gamma^2)}{2(4b^4-5b^2\gamma^2+\gamma^4)} < 0$, $\frac{\partial \pi_1^F}{\partial \mu} = \left(\frac{\partial w_1^F}{\partial \mu} - 1\right) q_1^F + (w_1^F - \mu)\frac{\partial q_1^F}{\partial \mu}$. Since $\frac{\partial w_1^F}{\partial \mu} - 1 = -\frac{2b^2-\gamma^2}{4b^2-\gamma^2} < 0$, then we obtain $\frac{\partial \pi_1^F}{\partial \mu} < 0$.

(iii) $\frac{\partial \pi_R^F}{\partial \mu} = \frac{\mu b^3(4b^2-3\gamma^2)-b\left[b^2(a_1+\kappa\phi)(4b^2-3\gamma^2)-a_2b\gamma^3-\phi(b+\gamma)^2(2b-\gamma)^2\right]}{2(b^2-\gamma^2)(4b^2-\gamma^2)^2}$. By solving the equation $\frac{\partial \pi_R^F}{\partial \mu} = 0$, we obtain $\mu_3 = (a_1+\phi\kappa) - \frac{a_2b\gamma^3+\phi(b+\gamma)^2(2b-\gamma)^2}{b^2(4b^2-3\gamma^2)}$. Thus, when $0 < \mu < \mu_3$, $\frac{\partial \pi_R^F}{\partial \mu} > 0$; when $\mu > \mu_3$, $\frac{\partial \pi_R^F}{\partial \mu} < 0$.

(iv) $\frac{\partial \pi_{SC}^F}{\partial \mu} = \frac{b\left[\mu(12b^4-9b^2\gamma^2+2\gamma^4)-\phi(2b-\gamma)^2(b+\gamma)^2+a_2(8b^3\gamma-3b\gamma^3)-(a_1+\phi\kappa)(12b^4-9b^2\gamma^2+2\gamma^4)\right]}{2(b^2-\gamma^2)(4b^2-\gamma^2)^2}$, which increases with μ. By solving the equation $\frac{\partial \pi_{SC}^F}{\partial \mu} = 0$, we obtain $\mu_4 = a_1 + \phi\kappa - \frac{a_2(8b^3\gamma-3b\gamma^3)-\phi(2b-\gamma)^2(b+\gamma)^2}{12b^4-9b^2\gamma^2+2\gamma^4}$. Thus, when $0 < \mu \le \mu_4$, $\frac{\partial \pi_{SC}^F}{\partial \mu} \le 0$, when $\mu > \mu_4$, $\frac{\partial \pi_{SC}^F}{\partial \mu} > 0$. □

Proof of Proposition 9.

(i) $\frac{\partial w_1^F}{\partial \phi} = \frac{b\gamma+2b^2\kappa-\gamma^2\kappa+2b^2-\gamma^2}{4b^2-\gamma^2} > 0$, $\frac{\partial q_1^F}{\partial \phi} = \frac{b(b\gamma+2b^2\kappa-\gamma^2\kappa+2b^2-\gamma^2)}{2(4b^4-5b^2\gamma^2+\gamma^4)} > 0$.

(ii) $\frac{\partial w_2^F}{\partial \phi} = -\frac{b\gamma+2b^2-\gamma^2+b\gamma\kappa}{4b^2-\gamma^2} < 0$, $\frac{\partial q_2^F}{\partial \phi} = -\frac{b\left(b\gamma+2b^2-\gamma^2+b\gamma\kappa\right)}{2\left(4b^4-5b^2\gamma^2+\gamma^4\right)} < 0$, $\frac{\partial \pi_2^F}{\partial \phi} = \frac{\partial w_2^F}{\partial \mu} q_2^F + w_2^F \frac{\partial q_2^F}{\partial \mu} < 0$.

(iii) $\frac{\partial p_1^F}{\partial \phi} = \frac{b\gamma+6b^2\kappa-2\gamma^2\kappa-2b^2}{8b^2-2\gamma^2}$. Thus, when $\kappa > (<)\frac{2b^2-b\gamma}{6b^2-2\gamma^2}$, $\frac{\partial p_1^F}{\partial \phi} > (<)0$, $\frac{\partial p_2^F}{\partial \phi} = -\frac{b(\gamma-2b+\gamma\kappa)}{8b^2-2\gamma^2}$. Thus, when $\kappa > (<)\frac{2b-\gamma}{\gamma}$, $\frac{\partial p_2^F}{\partial \phi} > 0$.

$\frac{\partial \pi_1^F}{\partial \phi} = \left(\frac{\partial w_1^F}{\partial \phi} - 1\right) q_1^F + \left(w_1^F - \mu\right)\frac{\partial q_1^F}{\partial \phi}$. Since $\frac{\partial w_1^F}{\partial \phi} - 1 = \frac{b\gamma+2b^2\kappa-\gamma^2\kappa-2b^2}{4b^2-\gamma^2}$. Thus, when $\kappa > \frac{2b^2-b\gamma}{2b^2-\gamma^2}$, $\frac{\partial \pi_1^F}{\partial \phi} > 0$. □

Proof of Proposition 10. The proof of Proposition 10 is similar to Proposition 4, so we omit it here. □

Proof of Proposition 11. The proof of Proposition 11 can be obtained by simple calculation, so we omit it here. □

Proof of Proposition 12.

(i) $\pi_1^S - \pi_1^N = \frac{b\kappa\tau\left(2b^2-\gamma^2\right)\left(4a_1b^2-2a_1\gamma^2-4b^2\mu+2\gamma^2\mu-2a_2b\gamma+2b^2\kappa\tau-\gamma^2\kappa\tau\right)}{(2b^2-2\gamma^2)(4b^2-\gamma^2)^2}$. Since $q_1^N > 0$ and $q_2^N > 0$, then we have $4a_1b^2 - 2a_1\gamma^2 - 4b^2\mu + 2\gamma^2\mu - 2a_2b\gamma + 2b^2\kappa\tau - \gamma^2\kappa\tau > 0$, which means that $\pi_1^S > \pi_1^N$. The rest of the proof is similar, so we omit it here.

(ii) $\pi_1^F - \pi_1^A = \frac{b\left(2a_1b^2-a_1\gamma^2-2b^2\mu+\gamma^2\mu-a_2b\gamma+2b^2\kappa\tau-\gamma^2\kappa\tau+2b^2\tau-\gamma^2\tau+b\gamma\tau\right)^2}{2(b^2-\gamma^2)(4b^2-\gamma^2)^2} - \frac{b\left(a_1\gamma^2-2a_1b^2+2b^2\mu-\gamma^2\mu+a_2b\gamma-2b^2\kappa\tau+\gamma^2\kappa\tau\right)^2}{2(b^2-\gamma^2)(4b^2-\gamma^2)^2(1-\tau)}$. Thus, there exists a τ_1 whereby, when $0 < \tau \leq \tau_1$, $\pi_1^F \geq \pi_1^A$, and when $\tau_1 < \tau < 1$, $\pi_1^F < \pi_1^A$. □

Proof of Proposition 13. The proof of Proposition 13 is similar to Proposition 12, so we omit it here. □

Proof of Proposition 14. The proof of Proposition 14 is similar to Proposition 12, so we omit it here. □

Proof of Proposition 15. (i)

$$\pi_{sc}^S - \pi_{sc}^N = \frac{b\kappa\tau\left(24a_1b^4+4a_1\gamma^4-18a_1b^2\gamma^2+6a_2b\gamma^3-16a_2b^3\gamma-24b^4\mu-4\gamma^4\mu+18b^2\gamma^2\mu+12b^4\kappa\tau+2\gamma^4\kappa\tau-9b^2\gamma^2\kappa\tau\right)}{(4b^2-\gamma^2)^2(4b^2-4\gamma^2)}$$

We can observe that $\pi_{sc}^S - \pi_{sc}^N$ decreases with μ.

By solving the equation $\pi_{sc}^S - \pi_{sc}^N$, we obtain $\mu_5 = \frac{24a_1b^4+4a_1\gamma^4+12b^4\kappa\tau+2\gamma^4\kappa\tau-18a_1b^2\gamma^2+6a_2b\gamma^3-16a_2b^3\gamma-9b^2\gamma^2\kappa\tau}{24b^4-18b^2\gamma^2+4\gamma^4}$.

Thus, if $a_1\left(24b^4 + 4\gamma^4 - 18b^2\gamma^2\right) + \tau\kappa\left(12b^4 + 2\gamma^4 - 9b^2\gamma^2\right) > a_2\left(16b^3\gamma - 6b\gamma^3\right)$, then when $0 < \mu \leq \mu_5$, $\pi_{sc}^S - \pi_{sc}^N \geq 0$, when $\mu > \mu_5$, $\pi_{sc}^S - \pi_{sc}^N < 0$.

The remaining proofs of Proposition 15 are similar, so we omit them here. □

References

1. An, J.; Cho, S.-H.; Tang, C.S. Aggregating Smallholder Farmers in Emerging Economies. *Prod. Oper. Manag.* **2015**, *24*, 1414–1429. [CrossRef]
2. Zasada, I. Multifunctional Peri-Urban Agriculture—A Review of Societal Demands and the Provision of Goods and Services by Farming. *Land Use Policy* **2011**, *28*, 639–648. [CrossRef]
3. Orsini, F.; Kahane, R.; Nono-Womdim, R.; Gianquinto, G. Urban agriculture in the developing world: A review. *Agron. Sustain. Dev.* **2013**, *33*, 695–720. [CrossRef]
4. Kao, T.-Y.; Chen, J.C.H.; Wu, J.-T.B.; Yang, M.-H. Poverty Reduction through Empowerment for Sustainable Development: A Proactive Strategy of Corporate Social Responsibility. *Corp. Soc. Responsib. Environ. Manag.* **2016**, *23*, 140–149. [CrossRef]

5. Yonghui. 2019 Corporate Social Responsibility Report of Yonghui Superstores. 2020. Available online: http://www.yonghui.com.cn/upload/Inv/6190375.PDF (accessed on 2 April 2020).
6. English.gov.cn. China to Promote Poverty Alleviation through Consumption. 2019. Available online: http://english.www.gov.cn/policies/latest_releases/2019/01/14/content_281476479024726.htm (accessed on 14 January 2019).
7. En.people.cn. Sales of China's Anti-Poverty Products Exceed 330 Billion Yuan in First 10 Months. 2020. Available online: http://english.www.gov.cn/news/topnews/202011/29/content_WS5fc38b93c6d0f72576940db9.html (accessed on 29 November 2020).
8. Yu, J.J.; Tang, C.S.; Sodhi, M.S.; Knuckles, J. Optimal Subsidies for Development Supply Chains. *Manuf. Serv. Oper. Manag.* **2020**, *22*, 1131–1147. [CrossRef]
9. Sodhi, M.S.; Tang, C.S. Supply-Chain Research Opportunities with the Poor as Suppliers or Distributors in Developing Countries. *Prod. Oper. Manag.* **2014**, *23*, 1483–1494. [CrossRef]
10. ChinaDaily. China's Poverty Reduction Offers Lessons for Africa. 2020. Available online: http://www.chinadaily.com.cn/a/202012/29/WS5fea88a5a31024ad0ba9f2bf.html (accessed on 29 December 2020).
11. Edmond Yeboah, N.; Feng, Y.; Oppong-Sekyere, D.; Boadi Nyamaah, J. Agricultural Supply Chain Risk Identification—A Case Finding from Ghana. *J. Manag. Strategy* **2014**, *5*, 31–48. [CrossRef]
12. Wang, H.H.; Wang, Y.; Delgado, M.S. The Transition to Modern Agriculture: Contract Farming in Developing Economies. *Am. J. Agric. Econ.* **2014**, *96*, 1257–1271. [CrossRef]
13. Fang, L.; Huang, C.-C. Targeted poverty alleviation in China: Evidence from Jingdong e-commerce poverty alleviation. *Poverty Public Policy* **2020**, *12*, 386–396. [CrossRef]
14. Kolk, A.; van Tulder, R. Poverty alleviation as business strategy? Evaluating commitments of frontrunner Multinational Corporations. *World Dev.* **2006**, *34*, 789–801. [CrossRef]
15. Zhou, J.; Fan, X.; Chen, Y.-J.; Tang, C.S. Information Provision and Farmer Welfare in Developing Economies. *Manuf. Serv. Oper. Manag.* **2020**, *23*, 230–245. [CrossRef]
16. Chen, Y.-J.; JGeorge, S.; Zuo-Jun Max, S. Training, Production, and Channel Separation in ITC's E-Choupal Network. *Prod. Oper. Manag.* **2013**, *22*, 348–364. [CrossRef]
17. Yu, J.J.; Tang, C.S.; Shen, Z.-J.M. Improving Consumer Welfare and Manufacturer Profit via Government Subsidy Programs: Subsidizing Consumers or Manufacturers? *Manuf. Serv. Oper. Manag.* **2018**, *20*, 752–766. [CrossRef]
18. Tang, C.S.; Sodhi, M.S.; Formentini, M. An analysis of partially-guaranteed-price contracts between farmers and agri-food companies. *Eur. J. Oper. Res.* **2016**, *254*, 1063–1073. [CrossRef]
19. García-Martínez, G.; Guijarro, F.; Poyatos, J.A. Measuring the social responsibility of European companies: A goal programming approach. *Int. Trans. Oper. Res.* **2019**, *26*, 1074–1095. [CrossRef]
20. Zhou, Y.-W.; Guo, J.; Zhou, W. Pricing/service strategies for a dual-channel supply chain with free riding and service-cost sharing. *Int. J. Prod. Econ.* **2018**, *196*, 198–210. [CrossRef]
21. Xie, J.; Zhang, W.; Liang, L.; Xia, Y.; Yin, J.; Yang, G. The revenue and cost sharing contract of pricing and servicing policies in a dual-channel closed-loop supply chain. *J. Clean. Prod.* **2018**, *191*, 361–383. [CrossRef]
22. Li, T.; Zhang, R.; Zhao, S.; Liu, B. Low carbon strategy analysis under revenue-sharing and cost-sharing contracts. *J. Clean. Prod.* **2019**, *212*, 1462–1477. [CrossRef]
23. Hosseini-Motlagh, S.-M.; Pazari, P.; Nouri, M. Coordination of promotional effort, corporate social responsibility and periodic review replenishment decisions in a two-echelon socially responsible supply chain. *J. Ind. Syst. Eng.* **2018**, *11*, 60–83.
24. Wang, Y.; Fan, R.; Shen, L.; Miller, W. Recycling decisions of low-carbon e-commerce closed-loop supply chain under government subsidy mechanism and altruistic preference. *J. Clean. Prod.* **2020**, *259*, 120883. [CrossRef]
25. Wan, X.; Jiang, B.; Li, Q.; Hou, X. Dual-channel environmental hotel supply chain network equilibrium decision under altruism preference and demand uncertainty. *J. Clean. Prod.* **2020**, *271*, 122595. [CrossRef]
26. Xu, F.; Wang, H. Competitive–Cooperative Strategy Based on Altruistic Behavior for Dual-Channel Supply Chains. *Sustainability* **2018**, *10*, 2103. [CrossRef]
27. Wang, Y.; Yu, Z.; Shen, L.; Dong, W. E-Commerce Supply Chain Models under Altruistic Preference. *Mathematics* **2021**, *9*, 632. [CrossRef]
28. Wang, Y.; Yu, Z.; Shen, L.; Fan, R.; Tang, R. Decisions and Coordination in E-Commerce Supply Chain under Logistics Outsourcing and Altruistic Preferences. *Mathematics* **2021**, *9*, 253. [CrossRef]
29. Guan, Z.; Ye, T.; Yin, R. Channel coordination under Nash bargaining fairness concerns in differential games of goodwill accumulation. *Eur. J. Oper. Res.* **2020**, *285*, 916–930. [CrossRef]
30. Caliskan-Demirag, O.; Chen, Y.F.; Li, J. Channel coordination under fairness concerns and nonlinear demand. *Eur. J. Oper. Res.* **2010**, *207*, 1321–1326. [CrossRef]
31. Cui, T.H.; Raju, J.S.; Zhang, Z.J. Fairness and Channel Coordination. *Manag. Sci.* **2007**, *53*, 1303–1314. [CrossRef]
32. Yang, J.; Xie, J.; Deng, X.; Xiong, H. Cooperative advertising in a distribution channel with fairness concerns. *Eur. J. Oper. Res.* **2013**, *227*, 401–407. [CrossRef]
33. Zheng, X.-X.; Liu, Z.; Li, K.W.; Huang, J.; Chen, J. Cooperative game approaches to coordinating a three-echelon closed-loop supply chain with fairness concerns. *Int. J. Prod. Econ.* **2019**, *212*, 92–110. [CrossRef]
34. Zhou, Y.; Bao, M.; Chen, X.; Xu, X. Co-op advertising and emission reduction cost sharing contracts and coordination in low-carbon supply chain based on fairness concerns. *J. Clean. Prod.* **2016**, *133*, 402–413. [CrossRef]

35. Shen, L.; Lin, F.; Wang, Y.; Su, X.; Li, H.; Zhou, R. Advertising Decisions of Platform Supply Chains Considering Network Externalities and Fairness Concerns. *Mathematics* **2022**, *10*, 2359. [CrossRef]
36. Arslan, M.C.; Turkay, M. EOQ Revisited with Sustainability Considerations. *Found. Comput. Decis. Sci.* **2013**, *38*, 223–249. [CrossRef]
37. Modak, N.M.; Kazemi, N.; Cárdenas-Barrón, L.E. Investigating structure of a two-echelon closed-loop supply chain using social work donation as a Corporate Social Responsibility practice. *Int. J. Prod. Econ.* **2019**, *207*, 19–33. [CrossRef]
38. Liu, Y.; Liu, Z.; Ren, W.; Lin Forrest, J.Y. A coordination mechanism through relational contract in a two-echelon supply chain. *J. Retail. Consum. Serv.* **2020**, *56*, 102156. [CrossRef]
39. Sodhi, M.S.; Tang, C.S. Buttressing Supply Chains against Floods in Asia for Humanitarian Relief and Economic Recovery. *Prod. Oper. Manag.* **2014**, *23*, 938–950. [CrossRef]
40. Wu, C.-H. Collaboration and sharing mechanisms in improving corporate social responsibility. *Cent. Eur. J. Oper. Res.* **2016**, *24*, 681–707. [CrossRef]
41. Tang, C.S.; Wang, Y.; Zhao, M. The Implications of Utilizing Market Information and Adopting Agricultural Advice for Farmers in Developing Economies. *Prod. Oper. Manag.* **2015**, *24*, 1197–1215. [CrossRef]
42. Puggioni, D.; Stefanou, S.E. The value of being socially responsible: A primal-dual approach. *Eur. J. Oper. Res.* **2019**, *276*, 1090–1103. [CrossRef]
43. Xia, Y.; Zu, X.; Shi, C. A profit-driven approach to building a "people-responsible" supply chain. *Eur. J. Oper. Res.* **2015**, *241*, 348–360. [CrossRef]
44. Yuen, K.F.; Thai, V.V.; Wong, Y.D. Are customers willing to pay for corporate social responsibility? A study of individual-specific mediators. *Total Qual. Manag. Bus. Excell.* **2016**, *27*, 912–926. [CrossRef]
45. Singh, N.; Vives, X. Price and quantity competition in a differentiated duopoly. *Rand J. Econ.* **1984**, *15*, 546–554. [CrossRef]
46. Wang, Y.; Niu, B.; Guo, P. On the Advantage of Quantity Leadership When Outsourcing Production to a Competitive Contract Manufacturer. *Prod. Oper. Manag.* **2013**, *22*, 104–119. [CrossRef]
47. Jiang, X.; Zhou, J. The Impact of Rebate Distribution on Fairness Concerns in Supply Chains. *Mathematics* **2021**, *9*, 778. [CrossRef]
48. Fehr, E.; Schmidt, K.M. A theory of fairness, competition, and cooperation. *Q. J. Econ.* **1999**, *114*, 817–868. [CrossRef]

mathematics

MDPI

Review

Revenue Management in Airlines and External Factors Affecting Decisions: The Harmonic Oscillator Model

Ivan Arraut [1,*,†], Wilson Rosado [2,†] and Victor Leong [1,†]

1 Faculty of Business and Law, University of Saint Joseph, Estrada Marginal da Ilha Verde, 14-17, Macao, China; victor.leong@usj.edu.mo
2 Department of Physics, Universidad de Sucre, Cra. 28 No 5-267, Puerta Roja, Sincelejo-Sucre 700001, Colombia; wilson.rosado@unisucre.edu.co
* Correspondence: ivan.arraut@usj.edu.mo
† Ivan Arraut and Wilson Rosado contributed equally since the beginning. Victor Leong contributed on the final practical analysis.

Abstract: The Revenue Management (RM) problem in airlines for a fixed capacity, single resource and two classes has been solved before by using a standard formalism. In this paper we propose a model for RM by using the semi-classical approach of the Quantum Harmonic Oscillator. We then extend the model to include external factors affecting the people's decisions, particularly those where collective decisions emerge.

Keywords: revenue management; harmonic oscillator; collective decisions; spontaneous symmetry breaking

MSC: 91C99; 91B80; 81V99

1. Introduction

Revenue management (RM) is one of the most important concepts to be used inside a business or organization. It consists in maximizing the revenue from a fixed amount of perishable inventory by using techniques like "market segmentation" as well as "demand management" [1,2]. By using standard techniques, other authors have demonstrated that it is possible to increase the revenue in any organization if the market is segmented [1,2]. The market segmentation appears as a Revenue Function (RF) depending on the price (p_i) and the demand (d_i). In standard situations, the price and the demand are considered to be continuous stochastic variables [3]. In the case of airlines, an increase in ticket prices corresponds to a reduction in demand and vice versa. If we investigate in the literature a system with two variables, behaving in a way such that when one variable increases the other decreases and vice versa, we will find immediately that the harmonic oscillator performs well in this aspect. We can see the example of the pendulum, which for small oscillation angles follows the harmonic behavior [4]. For the pendulum, the largest amplitude of oscillation corresponds to a minimal velocity and vice versa, a maximal velocity corresponds to a minimal amplitude. Similar behavior can be found in several different systems, as far as they follow this harmonic behavior. Understanding that the harmonic oscillator behaves qualitatively as the RM system, we can then use it as a potential model for modelling the behavior of the price and demand. In this paper, we will consider the RM system as a semi-classical version of the Quantum Harmonic Oscillator (QHO) and then price and demand become operators related to each other at the quantum level through the uncertainty principle [5]. Remarkably, as it has been just mentioned, the Harmonic oscillator naturally reproduces some important features typically observed in the RM scenario in airlines; they are: (1) Higher price for one class means lower demand and vice versa. (2) A higher class, naturally being more expensive, has a lower demand, except when there is a promotion, such that the demand becomes much higher than for the case corresponding to the lower classes for the same price. (3) The different classes represent two dimensions of the classical phase space of the harmonic oscillator, and the highest possible revenue is proportional to the sum of the largest possible projected areas of the rectangles circumscribed inside the ellipses projected over the different planes of the ellipsoid representing the phase space covered by the harmonic oscillator. These three features make the harmonic oscillator scenario ideal for explaining the observed pattern of the RM problem in airlines. This also demonstrates how general the harmonic oscillator system is, given the wide range of applications it has [6]. The dynamics of this airline system is modelled by using a Hamiltonian function, with its corresponding free parameters. The basic RM model can be formulated with the simplest harmonic oscillator potential. However, when some external factors, such as the COVID-19, a hurricane or any other natural disaster enters in this scenario, then we have to introduce additional terms inside the potential function in the Hamiltonian of the oscillator, with their corresponding free parameters. In these situations, the purely harmonic behavior is completely lost. In this paper, besides illustrating how to implement the Quantum Harmonic Oscillator in the RM problem in airlines, we also illustrate one generic scenario where we model the collective decisions taken by a group of persons. The collective behavior scenario is modelled as a phase transition process emerging from the spontaneous symmetry breaking of a symmetry inside the system, in analogy to what happens in some scenarios in material science and particle physics [7,8]. In this way, once the symmetry of the vacuum (equilibrium condition) is broken, the system naturally selects an arbitrary condition which forces all the customers (related to the demand variable) to make the same decision, no matter what happens. The symmetry is in general broken due to some small fluctuation forcing the system to select a fixed vacuum state; this can only happen after the same system reaches certain conditions determined by its free parameters. We also analyze the consequences of the collective decisions on the total expected revenue, which will naturally differ from the standard situation. Note that the scenario illustrated in this paper is different to the one shown in [9], where standard probabilistic methods were mentioned. The paper is organized as follows: In Section 2, we formulate the standard RM problem in airlines and we explain in detail the two-class problem. In Section 3, we explain how the Quantum Harmonic Oscillator and particularly, its semi-classical approach, adapts naturally to the two-class problem. In Section 4, we explain how the collective decisions emerge by using the concepts of phase transitions due to the spontaneous symmetry breaking of a symmetry of the system. We also calculate the largest possible revenue when the vacuum symmetry is broken, which is naturally different from the standard case because the phase space volume covered by the system collapses dramatically when collective decisions are taken. We then compare the results emerging from collective decisions with those corresponding to the standard Harmonic oscillator case. In Section 5, we show a practical example where the formalism proposed here can be used, in order to analyze the revenue of the partnership between the airlines Hong Kong Express and Cathay Pacific. Finally, In Section 6, we conclude.

2. The Standard Revenue Management Problem in Airlines

The standard RM problem identifies different classes $i = 1, 2, \ldots, n$ with the corresponding prices $p_1 > p_2 > \cdots > p_n$ and demands defined in an analogue way. The demand is in general a stochastic variable and the revenue management objective function is defined as

$$max\Pi^s = \sum_{i=1}^{n} p_i d_i(p_i), \tag{1}$$

with the fixed capacity of the plane constrained as follows:

$$\sum_{i=1}^{n} p_i d_i(p_i) \leq P \times C, \tag{2}$$

with C representing the capacity of the plane and P representing an average price. This condition just suggests that the revenue cannot overpass the plane's capacity. Market studies have demonstrated that a good market segmentation can increase the total revenue [1,2]. When the single resource (round trip) problem is analyzed, and considering only two classes, the first class tickets are priority in the sense that the company in charge is allowed to sell as many first class tickets as possible. However, in the practical sense, we understand that the demand for the first class is not able to fill the whole plane. Then, the mathematical problem connected with RM is to know how many second class tickets can be sold before starting to lose money for overselling them.

The Two Class Problem

Here we consider the demand for each class as X_1 (class 1) and X_2 (class 2), with the corresponding prices p_1 and p_2. Both demand variables are considered to be independent and continuous, with a probability density $f_{X_1}(.)$ and $f_{X_2}(.)$. As it was just mentioned, we will sell S_2 tickets for the second class and the problem consists in knowing when the selling process for the second class should be stopped such that we start to sell the tickets for the first class. Each class has the booking limits S_1 and S_2. Here $S_1 = C$ means that the first class (ideally) could be sold until filling the whole plane, something which certainly never happens. Given S_2, the expected profit is defined as

$$E[\Pi(S_2)] = p_2 S_2 + p_1 E_{min}(X_1, C - S_2). \tag{3}$$

This expression can be expressed more explicitly in the form

$$E[\Pi(S_2)] = p_2 S_2 + p_1 \left(E[X_1] + \int_{C-S_2}^{\infty} (C - S_2 - x_1) f_{X_1}(x_1) dx_1 \right). \tag{4}$$

Here, the first term is the revenue obtained from the second class. Additionally, $E[X_1]$ is the expectation value and $f_{X_1}(x_1)$ is the probability density distribution corresponding to the demand variable X_1. It has been demonstrated before that the maximal revenue is obtained when

$$F_{X_1}(C - S_2^*) = \frac{p_1 - p_2}{p_1}, \tag{5}$$

with F_{X_1} representing a cumulative density function of X_1, defined in the standard form as $F_{X_1}(C - S_2^*) = \int_0^{C-S_2^*} f_{X_1}(x_1) dx_1$. From Equation (5) we can obtain the optimal point where the revenue is maximal. That point is defined as

$$p_2 = p_1 \bar{F}_{X_1}(C - S_2^*), \tag{6}$$

with $\bar{F}_{X_1}$ being defined as $\bar{F}_{X_1}(C - S_2^*) = 1 - F_{X_1}(C - S_2^*)$. In [9], this result was interpreted as a limit value, after which we should stop selling tickets to the second class. Then, we can sell class 2 tickets as long as the ticket price is greater than or equal to the expected marginal seat revenue (EMSR) for the class 1. In general, if $S_2 < S_2^*$, then $p_2 > EMSR_1(C - S_2)$.

3. The Quantum Harmonic Oscillator for the Single Resource Two-Class Problem

A large demand in seats for some specific class corresponds to a low price. In the same way, a low demand for a particular class, corresponds to a high price of the same class. If we promote the price and the demand to be operators, then we can define the demand operator as $\hat{X}_i$ and the price operator as $\hat{p}_i$. A Quantum Mechanical system behaving in the same way as the price–demand relation for the RM problem in airlines is the Quantum Harmonic Oscillator (QHO) [6], although in this paper we will use its semi-classical version. If we consider first the quantum version of this system, then the operators corresponding to the price and the demand can be formulated such that they satisfy the standard commutation relations defined as

$$[\hat{p}_i, \hat{X}_j] = i\delta_{ij}, \quad [\hat{p}_i, \hat{p}_j] = [\hat{X}_i, \hat{X}_j] = 0. \tag{7}$$

Then, when we explore the RM problem from the perspective of Quantum Mechanics (QM), we have to select which space we want to use for analyzing the system. This means that we can work either in the demand space or in the price space. They are both related to each other by the standard Fourier transformation defined as

$$\psi(p_i) = \frac{1}{2\pi} \int_{-\infty}^{\infty} e^{ip_i x_i} \psi(x_i) dx_i. \tag{8}$$

For the customers, it is convenient to select the price space for analysis, because they are naturally interested only in the prices of the tickets. However, for the airline companies, since they are more concerned about the demand for the products which they offer, the demand space is the most convenient one for analyzing the system. Following the standard language of QM, we can interpret the price operator as the generator of the changes in the demand and vice versa; the demand operator for a product can be interpreted as the generator of the changes in prices. This is mathematically defined as

$$|x_i - a >= e^{-ip_i a/\hbar}|x_i >, \qquad |p_i + z >= e^{ix_i a/\hbar}|p_i > . \tag{9}$$

Now we can proceed to analyze the most basic situation. For practical calculations, however, we will consider the semi-classical approach to the harmonic oscillator problem.

The Two-Dimensional Case: Single Resource Two-Class Problem

If we analyze the two-dimensional version of the Quantum Harmonic Oscillator from the perspective of the RM problem in airlines, then each class corresponds to one dimension for the system. The Hamiltonian of the system is then defined as

$$\hat{H} = \sum_{i=1}^{2} \left(\frac{\hat{p}_i^2}{2} + \frac{1}{2}\omega_i^2 \hat{X}_i^2 \right). \tag{10}$$

The expected profit from the perspective of the Quantum Harmonic formulation can be derived in analogy with the result obtained in Equation (3). Evidently, instead of stochastic variables, the quantum formulation works with operators. However, for the analysis of the RM problem, the semi-classical approach of the oscillator is superior because the problem can be visualized graphically through the phase space. The Hamiltonian (10) has naturally two eigenvalues $E_{1,2}$, corresponding to each class. Since the first class is more expensive than the second class (for the same demand), then naturally it is expected that $E_1 > E_2$ in general. The Hamiltonian (10) reflects the typical behavior of the RM problem. Plotted in the phase space, the Hamiltonian for the two-class problem is a four-dimensional ellipsoid. As an illustration, we can observe Figure 1 [10]. Note that this figure illustrates a three-dimensional surface for the purpose of illustration; however, we have to take into account that here we are considering a four-dimensional ellipsoid.

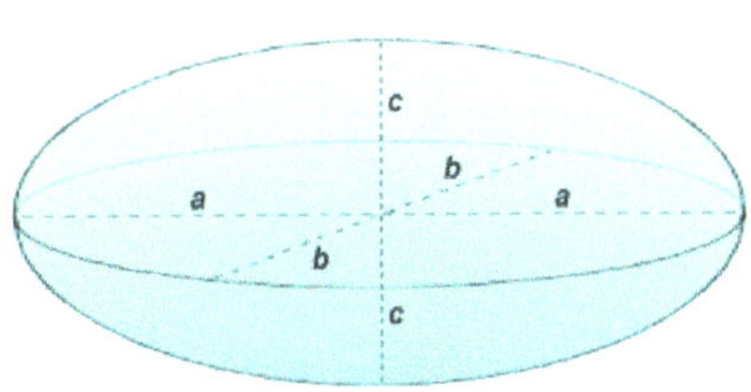

Figure 1. Ellipsoid illustrating the phase space covered by the Quantum Harmonic Oscillator. Without loss of generality, we can take $a = p_{max\,1}$ as the largest possible price for the first class and $b = X_{max\,1}$ as the largest possible demand for the first class. $c = p_{max\,2}$ might represent, for example, the largest possible price for the second class, with the corresponding largest demand for the second class given by a_{max}. Although we are talking about a four-dimensional ellipsoid, this figure illustrates a three-dimensional example.

Since $E_1 > E_2$, then the projected area of the two-dimensional ellipsoid corresponding to the first class is larger than the projected area corresponding to the second class. In general, the more expensive a class is, the taller will be the height of the projected rectangle. The harmonic oscillator model is so perfect for being adapted in the RM case that the phase space in Figure 1 illustrates the most interesting behavior related to this model, namely, in the hypothetical case when the first class is cheaper than second class, which can certainly happen under some short-time promotion or so, the demand becomes much larger than the one corresponding to the second class. This is something which certainly happens the whole time in reality. For example, at a supermarket during a day when expensive products are offered at a cheaper price, you will naturally see a huge amount of people, more than the usual, buying the most expensive products and probably ignoring the cheapest ones. This only happens because the customers will not see in the short term another chance of obtaining such expensive products at prices lower than the cheapest products. The same principle applies to the RM problem in airlines. As we have mentioned before, the Hamiltonian (10) obeys the eigenvalue problem

$$\hat{H}\psi(x_i,t) = E_i\psi(x_i,t). \tag{11}$$

The ellipsoid equation emerging from the semi-classical scenario is

$$1 = \frac{1}{2E}\omega^2 X^2 + \frac{1}{2E}p^2, \tag{12}$$

with $p^2 = p_1^2 + p_2^2$ and $X^2 = X_1^2 + X_2^2$. The most important quantity found by then was the largest possible revenue, which from the semi-classical perspective becomes a geometrical optimization problem for finding the largest possible rectangle which can be circumscribed inside the projected ellipse for each class. The projected ellipses are defined by the standard equation

$$1 = \frac{1}{2E_{1,2}}\omega^2 X_{1,2}^2 + \frac{1}{2E_{1,2}}p_{1,2}^2. \tag{13}$$

The optimization result is

$$E[\Pi] = A_1 + A_2 = \frac{2E_1^2}{\omega_1^2} + \frac{2E_2^2}{\omega_2^2}, \tag{14}$$

which for the case of an isotropic harmonic oscillator, becomes

$$E[\Pi] = \frac{2}{\omega^2}\left(E_1^2 + E_2^2\right). \tag{15}$$

If we use the standard expression for the eigenvalues of the QHO, we have

$$E_{1,2} = \omega\left(n_{1,2} + \frac{1}{2}\right). \tag{16}$$

Replacing this expression in Equation (15), we obtain

$$E[\Pi] = 2\omega\left(\left(n_1 + \frac{1}{2}\right)^2 + \left(n_2 + \frac{1}{2}\right)^2\right). \tag{17}$$

The total "energy" of the system $E = E_1 + E_2$, or more explicitly

$$E = \omega(n_1 + n_2 + 1), \tag{18}$$

is the constraint which complements (15). Equation (18) is just a way to express the values taken by the Hamiltonians of each class, $\hat{H}_1$ and $\hat{H}_2$. Although we could analyze practical examples by using the numbers n_1 and n_2, we will still use the standard Hamiltonian and its phase space in order to analyze practical problems. The number n_i just fixes the value of the Hamiltonian. Then, we can write, for example,

$$\hat{H} = \omega(n_1 + 1/2) = \frac{1}{2}\omega^2 X^2 + \frac{1}{2}p^2, \quad (19)$$

for one class. Then, n_1 just fixes the value taken by $\hat{H} = E$. Please see the Section 5 for the practical applications of the present formalism.

4. Collective Decisions in the RM Problem: Spontaneous Symmetry Breaking

We can extend the QHO model for including additional terms in the potential. These additional terms are related to the decisions taken by the passengers, influenced by some external factors. Under some special circumstances, these decisions turn out to be collective. Collective decisions can be interpreted as spontaneous symmetry breaking of the system if we consider the price and the demand as Quantum Fields. Similar situations emerge in finance and other research areas where collective decisions emerge naturally [11,12]. Here we will analyze the standard Hamiltonian of the form

$$\hat{H} = \sum_i \left(\frac{\hat{p}_i^2}{2m} + \frac{1}{2}\omega^2 \hat{X}_i^2 + \lambda \frac{\hat{X}_i^4}{4} \right), \quad (20)$$

where we can perceive that the potential term has the form

$$V(\hat{X}) = \frac{1}{2}\omega^2 \hat{X}^2 + \lambda \frac{\hat{X}^4}{4}. \quad (21)$$

This is a well-understood potential term in the research area of Quantum Field Theory [7,8]. In this part of the paper, we will not take into account the different classes because when the symmetry is broken, the distinction between classes is completely lost. Then, we will omit the subindex i from Equation (20) for the subsequent analysis. The potential (21) is a function where some combinations of the parameters ω and λ create the ideal scenario for reproducing spontaneous symmetry breaking after a small fluctuation affects the system [13–19]. Here we will interpret this symmetry breaking pattern in the language of the RM problem. We will also understand how the phase space is deformed due to the presence of the additional potential term $\lambda \frac{\hat{X}^4}{4!}$. The form of the potential (21) can be illustrated in Figure 2.

The figure shows that the spontaneous symmetry breaking occurs when $\omega^2 < 0$, such that the system develops more than one vacuum condition. If the demand operator has infinite components, then the system would develop infinite vacuum states, all of them degenerate. In the thermodynamic limit, any small fluctuation will force the system to select one of these vacuum states arbitrarily, defining then the collective decision pattern taken by all the persons affected by some particular event. The vacuum states appearing in Figure 2, can be obtained mathematically from Equation (21) if we calculate the condition $\partial V/\partial x = 0$, obtaining then the vacuum conditions $x_0^2 = \omega^2/\lambda$, valid when $\omega^2 < 0$. Here we understand that $x = \hat{X}\psi(x,t)$ and we take it as a Quantum Field. Note from the figure that when the symmetry is spontaneously broken, the usual vacuum state with $x_0 = 0$ becomes unstable and then the most stable condition turns out to be the one corresponding to one of the possible states breaking the symmetry spontaneously. Note that in the thermodynamic limit, a small fluctuation makes the system select from one of the degenerate states. Such fluctuation in the practical sense, and for our purposes in this paper, corresponds to the rumor propagation due to some catastrophic event. Catastrophic events like bad weather, volcanic explosions blocking air traffic, or even a pandemic disease affecting a region or the world, just create the ideal scenario for the standard vacuum state to become unstable. This means that the information of the catastrophic events is stored inside the free parameters of the potential defined in Equation (21). In the light of the

COVID-19 crisis, perhaps the worst pandemic crisis in this century [20], this kind of analysis became important. Due to the COVID-19 pandemic, for example, all the airlines had to stop most of the services. The demand for tickets then became low and the corresponding prices became too high, as it could be verified at the time by any airline website. Low-fare airlines stopped services completely and big airlines were only providing partial services. In this way, the offer is limited because the demand fell catastrophically due to the quarantine measures, social distance measures and any other measures affecting travel taken by different countries. By keeping inside the harmonic oscillator model, we would say that the phase space has been reduced considerably due to a catastrophic event. Spontaneous symmetry breaking is the most natural explanation to what was happening during the COVID-19 period. Basically, the pandemic itself (together with all the measures) created the symmetry-breaking scenario by changing the value of the free parameters in the system, provoking then the demand field to be fixed at some small value defined by some specific vacuum state ($x_0 = \omega/\lambda^{1/2}$ if we only take the positive branch of the solution). Note that the new Hamitonian (20) defines a phase space in agreement with Figure 3.

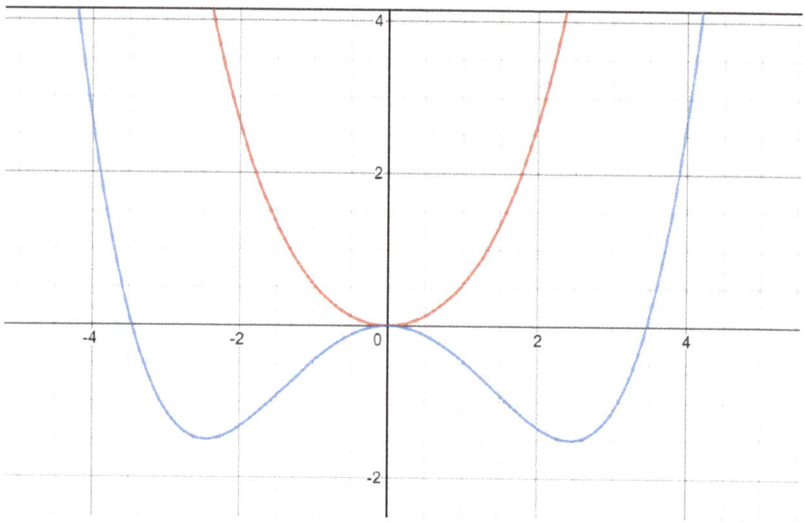

Figure 2. Symmetry breaking pattern with the x-axis representing the demand and the y-axis representing the price. The red line corresponds to the standard situation where both ω^2 and λ are positive. In this case, we only have one well-defined vacuum state. On the other hand, the blue curve corresponds to the case where $\omega^2 < 0$. This situation forces the system to develop two (or infinite) different vacuums as illustrated in the figure. The most stable state when $\omega^2 < 0$ corresponds to the one where the symmetry is spontaneously broken.

Note in addition that although the phase space looks larger for the new Hamiltonian (20), in comparison with the standard Harmonic oscillator case (10), the total expected revenue falls considerably. The total phase space volume is naturally larger when the symmetry is spontaneously broken because the larger possible prices also become catastrophically high because the airline companies cannot exploit fully the whole available phase space. This is because each portion of it now lives in a different Hilbert space, disjoint from each other in the thermodynamic limit [13,21]. Then, although the plot suggests larger available phase space, the small fluctuation provoked by the border measures due to the pandemic or any other disaster makes the system focus only on one single line of the phase space. Calculating the total expected revenue is not a problem in this case where there is no distinction between first or second class. By taking $x_0^2 = \omega^2/\lambda$ (with $\omega^2 < 0$), and if we replace this result in Equation (20), after taking

$$E = \frac{p^2}{2m} - \frac{1}{2}\omega^2 x^2 + \lambda \frac{x^4}{4}, \tag{22}$$

as the result of $\hat{H}\psi(x,t)$, we obtain the corresponding value for the price defined on the symmetry breaking vacuum state as

$$p_0 = \sqrt{2mE + \frac{m\omega^4}{2\lambda}}. \tag{23}$$

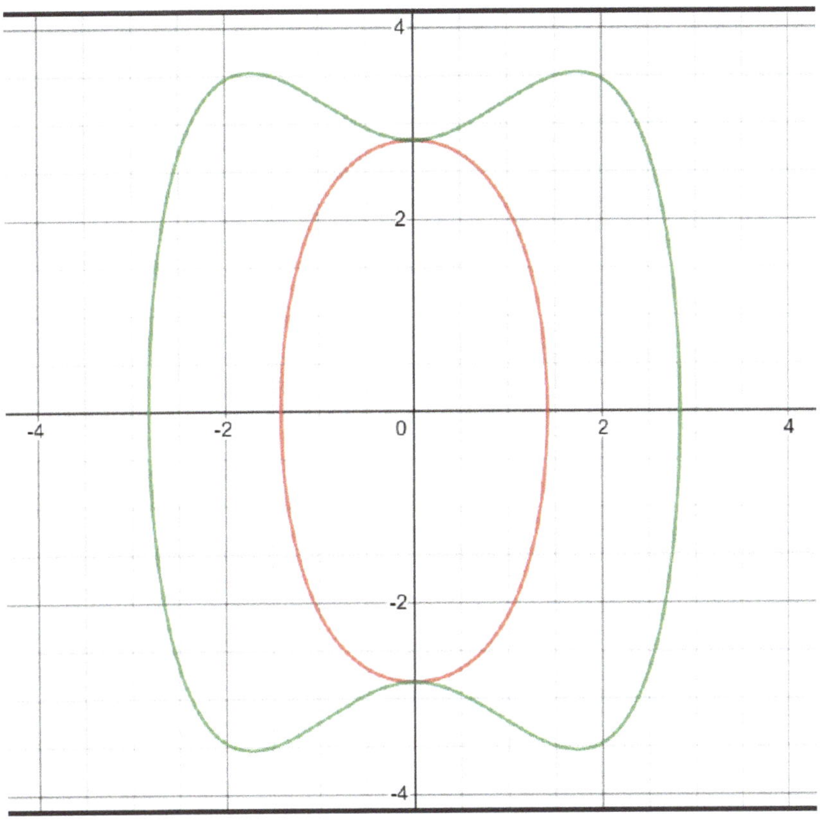

Figure 3. Changes on the phase space due to the presence of the quartic-order term in the Hamiltonian (20). The x-axis represents the demand and the y-axis represents the price. Note that the available phase space looks bigger when the quartic term appears. However, due to spontaneous symmetry breaking, the phase space is not fully available and it collapses to a line.

Here E is assumed to be fixed for each class. However, under spontaneous symmetry breaking, there is no distinction between classes and for that reason we will not be concerned about this classification. The result (23) represents the price of the ticket when the system has selected a particular vacuum state. Then, the expected revenue under spontaneous symmetry breaking (SSB) would be

$$E[\Pi]_{SSB} = p_0 x_0 = \omega\left(\frac{m}{\lambda}\right)^{1/2}\sqrt{2E + \frac{\omega^4}{2\lambda}}. \tag{24}$$

Then, for a fixed E, the expected revenue depends on the parameters m, ω and λ. By taking the semi-classical results already defined in Equation (18), we obtain

$$E[\Pi]_{SSB} = \omega^{3/2}\left(\frac{m}{\lambda}\right)^{1/2}\sqrt{(2n+1)+\frac{\omega^3}{2\lambda}}, \tag{25}$$

where we have taken the positive branch of the product between the price and the demand and we are considering only one class because the distinction already disappears at this level. If we compare this result with the expected revenue obtained from the harmonic oscillator model in Equation (17), by considering only one class, and considering $n = 0$, we obtain the condition such that the largest expected revenue in the standard harmonic oscillator system defined in Equation (10) is larger than the expected revenue in the present case. Such condition is defined by

$$\frac{\lambda^2}{2m\omega} \geq 2\lambda + \omega^3. \tag{26}$$

This condition is naturally satisfied for small frequencies, which typically happens in these situations. We can solve Equation (26) for λ, obtaining then

$$\lambda \geq 2m\omega\left(1+\sqrt{1+\frac{\omega^2}{2m}}\right), \tag{27}$$

which for small frequencies becomes approximately $\lambda \geq 4m\omega$. In this way, we have demonstrated that the expected revenue during crisis periods where the symmetry is spontaneously broken due to small external fluctuations, after the system reaches certain conditions, is much lower than the expected revenue under standard circumstances in most of the situations. We have also identified the parameters responsible of this behavior. The most important parameter is evidently the frequency of the system ω, which not only controls the vacuum behavior and whether we have symmetry breaking or not, but in addition, its absolute value marks the limits where the expected revenue falls with respect to the standard case. Note that we can take ω as the usual frequency, ignoring the fact that $\omega^2 < 0$ because we have already made the corresponding change of sign in Equation (22).

5. Numerical Example

In this section, we will work over some numerical examples for testing our model. We must remark that the QHO is a simplification of a general tendency. For the simplest case, the most general form for the Hamiltonian explaining the RM problem in airlines is

$$\hat{H} = a_0\hat{p} + b_0\hat{x} + a_1\hat{p}^2 + b_1\hat{x}^2 + a_2\hat{p}^3 + b_2\hat{x}^3 + a_3\hat{p}^4 + b_3\hat{x}^4 + \ldots \tag{28}$$

The degree of the polynomial expansion for the price and demand changes for different cases. For simplicity, in this article we have used the standard harmonic oscillator as a starting point for explaining the dynamical relation between price and demand in the airlines. In what follows, we will illustrate how the formulation proposed in this paper can be applied to a real-life situation. We will use as example the airline Hong Kong Express. The authors checked the website for Hong Kong Express on Thursday 15 February, and found that the lowest price for the tickets for going from Hong Kong to Tokyo was HKD 580. The same price repeated for several days and there were fluctuations over this price, with a peak in price of HKD 2888 on Sunday 25 February. In a non-hot season, a pattern of having more expensive tickets on Fridays was constant, with a price of HKD 608. The hot season by the end of March showed a tendency to have very high prices. The following table shows the values for the prices for different dates corresponding to the period between 16 February and 31 March. Based on the simplest model with the semi-classical approach of the Quantum Harmonic Oscillator, we made the predictions for the demand values. Although knowing the exact demand is impossible (it must be always conjectured), the model shows consistency in the sense that by controlling the price, the airlines can control the demand. In other words, in hot seasons, understanding that more people are willing to travel, the airlines increase the prices of the tickets in order to reduce the demand. We can then say that the airlines operate over the price space, and are able to control this variable. On the contrary, the customers operate over the demand space, because they have the power of deciding whether or not they buy a ticket.

Since Hong Kong Express is a cheap airline, considered to be a class-less airline in the sense that there is only one class, for this system we have a two-dimensional harmonic oscillator behavior. If we want to extend this system to include one more class, we have to take Hong Kong Express as a second class section and its partner airline, namely, Cathay Pacific, would correspond to the first class. In other words, we will take the economy class of Cathay Pacific as the first class for Hong Kong Express. This analysis is valid since both airlines correspond to the same company. We could notice that the price for the same mentioned period for Cathay Pacific represents a constant price for the economy class of HKD 5482. The reason for this price not to change is that the demand for this ticket is always low no matter the season; in other words, there is never enough demand for the airline (Cathay Pacific) to force it to decrease by increasing the price. Additionally, the proposed price is the minimal possible price for this airline. Evidently, most of the customers will select Hong Kong Express for this itinerary and this explains why the prices fluctuate season by season for this case. The explicit Hamiltonian for this system is

$$\hat{H} = a_0\hat{p}_{CX}^2 + b_0\hat{x}_{CX}^2 + a_1\hat{p}_{HKE}^2 + b_1\hat{x}_{HKE}^2. \tag{29}$$

This Hamiltonian is just a special case for the most general situation described in (28). It can be expressed in a compact way as

$$\hat{H} = \hat{H}_{CX} + \hat{H}_{HKE}. \tag{30}$$

We can define the parameters for each Hamiltonian $\hat{H}_{CX}$ and $\hat{H}_{HKE}$ separately. Ideally, in a single day, when the estimated demand increases so much, the airline then increases the price of the ticket until the demand decreases to a value equal to the plane capacity C (number of available seats). Let us define C_{CX} and C_{HKE} as the plane capacities for the corresponding airlines. Then, for example, the price HKD 2888 in the Table 1 corresponds to $x_{HKE} = C_{HKE}$. Then, we have the following information for the case of February 25th for Hong Kong Express:

$$\begin{aligned} \hat{H}_{HKE} &= (236)^2 + a_1(2,888)^2, \\ \hat{H}_{HKE} &= (4 \times 236)^2 + a_1(588)^2. \end{aligned} \tag{31}$$

In this previous expression, we have set $b_1 = 1$ in Equation (29) and we have left a_1 to be fixed by the initial conditions established in Equation (31). After solving the system, we find that $a_1 = 47.8$ with the corresponding units. Note that in Equation (31) we have fixed the demand at the largest price p_{HKE} = 2888 HKD with the plane capacity. After investigating on the websites, we found that Hong Kong Express' largest plane is an Airbus A321neo with a capacity of $C_{HKE} = 236$ passengers [22]. Then, we assume that the airline is using its largest plane for this route. Finally, if the price offered to the market is p_{HKE} = HKD 588, namely, the lowest price found during ordinary seasons, then we can assume that the demand related to this price in a day like 25 February would be four times the plane capacity, namely $x_{HKE} = 4 \times 236$ passengers for the lowest price. This is the information contained in Equation (31). Here we have two assumptions and they are: (1) The demand is four times the capacity of the plane when the price is the lowest during a hot day. (2) The demand is just equal to the plane capacity when the airline fixes the price to a largest possible value. In this way, we can say that on February 25th, the Hamiltonian for Hong Kong Express is

$$\hat{H}_{HKE} = 47.8p_{HKE}^2 + x_{HKE}^2. \tag{32}$$

For CX (Cathay Pacific), we have to fix the Hamiltonian in a similar way. However, we did not detect any fluctuation in the prices for the lowest fare of CX, which was fixed to $p_{CX} = 5482$, which is also the lowest possible fare. This fare is also special because it offers a free return ticket if the round trip option is selected. Yet HKE (Hong Kong Express) is still cheaper if a round trip is selected. The only advantage of CX is the free food service plus the commodity of the seats. However, the itinerary Hong Kong–Tokyo is only a 4 h trip more or less and many people decide not to take any meals during this flight when they use HKE. This makes us think that the demand for the CX tickets is so low that the plane is probably filled with a quarter or lower of its capacity. Then, we will assume $x_{CX} = \frac{1}{4}C_{CX}$. The most expensive ticket for CX during the same itinerary costs around HKD 9500. For this particular case, this price would correspond to a zero value for the demand, namely $x_{CX} \approx 0$ when p_{CX} = HKD 9500. Then, the conditions for the Hamiltonian of CX are

$$\hat{H}_{CX} = a_0(9500)^2,$$
$$\hat{H}_{CX} = a_0(5482)^2 + b_0(202/4)^2. \tag{33}$$

Here we are taking $C_{CX} = 202$ passengers, considering that the smallest plane of CX is the AirbusA321neo [23]. Here again, considering $b_0 = 1$, we obtain $a_0 = 4.23 \times 10^{-5}$. Then, the Hamiltonian for 25 February for CX is

$$\hat{H}_{CX} = 4.25 \times 10^{-5} p_{CX}^2 + x_{CX}^2. \tag{34}$$

Then, the Hamiltonian (29) is defined as

$$\hat{H} = 4.25 \times 10^{-5} p_{CX}^2 + x_{CX}^2 + 47.8 p_{HKE}^2 + x_{HKE}^2. \tag{35}$$

We can perceive the difference in price and demand scales, for this itinerary, between CX and HKE. The differences are reflected through the parameters b_0 and b_1, which differ from each other by six orders of magnitude. With the Hamiltonian (35), CX and HKE in partnership can decide if it is necessary to increase or decrease the prices for certain itinerary. Note that the Hamiltonian (35) is valid for one day. However, by changing the parameters b_0 and b_1, we can adapt the same Hamiltonian to different days. In other words, the parameters b_0 and b_1 change day by day in agreement with the maximum and minimum prices fixed by the airlines. Finally, we must remark that the quadratic dependence of the prices and demands for $\hat{H}_{CX}$ and $\hat{H}_{HKE}$ are shown for illustration purposes. However, for more general cases, the dependence might change. Yet still, the quadratic dependence represents a good initial ansatz. Our approach is just limited for the lack of demand date, something which is not provided by the airlines. Finally, we can evaluate the revenue for CX and HKE if we know the values of $\hat{H}_{CX}$ and $\hat{H}_{HKE}$. These values can be found by evaluating any of the pairs of price–demand values mentioned previously. The result is $\hat{H}_{CX} = 3.84 \times 10^3$ and $\hat{H}_{HKE} = 3.99 \times 10^8$. The basic optimization helping us to find the largest possible area of a square circumscribed inside an ellipse gives us the largest possible revenues for CX and HKE for the analyzed itinerary as

$$E[\Pi]_{CX} = 9.25 \times 10^5 \text{ HKD},$$
$$E[\Pi]_{HKE} = 2.75 \times 10^7 \text{ HKD}, \tag{36}$$

where we have divided the optimal result by four in order to take into account that the price and the demand are both positive numbers. The result (36) means that for the same itinerary and the same date, the cheapest fares offered by HKE give higher profits than the expensive fares offered by CX. This is a demonstration of the importance of segmentation and the emphasis on massification at the moment of offering a product. If there is a pandemic, only the expensive airlines would operate (and this was actually the case during COVID-19), and this represents significant losses, considering that HKE gives two order of magnitude more revenue than CX in the present example. During the COVID-19 pandemic, the demand was so low that in such a case, only the price p_{CX} = HKD 5482 would be available, no matter the demand. This is the case because the demand will be always much smaller than the plane's capacity. This catastrophic scenario is what generated a huge crisis in the airline industry and it is equivalent to a phase transition of the system, looking for a new ground state. This was the scenario illustrated in the Section 4.

Table 1. Special values for the prices for Hong Kong Express. The itinerary corresponds to the trip Hong Kong–Tokyo. The period under analysis was 15 February 2024–5 April 2024. All the prices during this period are equal to HKD 588, except for most of the dates appearing in the table.

Key Dates P	Price HKE (HKD)	Day of the Week
16 February	608	Friday
23 February	708	Friday
24 February	1508	Saturday
25 February	2888	Sunday
1 March	608	Friday
3 March	658	Sunday
4 March	658	Monday
5 March	658	Tuesday
8 March	608	Friday
9 March	652	Saturday
10 March	**588**	Sunday
14 March	658	Thursday
15 March	708	Friday
20 March	728	Wednesday
21 March	1068	Thursday
22 March	938	Friday
23 March	918	Saturday
24 March	918	Sunday
25 March	788	Monday
26 March	1268	Tuesday
27 March	1488	Wednesday
28 March	2018	Thursday
29 March	1588	Friday
30 March	1318	Saturday
31 March st	818	Sunday
5 April	608	Friday

Comparison with Other Methods

The standard methods used for analyzing the RM problem in airlines are based on the formalism illustrated in the Section 2. In such a case, the relation (2) already indicates that a large price for a ticket is related to a low demand and vice versa. The expected revenue is indicated by Equation (4) and the maximum revenue condition is indicated by Equation (5). However, the problem with the traditional method proposed in Section 2 is that the maximum revenue condition does not indicate an explicit relation between price and demand as can be observed from Equation (5); instead, this dependence appears via a probability density function $f_{X_1}(x_1)$, which must be pre-established from previous data analysis. This necessity of knowing the exact density distribution of the data is a significant disadvantage with respect to the harmonic oscillator method, which avoids this and other complications by providing analytic dependencies between price and demand via the phase space through the Hamiltonian formulation. Then, our method gives an initial good estimation of the price–demand relation for airlines without invoking statistical methods. Subsequently, the statistical methods are just used for adjusting the free parameters of the theory, here taken as those appearing in Equation (28). In other words, our method suggests in advance a functional dependence between price and demand, leaving the statistical methods just for refining the values of the free parameters involved in the analysis. Then, even without involving any statistical method, the proposed formalism gives us a good qualitative behavior of the price–demand relation when different classes are involved. Additionally, the related optimization problem, illustrated in Equation (14) and used for calculating the maximum revenue, is just a standard optimization problem from ordinary calculus, which is much simpler than the way the standard methods are able to calculate the ideal revenue via statistical probability density function. In summary, our method is mainly analytical and it just uses statistics as an auxiliary tool, while the other methods present a full statistical dependence.

6. Conclusions

In this paper we have developed an RM model based on an extension of the harmonic oscillator case. We have included a quartic-order term which is related to the interaction between consumers taking decisions based on external events. If $\omega^2 > 0$, then we have the usual standard harmonic oscillator case. This means that the quartic-order term does not affect the harmonic behavior so much in such a case. However, when $\omega^2 < 0$, the conditions of the system change completely. In such a case, what is considered to be the most stable condition becomes unstable. Such instability can be due to a catastrophic event, such as the current COVID-19 pandemic situation, for example. Then, a small fluctuation (rumors of people, news or information) of the system, forces it to select some vacuum state, which corresponds to a fixed price and demand. In this way, the whole phase space collapses to a single line. The product of the price and the demand, evaluated both on the symmetry breaking vacuum state, gives us the expected revenue for this case. We have demonstrated in Equation (26) that the expected revenue for the case where the symmetry is broken is lower than for the case where we have the standard behavior based on the semi-classical approach of the Quantum Harmonic Oscillator (semi-classical model). This is valid when the frequency ω is small under the standard units taken in the system. Here, evidently, we assume the same frequency in both cases (the standard case and the symmetry breaking case) for the purposes of comparison. Finally, we have illustrated an example where the Hamiltonian approach can be used for analyzing the price and demand for the partnership between CX (Cathay Pacific) and HKE (Hong Kong Express). We have proved that the segmentation of the airline's fare by including a cheaper option, generating massification, is the key ingredient for increasing the revenue of an airline company. Our approach also shows how to deal with the cases where there are partnerships between organizations. The proposed method has also a significant advantage with respect to the traditional way of calculating the RM in airlines, because it does not depend explicitly on prior statistical analysis. Instead, the statistics are reduced to the analysis of some free parameters which can be adjusted with the observations correspondingly.

Funding: This research received no external funding

Data Availability Statement: Data are contained within the article.

Conflicts of Interest: The authors declare no conflict of interest

References

1. Bish, D.R.; Bish, E.K.; Maddah, B. *Revenue Management and Capacity Planning;* Taylor, G.D., Ed.; CRC Press: Boca Raton, FL, USA, 2008.
2. Boyd, E.A.; Bilegan, I.C. Revenue Management and E-Commerce. *Manag. Sci.* **2003**, *49*, 1363. [CrossRef]
3. Littlewood, K. Forecasting and control of passenger bookings. *Proc. Agifors. Annu. Sympos.* **1972**, *12*, 95.
4. Serway, R.A.; Jewett, J.W. *Physics for Scientists and Engineers*, 9th ed.; Cengage Learning EMEA: Boston, MA, USA, 1 January 2013.
5. Heisenberg, W. Uber den anschaulichen Inhalt der quantentheoretischen Kinematik und Mechanik. *Z. Phys.* **1927**, *43*, 172198. [CrossRef]
6. Griffiths, D. J.*Introduction to Quantum Mechanics*, 2nd ed.; Prentice Hall: Hoboken, NJ, USA, 2004.
7. Peskin, M.E.; Schroeder, D.V. *An Introduction to Quantum Field Theory;* CRC Press: Boca Raton, FL, USA, 2018.
8. Ryder, L.H. *Quantum Field Theory;* Cambridge University Press: Cambridge, UK; ISBN 978-0-521-472423.
9. Belobaba, P.P. Applications of a probabilistic decision model to airline seat inventory control. *Opns. Res.* **1989**, *37*, 183. [CrossRef]
10. Available online: https://www.technologyuk.net/mathematics/geometry/ellipsoids.shtml (accessed on 1 September 2023).
11. Arraut, I.; Au, A.; Ching-biu Tse, A. Spontaneous symmetry breaking in quantum finance. *EPL* **2020**, *6*, 68003. [CrossRef]

12. Calmet, X.; Wiesendanger, S. N. An analytical perturbative solution to the Merton–Garman model using symmetries. *J. Futur. Mark.* **2020**, *40*, 3–22. [CrossRef]
13. Goldstone, J.; Salam, A.; Weinberg, S. Broken Symmetries. *Phys. Rev.* **1962**, *127*, 965. [CrossRef]
14. Bludman, S.A.; Klein, A. Broken Symmetries and Massless Particles. *Phys. Rev.* **1963**, *131*, 2364. [CrossRef]
15. Goldstone, J. Field theories with Superconductor solutions. *Nuovo. Cim.* **1961**, *19*, 154. [CrossRef]
16. Nambu, Y.; Jona-Lasinio, G. Dynamical Model of Elementary Particles Based on an Analogy with Superconductivity. I. *Phys. Rev.* **1961**, *122*, 345. [CrossRef]
17. Arraut, I., The Quantum Yang-Baxter Conditions: The Fundamental Relations behind the Nambu-Goldstone Theorem. *Symmetry* **2019**, *11*, 803. [CrossRef]
18. Arraut, I. The origin of the mass of the Nambu–Goldstone bosons. *Int. J. Mod. Phys. A* **2018**, *33*, 1850041. [CrossRef]
19. Arraut, I. The Nambu-Goldstone theorem in non-relativistic systems. *Int. J. Mod. Phys. A* **2017**, *32*, 1750127. [CrossRef]
20. Available online: https://www.worldometers.info/coronavirus/ (accessed on 1 December 2023).
21. Brauner, T. Spontaneous Symmetry Breaking and Nambu–Goldstone Bosons in Quantum Many-Body Systems. *Symmetry* **2010**, *2*, 609–657. [CrossRef]
22. Available online: https://en.wikipedia.org/wiki/HK_Express (accessed on 1 March 2024).
23. Available online: https://en.wikipedia.org/wiki/Cathay_Pacific_fleet (accessed on 1 March 2024).

MDPI AG
Grosspeteranlage 5
4052 Basel
Switzerland
Tel.: +41 61 683 77 34

Mathematics Editorial Office
E-mail: mathematics@mdpi.com
www.mdpi.com/journal/mathematics

www.ingramcontent.com/pod-product-compliance
Lightning Source LLC
Chambersburg PA
CBHW061548210926
51850CB00009B/2033
9783725857012